Springer Theses

Recognizing Outstanding Ph.D. Research

Aims and Scope

The series "Springer Theses" brings together a selection of the very best Ph.D. theses from around the world and across the physical sciences. Nominated and endorsed by two recognized specialists, each published volume has been selected for its scientific excellence and the high impact of its contents for the pertinent field of research. For greater accessibility to non-specialists, the published versions include an extended introduction, as well as a foreword by the student's supervisor explaining the special relevance of the work for the field. As a whole, the series will provide a valuable resource both for newcomers to the research fields described, and for other scientists seeking detailed background information on special questions. Finally, it provides an accredited documentation of the valuable contributions made by today's younger generation of scientists.

Theses are accepted into the series by invited nomination only and must fulfill all of the following criteria

- They must be written in good English.
- The topic should fall within the confines of Chemistry, Physics, Earth Sciences, Engineering and related interdisciplinary fields such as Materials, Nanoscience, Chemical Engineering, Complex Systems and Biophysics.
- The work reported in the thesis must represent a significant scientific advance.
- If the thesis includes previously published material, permission to reproduce this must be gained from the respective copyright holder.
- They must have been examined and passed during the 12 months prior to nomination.
- Each thesis should include a foreword by the supervisor outlining the significance of its content.
- The theses should have a clearly defined structure including an introduction accessible to scientists not expert in that particular field.

More information about this series at http://www.springer.com/series/8790

Yuto Ashida

Quantum Many-Body Physics in Open Systems: Measurement and Strong Correlations

Doctoral Thesis accepted by
The University of Tokyo, Tokyo, Japan

Author
Dr. Yuto Ashida
Department of Applied Physics
The University of Tokyo
Tokyo, Japan

Supervisor
Prof. Masahito Ueda
Department of Physics
The University of Tokyo
Tokyo, Japan

ISSN 2190-5053 ISSN 2190-5061 (electronic)
Springer Theses
ISBN 978-981-15-2582-7 ISBN 978-981-15-2580-3 (eBook)
https://doi.org/10.1007/978-981-15-2580-3

© Springer Nature Singapore Pte Ltd. 2020
This work is subject to copyright. All rights are reserved by the Publisher, whether the whole or part of the material is concerned, specifically the rights of translation, reprinting, reuse of illustrations, recitation, broadcasting, reproduction on microfilms or in any other physical way, and transmission or information storage and retrieval, electronic adaptation, computer software, or by similar or dissimilar methodology now known or hereafter developed.
The use of general descriptive names, registered names, trademarks, service marks, etc. in this publication does not imply, even in the absence of a specific statement, that such names are exempt from the relevant protective laws and regulations and therefore free for general use.
The publisher, the authors and the editors are safe to assume that the advice and information in this book are believed to be true and accurate at the date of publication. Neither the publisher nor the authors or the editors give a warranty, expressed or implied, with respect to the material contained herein or for any errors or omissions that may have been made. The publisher remains neutral with regard to jurisdictional claims in published maps and institutional affiliations.

This Springer imprint is published by the registered company Springer Nature Singapore Pte Ltd.
The registered company address is: 152 Beach Road, #21-01/04 Gateway East, Singapore 189721, Singapore

To my parents.

Supervisor's Foreword

A grand challenge in modern physics is to understand out-of-equilibrium systems subject to quantum measurement or an interaction with an environment. With rapid developments in atomic, molecular and optical (AMO) physics, out-of-equilibrium quantum physics has attracted increasing attention. In this thesis, Yuto Ashida combined techniques in quantum optics and condensed matter physics to explore several fundamental aspects of open quantum physics with special emphasis on quantum criticality, many-body dynamics, and system-environment entanglement.

Recent advances in AMO experiments have made it possible to manipulate quantum many-body systems at the single-quantum level. At such high resolution, Heisenberg's uncertainty relation places fundamental constraints on observable quantities through measurement backaction. Here, a crucial question is how the conventional quantum many-body physics should be altered under such atomic-level observation. Combining theoretical techniques from quantum optics, many-body physics, and statistical mechanics, Yuto has demonstrated that quantum measurement and an interaction with an environment fundamentally alters the dynamics of a system, giving rise to unique quantum many-body phenomena due to the nonunitary nature of quantum measurement.

First, Yuto combined the renormalization-group method with the cutting-edge numerical technique (iTEBD) to extend the Berezinskii-Kosterlitz-Thouless paradigm to the realm of open quantum systems. Second, Yuto studied how the measurement backaction qualitatively alters out-of-equilibrium many-body dynamics and found that correlations can propagate faster than the conventional speed limit known as the Lieb-Robinson bound at the expense of the probabilistic nature of quantum measurement. Third, Yuto employed the eigenstate thermalization hypothesis to address single-trajectory thermalization and heating in generic many-body systems under quantum measurement. Fourth, Yuto devised a new canonical transformation to solve generic quantum impurity problems in and out of equilibrium and developed a powerful variational approach to the nonequilibrium Kondo models and many-body systems recently realized in Rydberg gases.

Understanding physics in open and out-of-equilibrium quantum systems is still in its infancy and largely unexplored especially in the context of many-body physics. I believe that the results obtained by Yuto in this Thesis should also be applicable to other systems in AMO physics and condensed matter physics.

Tokyo, Japan
December 2019

Prof. Masahito Ueda

Abstract

This Thesis studies the fundamental aspects of many-body physics in quantum systems open to an external world. The ability to observe and manipulate quantum matter at the single-quantum level has revolutionized our approach to many-body physics. At such an ultimate resolution, a quantum system exhibits an unavoidable change due to the measurement backaction from an external observer. Moreover, various types of controlled couplings between quantum systems and external environments have been experimentally realized. These remarkable developments thus point to a new research arena of *open* many-body systems, where an interaction with an external observer and environment plays a major role. The first part of this Thesis is devoted to elucidating the influence of measurement backaction from an external observer on quantum critical phenomena and out-of-equilibrium many-body dynamics. The second part of this Thesis is devoted to revealing in- and out-of-equilibrium physics of an open quantum system strongly correlated with an external environment, where the entanglement between the system and the environment plays an essential role. The results obtained in this Thesis should serve as pivotal roles in understanding many-body physics of quantum systems open to an external world, and they are applicable to experimental systems in atomic, molecular and optical physics, quantum information science, and condensed matter physics.

Parts of this thesis have been published in the following journal articles:

1. *Efficient variational approach to dynamics of a spatially extended bosonic Kondo model,* **Yuto Ashida**, Tao Shi, Richard Schmidt, H. R. Sadeghpour, J. Ignacio Cirac and Eugene Demler, https://journals.aps.org/pra/abstract/10.1103/PhysRevA.100.043618 Physical Review A **100**, 043618 (2019). [Selected for Editors' Suggestion.]
2. *Quantum Rydberg Central Spin Model,* **Yuto Ashida**, Tao Shi, Richard Schmidt, H. R. Sadeghpour, J. Ignacio Cirac and Eugene Demler, https://journals.aps.org/prl/abstract/10.1103/PhysRevLett.123.183001 Physical Review Letters **123**, 183001 (2019). [Selected for Editors' Suggestion.]
3. *Thermalization and Heating Dynamics in Open Generic Many-Body Systems,* **Yuto Ashida**, Keiji Saito and Masahito Ueda, https://journals.aps.org/prl/abstract/10.1103/PhysRevLett.121.170402 Physical Review Letters **121**, 170402 (2018).
4. *Variational principle for quantum impurity systems in and out of equilibrium: Application to Kondo problems,* **Yuto Ashida**, Tao Shi, Mari C. Bañuls, J. Ignacio Cirac and Eugene Demler, https://journals.aps.org/prb/abstract/10.1103/PhysRevB.98.024103 Physical Review B **98**, 024103 (2018).
5. *Solving Quantum Impurity Problems in and out of Equilibrium with the Variational Approach,* **Yuto Ashida**, Tao Shi, Mari C. Bañuls, J. Ignacio Cirac and Eugene Demler, https://journals.aps.org/prl/abstract/10.1103/PhysRevLett.121.026805 Physical Review Letters **121**, 026805 (2018).
6. *Exploring the anisotropic Kondo model in and out of equilibrium with alkaline-earth atoms,* Marton Kanász-Nagy, **Yuto Ashida**, Tao Shi, Catalin P. Moca, Tatsuhiko N. Ikeda, Simon Fölling, J. Ignacio Cirac, Gergely Zaránd and Eugene Demler, https://journals.aps.org/prb/abstract/10.1103/PhysRevB.97.155156 Physical Review B **97**, 155156 (2018).
7. *Full-Counting Many-Particle Dynamics: Nonlocal and Chiral Propagation of Correlations,* **Yuto Ashida** and Masahito Ueda, https://journals.aps.org/prl/abstract/10.1103/PhysRevLett.120.185301 Physical Review Letters **120**, 185301 (2018).
8. *Many-body interferometry of magnetic polaron dynamics,* **Yuto Ashida**, Richard Schmidt, Leticia Tarruell and Eugene Demler, https://journals.aps.org/prb/abstract/10.1103/PhysRevB.97.060302 Physical Review B **97**, 060302(R) (2018) [Rapid Communication]. (Featured on https://www.nature.com/articles/s41567-018-0088-x Research highlight in Nature Physics **14**, 206 (2018)).
9. *Parity-time-symmetric quantum critical phenomena,* **Yuto Ashida**, Shunsuke Furukawa and Masahito Ueda, https://www.nature.com/articles/ncomms15791 Nature Communications **8**, 15791 (2017).

Parts of this thesis have been published in the following journal articles:

10. *Multiparticle quantum dynamics under real-time observation*, **Yuto Ashida** and Masahito Ueda, https://journals.aps.org/pra/abstract/10.1103/PhysRevA.95.022124 Physical Review A **95**, 022124 (2017).
11. *Quantum critical behavior influenced by measurement backaction in ultracold gases*, **Yuto Ashida**, Shunsuke Furukawa and Masahito Ueda, https://journals.aps.org/pra/abstract/10.1103/PhysRevA.94.053615 Physical Review A **94**, 053615 (2016).

Acknowledgements

First and foremost, I would like to thank my supervisor, Prof. Masahito Ueda for the tremendous support he has provided all over my Ph.D. work. He is always an inspiring scientist giving to me his fascination and joy in doing physics, and is also a proficient advisor guiding me to make efforts in the right direction. I have learned a lot from his fearless attitude toward jumping in uncharted terrain.

I am also deeply thankful to Prof. Eugene Demler for hosting me at Harvard University, providing a lot of valuable advice and making many stimulating discussions throughout the work presented in the second part of this Thesis. I am also thankful to my collaborators, Prof. Shunsuke Furukawa for work presented in Chap. 3, Prof. Keiji Saito for a part of work presented in Chap. 4, Prof. J. Ignacio Cirac, Prof. Tao Shi, Dr. Mari Carmen Bañuls, and Prof. H. R. Sadeghpour for work presented in Chap. 5, and Dr. Richard Schmidt and Prof. Leticia Tarruell for work presented in Chap. 6.

I am also thankful to members of Ueda group in the University of Tokyo for their insightful comments on my study. They have also made the lab a delightful place and provided countless humorous moments.

It has been my great pleasure to have discussions with many researchers, including Dr. Olalla A. Castro-Alvaredo, Prof. Ippei Danshita, Dr. Takeshi Fukuhara, Prof. Leonid Glazman, Dr. Fabian Grusdt, Prof. Naomichi Hatano, Prof. Yogesh Joglekar, Dr. Márton Kanász-Nagy, Prof. Hosho Katsura, Prof. Masato Koashi, Prof. Vladimir Konotop, Dr. Zala Lenarcic, Prof. Jesper Levinsen, Prof. Igor Mekhov, Prof. Takashi Mori, Dr. Masaya Nakagawa, Prof. Meera Parish, Prof. Tilman Pfau, Dr. Hannes Pichler, Prof. Peter Rabl, Prof. Jörg Schmiedmayer, Dr. Yulia Shchadilova, Prof. Yoshiro Takahashi, Mr. Takafumi Tomita, and Prof. Gergely Zaránd.

I wish to thank my parents and wife for their supports, understandings, and encouragements.

Finally, I acknowledge the financial support from the Japan Society for the Promotion of Science (JSPS) through the Program for Leading Graduate Schools (ALPS) and Grant-in-Aid for JSPS Research Fellow (JPSJ KAKENHI Grant No. JP16J03613).

Contents

1 **Motivation and Outline** 1
   References 7

2 **Continuous Observation of Quantum Systems** 13
   2.1 General Theory of the Nonunitary Evolution 13
      2.1.1 Indirect Measurement Model 13
      2.1.2 Mathematical Characterization of Measurement
            Processes 15
   2.2 Continuous Observation of Quantum Systems 17
      2.2.1 Repeated Indirect Measurements 17
      2.2.2 Quantum Jump Process 20
      2.2.3 Diffusive Limit 24
      2.2.4 Physical Example: Site-Resolved Measurement
            of Atoms 25
   References 27

3 **Quantum Critical Phenomena** 29
   3.1 Introduction 29
      3.1.1 Motivation: Measurement Backaction on Strongly
            Correlated Systems 29
      3.1.2 Universal Low-Energy Behavior in One-Dimensional
            Quantum Systems 32
   3.2 Backaction on Quantum Criticality I: The Quadratic Term 39
      3.2.1 Model 39
      3.2.2 Correlation Functions: Bifurcating Critical Exponents 42
      3.2.3 Realization in a 1D Ultracold Bose Gas
            with a Two-Body Loss 45
   3.3 Backaction on Quantum Criticality II: The Potential Term 47
      3.3.1 Model and Its Symmetry 47
      3.3.2 Perturbative Renormalization Group Analysis 48

xv

		3.3.3	Numerical Demonstration in a Non-Hermitian Spin-Chain Model	51
		3.3.4	Nonperturbative Renormalization Group Analysis	55
		3.3.5	Realization in Ultracold Gases with a One-Body Loss	59
		3.3.6	Short Summary	64
	3.4	Backaction on Quantum Phase Transitions		65
		3.4.1	Model	65
		3.4.2	Measurement-Induced Shift of the Quantum Critical Point	66
		3.4.3	Realization with Ultracold Gases in an Optical Lattice	69
	3.5	Experimental Realizations in Ultracold Gases		70
	3.6	Conclusions and Outlook		73
	References			80
4	**Out-of-Equilibrium Quantum Dynamics**			87
	4.1	Propagation of Correlations and Entanglement Under Measurement		88
		4.1.1	Introduction	88
		4.1.2	General Idea: The Full-Counting Many-Particle Dynamics	89
		4.1.3	System: Atoms Subject to a Spatially Modulated Loss	90
		4.1.4	Nonequilibrium Dynamics of Correlations and Entanglement	93
	4.2	Thermalization and Heating Dynamics Under Measurement		100
		4.2.1	Introduction	100
		4.2.2	Statistical Ensemble Under Minimally Destructive Observation	102
		4.2.3	Numerical Simulations in Nonintegrable Open Many-Body Systems	105
		4.2.4	Application to Many-Body Lindblad Dynamics	111
	4.3	Diffusive Quantum Dynamics Under Measurement		112
		4.3.1	Introduction	112
		4.3.2	System: Atoms Under Spatial Observation	113
		4.3.3	Minimally Destructive Spatial Observation	115
		4.3.4	Many-Body Stochastic Schrödinger Equations	116
		4.3.5	Numerical Demonstrations	119
	4.4	Experimental Situations in Ultracold Gases		121
	4.5	Conclusions and Outlook		123
	References			136
5	**Quantum Spin in an Environment**			145
	5.1	Introduction		145
	5.2	Disentangling Canonical Transformation		149
		5.2.1	Disentangling a Single Spin-1/2 Impurity and an Environment	150

		5.2.2	Disentangling Two Spin-1/2 Impurities and an Environment	152
		5.2.3	Disentangling the Single-Impurity Anderson Model	155
	5.3	Efficient Variational Approach to Generic Spin-Impurity Systems		157
		5.3.1	Fermionic Gaussian States	157
		5.3.2	Variational Time Evolution of the Covariance Matrix	158
		5.3.3	General Expression of the Functional Derivative	160
	5.4	Application to the Anisotropic Kondo Model		162
		5.4.1	Kondo Problem	162
		5.4.2	Entanglement Structure of the Variational Ground State	164
		5.4.3	Benchmark Tests with the Matrix-Product States	165
		5.4.4	Benchmark Test with the Bethe Ansatz Solution	170
		5.4.5	Tests of Nonperturbative Scaling and Universal Behavior	171
		5.4.6	Spatiotemporal Dynamics after the Quench	172
	5.5	Application to the Two-Lead Kondo Model		173
		5.5.1	Model	173
		5.5.2	Spatiotemporal Dynamics of the Environment after the Quench	174
		5.5.3	Transport Properties	176
	5.6	Experimental Implementation in Ultracold Gases		178
	5.7	Generalization to a Bosonic Environment		180
		5.7.1	General Formalism	180
		5.7.2	Application to the Rydberg Central Spin Problem	184
	5.8	Conclusions and Outlook		190
	References			196
6	**Quantum Particle in a Magnetic Environment**			205
	6.1	Introduction: New Frontiers in Polaron Physics		205
	6.2	System: A Mobile Particle in a Synthetic Magnetic Environment		207
	6.3	Many-Body Interferometry Acting on the Environment		210
	6.4	Out-of-Equilibrium Dynamics of the Magnetic Polaron		211
		6.4.1	Time-Dependent Variational Principle	211
		6.4.2	Quantum Dynamics of the Environment	212
		6.4.3	Many-Body Bound States	214
	6.5	Experimental Implementation in Ultracold Atomic Gases		216
	6.6	Conclusions and Outlook		220
	References			221
7	**Conclusions and Outlook**			225
Curriculum Vitae				227

Notations and Abbreviations

Notations

h	Planck's constant $h = 6.626 \times 10^{-34}$ J·s
$\hbar$	Reduced Planck's constant $\hbar = h/(2\pi) = 1.055 \times 10^{-34}$ J·s
k_B	Boltzmann constant $k_B = 1.381 \times 10^{-23}$ J/K
c	Speed of light $c = 299792458$ m/s
e	Electric charge $e = 1.602 \times 10^{-19}$ C
i	Imaginary unit $i = \sqrt{-1}$
π	Ratio of circumference of a circle to its diameter $\pi = 3.14159\ldots$
$\mathbb{R}$	Real number field
$\mathbb{C}$	Complex number field
$\mathbb{N}$	Natural number field
$\mathcal{H}$	Hilbert space
$|\psi\rangle \in \mathcal{H}$	State vector in $\mathcal{H}$
$\langle\psi|$	Dual vector of $|\psi\rangle$
$\langle\phi|\psi\rangle \in \mathbb{C}$	Inner product of $|\psi\rangle$ and $|\phi\rangle$
$\mathcal{L}(\mathcal{H})$	Set of linear operators on $\mathcal{H}$
$\hat{O} \in \mathcal{L}(\mathcal{H})$ (hatted symbols)	Linear operator
$\hat{I} \in \mathcal{L}(\mathcal{H})$	Identity operator
$\hat{\rho} \in \mathcal{L}(\mathcal{H})$	Density operator satisfying $\hat{\rho} \geq 0$ and $\text{Tr}[\hat{\rho}] = 1$
$[\hat{A}, \hat{B}]$	Commutator $\hat{A}\hat{B} - \hat{B}\hat{A}$ of linear operators
$\{\hat{A}, \hat{B}\}$	Anticommutator $\hat{A}\hat{B} + \hat{B}\hat{A}$ of linear operators
dN	Marked point stochastic process
dW	Wiener stochastic process
$E[\cdot]$	Ensemble average over a stochastic process
K	Tomonaga-Luttinger-liquid parameter
$\delta(x)$	Dirac's delta function

$\hat{\sigma}$	Vector of the Pauli matrices
I_d	$d \times d$ identity matrix
$\hat{\mathbb{P}}$	Parity operator satisfying $\hat{\mathbb{P}}^2 = 1$
Pf[A]	Pfaffian of a real antisymmetric matrix A
$\mathcal{A}[M]$	Matrix antisymmetrization $(M - M^{\mathrm{T}})/2$
$\mathcal{S}[M]$	Matrix symmetrization $(M + M^{\mathrm{T}})/2$

Abbreviations

Ads	Anderson
AFM	Antiferromagnetic
AMO	Atomic, Molecular and Optical
BA	Bethe Ansatz
BEC	Bose-Einstein Condensate
BKT	Berezinskii-Kosterlitz-Thouless
CDP	Cluster Decomposition Property
CFT	Conformal Field Theory
CMC	Center-of-Mass Coordinate
corr	Correlation
CPTP	Complete Positive and Trace Preserving
DFS	Decoherence Free Subspace
DMFT	Dynamical Mean-Field Theory
DMRG	Density-Matrix Renormalization Group
eff	Effective
EP	Exceptional Point
eq	Equilibrium
ETH	Eigenstate Thermalization Hypothesis
FM	Ferromagnetic
FRG	Functional Renormalization Group
FS	Fermi Sea
G (capital letter)	Gaussian
GS	Ground State
H.c.	Hermitian conjugate
imp	Impurity
int	Interaction
iTEBD	infinite Time-Evolving Block Decimation
K (capital letter)	Kondo
L (capital letter)	Left
LLP	Lee-Low-Pines
LR	Lieb-Robinson
mag	Magnetization
MI	Mott Insulator

MPO	Matrix-Product Operator
MPS	Matrix-Product States
MVPE	Matrix-Vector Product Ensemble
NGS	Non-Gaussian State
NRG	Numerical Renormalization Group
POVM	Positive Operator-Valued Measure
PT	Parity-Time
R (capital letter)	Right
RG	Renormalization Group
SIM	Spin-Impurity Model
SSCP	Spectral Singular Critical Point
tDMRG	time-dependent Density-Matrix Renormalization Group
TD-NRG	Time-Dependent-Numerical Renormalization Group
TEBD	Time-Evolving Block Decimation
TLL	Tomonaga-Luttinger Liquid

Chapter 1
Motivation and Outline

Abstract We review the background of the results presented in this thesis, which studies the fundamental aspects of many-body physics in quantum systems open to an external world. Firstly, we review recent remarkable experimental developments of observing and manipulating quantum matter at the single-quantum level. We discuss how they motivate the studies of elucidating the influence of an external observer on quantum many-body phenomena beyond the conventional paradigm of closed systems. Secondly, we review the physics of open quantum systems strongly correlated with an external environment and motivate the need of unveiling the entanglement between the system and the environment to understand genuine many-body effects such as the Kondo effect. Finally, we provide a short summary of each Chapter.

Keywords Open quantum systems · Quantum many-body systems · Ultracold atoms · Kondo effect

Ultimately, — in the great future — we can arrange the atoms the way we want; the very atoms, all the way down! What would happen if we could arrange the atoms one by one the way we want them?

Richard Feynman asked this visionary question in 1959 [1]. Once considered to be of purely academic interest, such microscopic manipulations and observations of genuine quantum systems are now routinely performed in the laboratory. In both theory and experiment, understanding of physics in controlled quantum systems has become one of the central problems in science.

The last two decades have witnessed remarkable developments in studies of in- and out-of-equilibrium many-body physics of an isolated, closed quantum system, as primarily promoted by rapid advances in atomic, molecular and optical (AMO) physics. Meanwhile, recent experimental advances have allowed one to *measure* and *manipulate* many-body systems at the single-quantum level, thus revolutionizing our approach to many-body physics. With such an ultimate resolution, the measurement backaction, which is the fundamental effect acted by an external observer, becomes significant. Moreover, microscopic manipulation capabilities of many-body systems have realized quantum systems strongly correlated with an external environment. These remarkable advances thus point to a new arena of many-body physics that is *open* to an external world, in which interactions with an external observer or an environment play a major role. The research of this Thesis is devoted to addressing the question of how the ability to measure and manipulate single quanta can create a new frontier of many-body physics beyond its conventional paradigm.

The first part of this Thesis elucidates how the influences of measurement backaction from an external observer trigger new types of many-body phenomena that have no analogues in closed systems (Fig. 1.1a). The time evolution of an isolated quantum system is described by a single Hermitian operator, i.e., the Hamiltonian. In contrast, under continuous observation, the dynamics becomes intrinsically nonunitary due to the measurement backaction and is characterized not only by the Hamiltonian but also by a measurement process. After reviewing a general theory of the nonunitary dynamics under continuous measurement in Chap. 2, we apply it to study effects of measurements on quantum many-body phenomena in Chaps. 3 and 4.

Our studies are mainly motivated by recent revolutionary developments in AMO physics. Already a number of groundbreaking experiments have been achieved over the past decade [2–31]. In particular, realizations of the technique known as quantum gas microscopy [2–13] have enabled one to detect a large number of atoms trapped in an optical lattice at the single-atom precision. Other examples include direct observations of quasi-long-range order predicted by the Tomonaga-Luttinger liquid (TLL) theory [14] and the Berezinskii-Kosterlitz-Thouless (BKT) transition [15, 16], microscopic observations of the superfluid-to-Mott insulator transition [17], quantum walk of single [18] and two atoms [19], light-cone propagation of correlations limited by the Lieb-Robinson (LR) velocity v_{LR} [20–22], the propagation of spin impurities [23] and magnons [24], and measurements of entanglement entropy [25] and antiferromagnetic fermionic correlations [26–29]. Similar microscopic observations have also been made in trapped ions [30, 31]. Further developments of in-situ imaging techniques will allow one to perform a nondestructive, real-time monitoring of the many-body dynamics [11, 32–39]. The aim of the first part of this Thesis is to show

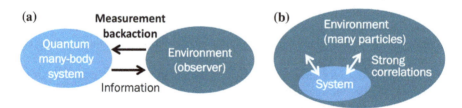

Fig. 1.1 Two distinct classes of open quantum many-body systems studied in this Thesis. **a** Quantum many-body systems under continuous observation discussed in the first part of Thesis. An external observer plays a role as an environment and extracts information about a quantum many-body system by performing continuous measurements. This causes an inevitable change of the measured system, i.e., the measurement backaction. **b** Quantum systems strongly correlated with environment discussed in the second part of Thesis. Strong correlations between a quantum system and a many-body environment invalidate the Born-Markov approximation; one has to explicitly take into account the system-environment entanglement

how one can utilize these revolutionary techniques to reveal previously unexplored effects of measurements on quantum many-body physics.

To achieve this aim, we focus on two fundamental aspects of many-body physics, namely, *quantum critical phenomena* and *out-of-equilibrium dynamics*. In Chap. 3, we analyze an effective non-Hermitian Hamiltonian that governs the nonunitary evolution under continuous observation and elucidate the influence of measurement backaction on quantum criticality. In particular, we identify two possible types of relevant non-Hermitian perturbations to one-dimensional quantum critical systems and show how they significantly alter the underlying critical phenomena and phase transitions beyond the conventional paradigms of the TLL theory and the BKT transition. We also analyze the influence of measurement backaction on quantum phase transitions in higher dimensions by focusing on the superfluid-to-Mott insulator transition in the Bose-Hubbard model. In Chap. 4, we discuss out-of-equilibrium dynamics influenced by the measurement backaction. We present three general theoretical frameworks and apply them to specific models to elucidate the underlying physics. Firstly, we introduce the notion of the full-counting dynamics that is open many-particle dynamics conditioned on certain measurement outcomes. Based on this formalism, we address propagation of correlations and that of information through many-body systems under measurement backaction. We consider an exactly solvable model and show that, by harnessing backaction due to observation of individual quanta, correlations can propagate beyond the LR bound at the cost of the probabilistic nature of quantum measurement. Secondly, we develop the framework of quantum thermalization in open many-body systems, where we consider couplings of quantum systems to generic Markovian environments permitted by quantum measurements and engineered dissipations. This gives yet another insight into why thermodynamics emerges so universally in our world. Finally, we formulate diffusive multiparticle dynamics subject to minimally destructive spatial observation. We derive a diffusive stochastic time-evolution equation to describe motions of indistinguishable particles under measurement by taking the limit of weak-spatial resolution and strong atom-light coupling. We demonstrate that the measurement backaction qualitatively alters the underlying dynamics depending on the distinguishability of particles. For all the theoretical considerations presented in this part, we propose possible experimental realizations with quantum gases.

The second part of this Thesis is devoted to studying many-body physics in quantum systems that are strongly correlated with an external environment, where the entanglement between the system and the environment plays a central role (Fig. 1.1b). In such situations, the strong system-bath correlations invalidate the Born-Markov approximation. We thus need to explicitly take into account the degrees of freedom of the environment rather than eliminating them as done in the master-equation approach. We focus on quantum impurity as the most fundamental paradigm of such a strongly correlated open quantum system. Historically, the physics of quantum impurity has originally been studied in the context of solid-state materials. The two most fundamental concepts developed there are the Kondo-singlet state [40], which is the many-body bound state formed by a localized spin impurity and the fermionic environment, and the polaron [41], which is a quasiparticle excitation formed by a mobile impurity dressed by surrounding phonon excitations. To this date, a broad class of problems that correspond to quantum impurity correlated with an external environment have been at the forefront of many subfields in physics. For instance, they have proven crucial to understanding thermodynamic properties of strongly correlated solid-state materials [42–46], decoherence [47–49] and transport phenomena [50–61] of nanodevices such as quantum dots and Nitrogen-vacancy center that are promising candidates for future quantum information technology, and lie at the heart of powerful numerical methods such as the dynamical mean-field theory [62]. The physics of a quantum impurity has recently attracted renewed interests owing to experimental developments of manipulation techniques in ultracold gases [63–69], molecular electronics [53], carbon nanotubes [54, 70, 71], and nanodevices [59, 72]. While equilibrium properties of a quantum impurity are well established, these new techniques motivate a surge of studies in an out-of-equilibrium regime which is still an area of active research with many open questions. The aim of the second part of this Thesis is to elucidate the role of strong correlations in the prototypical open quantum systems for both in- and out-of-equilibrium regimes.

In Chap. 5, we develop a versatile and efficient theoretical approach to solving generic quantum spin-impurity problems in and out of equilibrium and then apply it to reveal previously unexplored nonequilibrium dynamics. A quantum impurity in nonequilibrium regimes has been previously analyzed by a number of theoretical approaches [73–113]. In spite of the rich theoretical toolbox, analysis of the long-time dynamics remains very challenging. Previous approaches become increasingly costly at long times due to, for instance, artifacts of the logarithmic discretization in Wilson's numerical renormalization group [114] or large entanglement in the time-evolved state [115]. Another difficulty is to extend the previous approaches to generic spin-impurity models beyond the simplest Kondo models. These major challenges motivate a study to develop a new theoretical approach to quantum impurity systems. To overcome the challenges, we introduce a new canonical transformation that can completely decouple the impurity and the environmental degrees of freedom. We achieve this by employing the parity symmetry hidden in the total Hamiltonian of spin-impurity models. We combine the transformation with the fermionic Gaussian states and introduce a family of efficient variational many-body wavefunctions that can encode strong impurity-environment correlations. We benchmark our approach by demonstrating its successful application to the anisotropic [47] and two-lead [79] Kondo models and also by comparing it to results obtained by other methods such as the matrix-product state ansatz [116] and the Bethe ansatz [117–119]. We apply our method to reveal new types of nonequilibrium dynamics that are difficult to explore in the previous approaches. We propose a possible experiment in ultracold gases to test the predicted spatiotemporal dynamics by using quantum gas microscopy. We also extend our approach to a bosonic environment and apply it to study the strongly correlated system of spinful Rydberg molecules, which has been realized in state-of-the-art experiments [120–122]. In Chap. 6, we analyze out-of-equilibrium physics of yet another fundamental class of quantum impurities, that is, a mobile spinless impurity known as polaron [41]. As a concrete physical system, we study an impurity atom strongly interacting with a two-component Bose-Einstein condensate mimicking a synthetic magnetic environment. We show how the Ramsey interference technique acting on the environment can be used to directly measure novel out-of-equilibrium dynamics of magnetic polarons beyond the conventional paradigm of solid-state physics. We also discuss its concrete experimental realization in ultracold gases.

Outline

Let me summarize this introduction by providing an overview of the following Chapters. Each Chapter includes a detailed introduction to the presented results and their relations to previous works.

Chapter 2 reviews a general theory of continuous observation of quantum systems. We formulate a stochastic time-evolution equation of quantum systems subject to measurement backaction. Each stochastic realization of the dynamics is known as the quantum trajectory. We introduce several important classes of continuously monitored dynamics defined by a subensemble of quantum trajectories, which will play crucial roles in the subsequent discussions of the first part of this Thesis.

Chapter 3 addresses the influences of measurement backaction from continuous observation on quantum critical phenomena. We do this by introducing and analyzing an effective non-Hermitian Hamiltonian describing the underlying dynamics with continuous monitoring. We identify relevant non-Hermitian perturbations to the Tomonaga-Luttinger liquid and discuss anomalous one-dimensional critical phenomena triggered by the measurement backaction. We also discuss the influence of continuous observation on quantum phase transitions in higher dimensions by studying the Bose-Hubbard model as a concrete example. Possible experimental realizations with engineered dissipations and quantum gas microscopy are discussed. This Chapter is based on the publications [123, 124]:

- *Parity-time-symmetric quantum critical phenomena*, Yuto Ashida, Shunsuke Furukawa and Masahito Ueda, Nature Communications **8**, 15791 (2017).
- *Quantum critical behavior influenced by measurement backaction in ultracold gases*, Yuto Ashida, Shunsuke Furukawa and Masahito Ueda, Physical Review A **94**, 053615 (2016).

Chapter 4 analyzes out-of-equilibrium dynamics of many-particle systems under measurement backaction. We develop three general theoretical frameworks and apply them to specific models to elucidate the underlying physics. First, we introduce the notion of open many-particle dynamics conditioned on the number of quantum jump events, which we term the full-counting dynamics. We show the emergence of unique phenomena such as nonlocal and chiral propagation of correlations beyond the Lieb-Robinson bound. Second, we extend the framework of quantum thermalization to open generic many-body systems perturbed by measurements or engineered dissipations. Third, we formulate diffusive multiparticle dynamics under a minimally destructive spatial observation and demonstrate that the measurement indistinguishablity of particles can qualitatively alter the underlying transport dynamics. Possible experimental realizations in ultracold gases are discussed. This Chapter is based on the publications [125–127]:

- *Thermalization and Heating Dynamics in Open Generic Many-Body Systems*, Yuto Ashida, Keiji Saito and Masahito Ueda, Physical Review Letters **121,** 170402 (2018).
- *Full-Counting Many-Particle Dynamics: Nonlocal and Chiral Propagation of Correlations*, Yuto Ashida and Masahito Ueda, Physical Review Letters **120,** 185301 (2018).
- *Multiparticle quantum dynamics under real-time observation*, Yuto Ashida and Masahito Ueda, Physical Review A **95,** 022124 (2017).

Chapter 5 addresses in- and out-of-equilibrium physics of quantum spin impurities strongly correlated with an external many-body environment. Introducing new canonical transformations, we develop a versatile and efficient variational approach to solving generic quantum impurity problems. We apply our approach to the Kondo-type models and benchmark it by comparing to the results obtained from other numerical and analytical methods. We also reveal new types of out-of-equilibrium phenomena that are difficult to obtain in the previous approaches. We discuss possible experimental realizations using ultracold atoms to test the presented theoretical predictions. This Chapter is based on the publications [128–132]:

- *Solving Quantum Impurity Problems in and out of Equilibrium with the Variational Approach*, Yuto Ashida, Tao Shi, Mari C. Bañuls, J. Ignacio Cirac and Eugene Demler, Physical Review Letters **121,** 026805 (2018).
- *Variational principle for quantum impurity systems in and out of equilibrium: application to Kondo problems*, Yuto Ashida, Tao Shi, Mari C. Bañuls, J. Ignacio Cirac and Eugene Demler, Physical Review B **98,** 024103 (2018).
- *Exploring the anisotropic Kondo model in and out of equilibrium with alkaline-earth atoms*, Marton Kanász-Nagy, Yuto Ashida, Tao Shi, Catalin P. Moca, Tatsuhiko N. Ikeda, Simon Fölling, J. Ignacio Cirac, Gergely Zaránd and Eugene Demler, Physical Review B **97,** 155156 (2018).
- *Quantum Rydberg Central Spin Model*, Yuto Ashida, Tao Shi, Richard Schmidt, H. R. Sadeghpour, J. Ignacio Cirac and Eugene Demler, Physical Review Letters **123,** 183001 (2019).
- *Efficient variational approach to dynamics of a spatially extended bosonic Kondo model*, Yuto Ashida, Tao Shi, Richard Schmidt, H. R. Sadeghpour, J. Ignacio Cirac and Eugene Demler, Physical Review A **100,** 043618 (2019).

Chapter 6 studies out-of-equilibrium dynamics of a mobile particle strongly interacting with a magnetic environment. Analyzing the impurity atom coupled to a synthetic magnetic environemnt created by the two-component Bose-Einstein condensate, we demonstrate the emergence of unique dynamics in the strongly coupling regime that is not attainable in the conventional solid-state systems. We identify its origin as the nontrivial interplay between few- and many-body bound states. We discuss a concrete experimental realization of our consideration. This Chapter is based on the publication [133]:

- *Many-body interferometry of magnetic polaron dynamics*, Yuto Ashida, Richard Schmidt, Leticia Tarruell and Eugene Demler, Physical Review B **97,** 060302(R) (2018).

Chapter 7 concludes this Thesis with a future outlook.

References

1. Feynman RP (1960) There's plenty of room at the bottom. Caltech's Engineering and Science Magazine, Pasadena, USA
2. Bakr WS, Gillen JI, Peng A, Fölling S, Greiner M (2009) A quantum gas microscope for detecting single atoms in a Hubbard-regime optical lattice. Nature 462:74–77
3. Sherson JF, Weitenberg C, Endres M, Cheneau M, Bloch I, Kuhr S (2010) Single-atom-resolved fluorescence imaging of an atomic Mott insulator. Nature 467:68–72
4. Miranda M, Inoue R, Okuyama Y, Nakamoto A, Kozuma M (2015) Site-resolved imaging of ytterbium atoms in a two-dimensional optical lattice. Phys Rev A 91:063414
5. Cheuk LW, Nichols MA, Okan M, Gersdorf T, Ramasesh VV, Bakr WS, Lompe T, Zwierlein MW (2015) Quantum-gas microscope for fermionic atoms. Phys Rev Lett 114:193001
6. Parsons MF, Huber F, Mazurenko A, Chiu CS, Setiawan W, Wooley-Brown K, Blatt S, Greiner M (2015) Site-resolved imaging of fermionic ^{6}Li in an optical lattice. Phys Rev Lett 114:213002
7. Haller E, Hudson J, Kelly A, Cotta DA, Bruno P, Bruce GD, Kuhr S (2015) Single-atom imaging of fermions in a quantum-gas microscope. Nat Phys 11:738–742
8. Omran A, Boll M, Hilker TA, Kleinlein K, Salomon G, Bloch I, Gross C (2015) Microscopic observation of Pauli blocking in degenerate fermionic lattice gases. Phys Rev Lett 115:263001
9. Edge GJA, Anderson R, Jervis D, McKay DC, Day R, Trotzky S, Thywissen JH (2015) Imaging and addressing of individual fermionic atoms in an optical lattice. Phys Rev A 92:063406
10. Yamamoto R, Kobayashi J, Kuno T, Kato K, Takahashi Y (2016) An ytterbium quantum gas microscope with narrow-line laser cooling. New J Phys 18:023016
11. Alberti A, Robens C, Alt W, Brakhane S, Karski M, Reimann R, Widera A, Meschede D (2016) Super-resolution microscopy of single atoms in optical lattices. New J Phys 18:053010
12. Nelson KD, Li X, Weiss DS (2007) Imaging single atoms in a three-dimensional array. Nat Phys 3:556–560
13. Gericke T, Wurtz P, Reitz D, Langen T, Ott H (2008) High-resolution scanning electron microscopy of an ultracold quantum gas. Nat Phys 4:949–953
14. Hofferberth S, Lesanovsky I, Schumm T, Gritsev V, Demler E, Schmiedmayer J (2008) Probing quantum and thermal noise in an interacting many-body system. Nat Phys 4:489–495
15. Hadzibabic Z, Krüger P, Cheneau M, Battelier B, Dalibard J (2006) Berezinskii-Kosterlitz-Thouless crossover in a trapped atomic gas. Nature 441:1118–1121
16. Hung C-L, Zhang X, Gemelke N, Chin C (2011) Observation of scale invariance and universality in two-dimensional Bose gases. Nature 470:236–239
17. Bakr WS, Peng A, Tai ME, Ma R, Simon J, Gillen JI, Fölling S, Pollet L, Greiner M (2010) Probing the superfluid to Mott insulator transition at the single-atom level. Science 329:547–550
18. Weitenberg C, Endres M, Sherson JF, Cheneau M, Schauß P, Fukuhara T, Bloch I, Kuhr S (2011) Single-spin addressing in an atomic Mott insulator. Nature 471:319–324
19. Preiss PM, Ma R, Tai ME, Lukin A, Rispoli M, Zupancic P, Lahini Y, Islam R, Greiner M (2015) Strongly correlated quantum walks in optical lattices. Science 347:1229–1233
20. Cheneau M, Barmettler P, Poletti D, Endres M, Schauss P, Fukuhara T, Gross C, Bloch I, Kollath C, Kuhr S (2012) Light-cone-like spreading of correlations in a quantum many-body system. Nature 481:484–487

21. Langen T, Geiger R, Kuhnert M, Rauer B, Schmiedmayer J (2013) Local emergence of thermal correlations in an isolated quantum many-body system. Nat Phys 9:640–643
22. Fukuhara T, Hild S, Zeiher J, Schauß P, Bloch I, Endres M, Gross C (2015) Spatially resolved detection of a spin-entanglement wave in a Bose-Hubbard chain. Phys Rev Lett 115:035302
23. Fukuhara T, Kantian A, Endres M, Cheneau M, Schauß P, Hild S, Bellem D, Schollwöck U, Giamarchi T, Gross C, Bloch I, Kuhr S (2013) Quantum dynamics of a mobile spin impurity. Nat Phys 9:235–241
24. Fukuhara T, Schausz P, Endres M, Hild S, Cheneau M, Bloch I, Gross C (2013) Microscopic observation of magnon bound states and their dynamics. Nature 502:76–79
25. Islam R, Ma R, Preiss PM, Tai ME, Lukin A, Rispoli M, Greiner M (2015) Measuring entanglement entropy in a quantum many-body system. Nature 528:77–83
26. Parsons MF, Mazurenko A, Chiu CS, Ji G, Greif D, Greiner M (2016) Site-resolved measurement of the spin-correlation function in the Fermi-Hubbard model. Science 353:1253–1256
27. Boll M, Hilker TA, Salomon G, Omran A, Nespolo J, Pollet L, Bloch I, Gross C (2016) Spin- and density-resolved microscopy of antiferromagnetic correlations in fermi-hubbard chains. Science 353:1257–1260
28. Cheuk LW, Nichols MA, Lawrence KR, Okan M, Zhang H, Khatami E, Trivedi N, Paiva T, Rigol M, Zwierlein MW (2016) Observation of spatial charge and spin correlations in the 2D Fermi-Hubbard model. Science 353:1260–1264
29. Mazurenko A, Chiu CS, Ji G, Parsons MF, Kanász-Nagy M, Schmidt R, Grusdt F, Demler E, Greif D, Greiner M (2017) A cold-atom Fermi-Hubbard antiferromagnet. Nature 545:462–466
30. Jurcevic P, Lanyon BP, Hauke P, Hempel C, Zoller P, Blatt R, Roos CF (2014) Quasiparticle engineering and entanglement propagation in a quantum many-body system. Nature 511:202–205
31. Richerme P, Gong Z-X, Lee A, Senko C, Smith J, Moss-Feig M, Michalakis S, Gorshkov AV, Monroe C (2014) Non-local propagation of correlations in quantum systems with long-range interactions. Nature 511:198–201
32. Gemelke N, Zhang X, Hung C-L, Chin C (2009) In situ observation of incompressible Mott-insulating domains in ultracold atomic gases. Nature 460:995–998
33. Patil YS, Chakram S, Aycock LM, Vengalattore M (2014) Nondestructive imaging of an ultracold lattice gas. Phys Rev A 90:033422
34. Patil YS, Chakram S, Vengalattore M (2015) Measurement-induced localization of an ultra-cold lattice gas. Phys Rev Lett 115:140402
35. Preiss PM, Ma R, Tai ME, Simon J, Greiner M (2015) Quantum gas microscopy with spin, atom-number, and multilayer readout. Phys Rev A 91:041602
36. Ashida Y, Ueda M (2015) Diffraction-unlimited position measurement of ultracold atoms in an optical lattice. Phys Rev Lett 115:095301
37. Ashida Y, Ueda M (2016) Precise multi-emitter localization method for fast super-resolution imaging. Opt Lett 41:72–75
38. Mazzucchi G, Kozlowski W, Caballero-Benitez SF, Elliott TJ, Mekhov IB (2016) Quantum measurement-induced dynamics of many-body ultracold bosonic and fermionic systems in optical lattices. Phys Rev A 93:023632
39. Wigley PB, Everitt PJ, Hardman KS, Hush MR, Wei CH, Sooriyabandara MA, Manju P, Close JD, Robins NP, Kuhn CCN (2016) Non-destructive shadowgraph imaging of ultra-cold atoms. Opt Lett 41:4795–4798
40. Kondo J (1964) Resistance minimum in dilute magnetic alloys. Prog Theor Phys 32:37–49
41. Landau L, Pekar S (1948) Effective mass of the polaron. J Exp Theor Phys 423:71–74
42. Andres K, Graebner JE, Ott HR (1975) 4f-virtual-bound-state formation in Cea$_3$ at low temperatures. Phys Rev Lett 35:1779–1782
43. Hewson AC (1997) The Kondo problem to heavy fermions. Cambridge University Press, Cambridge New York
44. Löhneysen HV, Rosch A, Vojta M, Wölfle PW (2007) Fermi-liquid instabilities at magnetic quantum phase transitions. Rev Mod Phys 79, 1015

References

45. Gegenwart P, Si Q, Steglich F (2008) Quantum criticality in heavy-fermion metals. Nat Phys 4:186–197
46. Si Q, Steglich F (2010) Heavy fermions and quantum phase transitions. Science 329:1161–1166
47. Leggett AJ, Chakravarty S, Dorsey AT, Fisher MPA, Garg A, Zwerger W (1987) Dynamics of the dissipative two-state system. Rev Mod Phys 59:1–85
48. Loss D, DiVincenzo DP (1998) Quantum computation with quantum dots. Phys Rev A 57:120–126
49. Zhang W, Konstantinidis N, Al-Hassanieh KA, Dobrovitski VV (2007) Modelling decoherence in quantum spin systems. J Phys Cond Matt 19:083202
50. Glazman L, Raikh M (1988) Resonant Kondo transparency of a barrier with quasilocal impurity states. JETP Lett 47:452–455
51. Ng TK, Lee PA (1988) On-site coulomb repulsion and resonant tunneling. Phys Rev Lett 61:1768
52. Meir Y, Wingreen NS, Lee PA (1993) Low-temperature transport through a quantum dot: the Anderson model out of equilibrium. Phys Rev Lett 70:2601–2604
53. Liang W, Shores MP, Bockrath M, Long JR, Park H (2002) Kondo resonance in a single-molecule transistor. Nature 417:725–729
54. Yu LH, Natelson D (2004) The Kondo effect in C60 single-molecule transistors. Nano Lett 4:79–83
55. Goldhaber-Gordon D, Göres J, Kastner MA, Shtrikman H, Mahalu D, Meirav U (1998) From the Kondo regime to the mixed-valence regime in a single-electron transistor. Phys Rev Lett 81:5225–5228
56. Cronenwett SM, Oosterkamp TH, Kouwenhoven LP (1998) A tunable Kondo effect in quantum dots. Science 281:540–544
57. Simmel F, Blick RH, Kotthaus JP, Wegscheider W, Bichler M (1999) Anomalous Kondo effect in a quantum dot at nonzero bias. Phys Rev Lett 83:804–807
58. van der Wiel WG, Franceschi SD, Fujisawa T, Elzerman JM, Tarucha S, Kouwenhoven LP (2000) The Kondo effect in the unitary limit. Science 289:2105–2108
59. Potok RM, Rau IG, Shtrikman H, Oreg Y, Goldhaber-Gordon D (2007) Observation of the two-channel Kondo effect. Nature 446:167–171
60. Kretinin AV, Shtrikman H, Goldhaber-Gordon D, Hanl M, Weichselbaum A, von Delft J, Costi T, Mahalu D (2011) Spin-$\frac{1}{2}$ Kondo effect in an in as nanowire quantum dot: unitary limit, conductance scaling, and Zeeman splitting. Phys Rev B 84:245316
61. Kretinin AV, Shtrikman H, Mahalu D (2012) Universal line shape of the Kondo zero-bias anomaly in a quantum dot. Phys Rev B 85:201301
62. Georges A, Kotliar G, Krauth W, Rozenberg MJ (1996) Dynamical mean-field theory of strongly correlated fermion systems and the limit of infinite dimensions. Rev Mod Phys 68:13–125
63. Schirotzek A, Wu C-H, Sommer A, Zwierlein MW (2009) Observation of fermi polarons in a tunable fermi liquid of ultracold atoms. Phys Rev Lett 102:230402
64. Nascimbène S, Navon N, Jiang KJ, Tarruell L, Teichmann M, McKeever J, Chevy F, Salomon C (2009) Collective oscillations of an imbalanced fermi gas: Axial compression modes and polaron effective mass. Phys Rev Lett 103:170402
65. Koschorreck M, Pertot D, Vogt E, Frohlich B, Feld M, Köhl M (2012) Attractive and repulsive Fermi polarons in two dimensions. Nature 485:619–622
66. Kohstall C, Zaccanti M, Jag M, Trenkwalder A, Massignan P, Bruun GM, Schreck F, Grimm R (2012) Metastability and coherence of repulsive polarons in a strongly interacting Fermi mixture. Nature 485:615–618
67. Meinert F, Knap M, Kirilov E, Jag-Lauber K, Zvonarev MB, Demler E, Nägerl H-C (2017) Bloch oscillations in the absence of a lattice. Science 356:945–948
68. Hu M-G, Van de Graaff MJ, Kedar D, Corson JP, Cornell EA, Jin DS (2016) Bose polarons in the strongly interacting regime. Phys Rev Lett 117:055301

69. Jørgensen NB, Wacker L, Skalmstang KT, Parish MM, Levinsen J, Christensen RS, Bruun GM, Arlt JJ (2016) Observation of attractive and repulsive polarons in a Bose-Einstein condensate. Phys Rev Lett 117:055302
70. Makarovski A, Liu J, Finkelstein G (2007) Evolution of transport regimes in carbon nanotube quantum dots. Phys Rev Lett 99:066801
71. Chorley SJ, Galpin MR, Jayatilaka FW, Smith CG, Logan DE, Buitelaar MR (2012) Tunable Kondo physics in a carbon nanotube double quantum dot. Phys Rev Lett 109:156804
72. Iftikhar Z, Jezouin S, Anthore A, Gennser U, Parmentier FD, Cavanna A, Pierre F (2015) Two-channel Kondo effect and renormalization flow with macroscopic quantum charge states. Nature 526:233–236
73. Schmidt TL, Werner P, Mühlbacher L, Komnik A (2008) Transient dynamics of the Anderson impurity model out of equilibrium. Phys Rev B 78:235110
74. Werner P, Oka T, Millis AJ (2009) Diagrammatic Monte Carlo simulation of nonequilibrium systems. Phys Rev B 79:035320
75. Schiró M, Fabrizio M (2009) Real-time diagrammatic Monte Carlo for nonequilibrium quantum transport. Phys Rev B 79:153302
76. Werner P, Oka T, Eckstein M, Millis AJ (2010) Weak-coupling quantum Monte Carlo calculations on the Keldysh contour: theory and application to the current-voltage characteristics of the Anderson model. Phys Rev B 81:035108
77. Cohen G, Gull E, Reichman DR, Millis AJ, Rabani E (2013) Numerically exact long-time magnetization dynamics at the nonequilibrium Kondo crossover of the Anderson impurity model. Phys Rev B 87:195108
78. Nordlander P, Pustilnik M, Meir Y, Wingreen NS, Langreth DC (1999) How long does it take for the Kondo effect to develop? Phys Rev Lett 83:808–811
79. Kaminski A, Nazarov YV, Glazman LI (2000) Universality of the Kondo effect in a quantum dot out of equilibrium. Phys Rev B 62:8154–8170
80. Hackl A, Kehrein S (2008) Real time evolution in quantum many-body systems with unitary perturbation theory. Phys Rev B 78:092303
81. Keil M, Schoeller H (2001) Real-time renormalization-group analysis of the dynamics of the spin-boson model. Phys Rev B 63:180302
82. Pletyukhov M, Schuricht D, Schoeller H (2010) Relaxation versus decoherence: spin and current dynamics in the anisotropic Kondo model at finite bias and magnetic field. Phys Rev Lett 104:106801
83. Hackl A, Roosen D, Kehrein S, Hofstetter W (2009) Nonequilibrium spin dynamics in the ferromagnetic Kondo model. Phys Rev Lett 102:196601
84. Hackl A, Vojta M, Kehrein S (2009) Nonequilibrium magnetization dynamics of ferromagnetically coupled Kondo spins. Phys Rev B 80:195117
85. Tomaras C, Kehrein S (2011) Scaling approach for the time-dependent Kondo model. Europhys Lett 93:47011
86. Bera S, Nazir A, Chin AW, Baranger HU, Florens S (2014) Generalized multipolaron expansion for the spin-boson model: environmental entanglement and the biased two-state system. Phys Rev B 90:075110
87. Florens S, Snyman I (2015) Universal spatial correlations in the anisotropic Kondo screening cloud: analytical insights and numerically exact results from a coherent state expansion. Phys Rev B 92:195106
88. Blunden-Codd Z, Bera S, Bruognolo B, Linden N-O, Chin AW, von Delft J, Nazir A, Florens S (2017) Anatomy of quantum critical wave functions in dissipative impurity problems. Phys Rev B 95:085104
89. White SR, Feiguin AE (2004) Real-time evolution using the density matrix renormalization group. Phys Rev Lett 93:076401
90. Schmitteckert P (2004) Nonequilibrium electron transport using the density matrix renormalization group method. Phys Rev B 70:121302
91. Al-Hassanieh KA, Feiguin AE, Riera JA, Büsser CA, Dagotto E (2006) Adaptive time-dependent density-matrix renormalization-group technique for calculating the conductance of strongly correlated nanostructures. Phys Rev B 73:195304

References

92. Dias da Silva LGGV, Heidrich-Meisner F, Feiguin AE, Büsser CA, Martins GB, Anda EV, Dagotto E (2008) Transport properties and Kondo correlations in nanostructures: Time-dependent dmrg method applied to quantum dots coupled to Wilson chains. Phys Rev B 78:195317
93. Weichselbaum A, Verstraete F, Schollwöck U, Cirac JI, von Delft J (2009) Variational matrix-product-state approach to quantum impurity models. Phys Rev B 80:165117
94. Heidrich-Meisner F, Feiguin AE, Dagotto E (2009) Real-time simulations of nonequilibrium transport in the single-impurity Anderson model. Phys Rev B 79:235336
95. Heidrich-Meisner F, González I, Al-Hassanieh KA, Feiguin AE, Rozenberg MJ, Dagotto E (2010) Nonequilibrium electronic transport in a one-dimensional Mott insulator. Phys Rev B 82:205110
96. Nghiem HTM, Costi TA (2017) Time evolution of the Kondo resonance in response to a quench. Phys Rev Lett 119:156601
97. Anders FB, Schiller A (2005) Real-time dynamics in quantum-impurity systems: A time-dependent numerical renormalization-group approach. Phys Rev Lett 95:196801
98. Anders FB, Schiller A (2006) Spin precession and real-time dynamics in the Kondo model: time-dependent numerical renormalization-group study. Phys Rev B 74:245113
99. Anders FB, Bulla R, Vojta M (2007) Equilibrium and nonequilibrium dynamics of the sub-ohmic spin-boson model. Phys Rev Lett 98:210402
100. Anders FB (2008) Steady-state currents through nanodevices: a scattering-states numerical renormalization-group approach to open quantum systems. Phys Rev Lett 101:066804
101. Roosen D, Wegewijs MR, Hofstetter W (2008) Nonequilibrium dynamics of anisotropic large spins in the Kondo regime: time-dependent numerical renormalization group analysis. Phys Rev Lett 100:087201
102. Eckel J, Heidrich-Meisner F, Jakobs SG, Thorwart M, Pletyukhov M, Egger R (2010) Comparative study of theoretical methods for non-equilibrium quantum transport. New J Phys 12:043042
103. Lechtenberg B, Anders FB (2014) Spatial and temporal propagation of Kondo correlations. Phys Rev B 90:045117
104. Nuss M, Ganahl M, Arrigoni E, von der Linden W, Evertz HG (2015) Nonequilibrium spatiotemporal formation of the Kondo screening cloud on a lattice. Phys Rev B 91:085127
105. Dóra B, Werner MA, Moca CP (2017) Information scrambling at an impurity quantum critical point. Phys Rev B 96:155116
106. Lesage F, Saleur H, Skorik S (1996) Time correlations in 1D quantum impurity problems. Phys Rev Lett 76:3388–3391
107. Lesage F, Saleur H (1998) Boundary interaction changing operators and dynamical correlations in quantum impurity problems. Phys Rev Lett 80:4370–4373
108. Schiller A, Hershfield S (1998) Toulouse limit for the nonequilibrium Kondo impurity: Currents, noise spectra, and magnetic properties. Phys Rev B 58:14978–15010
109. Lobaskin D, Kehrein S (2005) Crossover from nonequilibrium to equilibrium behavior in the time-dependent Kondo model. Phys Rev B 71:193303
110. Vasseur R, Trinh K, Haas S, Saleur H (2013) Crossover physics in the nonequilibrium dynamics of quenched quantum impurity systems. Phys Rev Lett 110:240601
111. Ghosh S, Ribeiro P, Haque M (2014) Real-space structure of the impurity screening cloud in the resonant level model. J Stat Mech Theor Exp 2014:P04011
112. Medvedyeva M, Hoffmann A, Kehrein S (2013) Spatiotemporal buildup of the Kondo screening cloud. Phys Rev B 88:094306
113. Bolech CJ, Shah N (2016) Consistent bosonization-debosonization. ii. the two-lead Kondo problem and the fate of its nonequilibrium toulouse point. Phys Rev B 93:085441
114. Rosch A (2012) Wilson chains are not thermal reservoirs. Eur Phys J B 85:6
115. Schollwöck U (2011) The density-matrix renormalization group in the age of matrix product states. Ann Phys 326:96–192
116. Verstraete F, Murg V, Cirac JI (2008) Matrix product states, projected entangled pair states, and variational renormalization group methods for quantum spin systems. Adv Phys 57:143–224

117. Kawakami N, Okiji A (1981) Exact expression of the ground-state energy for the symmetric Anderson model. Phys Lett A 86:483–486
118. Andrei N, Furuya K, Lowenstein JH (1983) Solution of the Kondo problem. Rev Mod Phys 55:331–402
119. Schlottmann P (1989) Some exact results for dilute mixed-valent and heavy-fermion systems. Phys Rep 181:1–119
120. Bendkowsky V, Butscher B, Nipper J, Shaffer JP, Löw R, Pfau T (2009) Observation of ultralong-range Rydberg molecules. Nature 458:1005–1008
121. Gaj A, Krupp AT, Balewski JB, Löw R, Hofferberth S, Pfau T (2014) From molecular spectra to a density shift in dense Rydberg gases. Nat Commun 5:4546
122. Böttcher F, Gaj A, Westphal KM, Schlagmüller M, Kleinbach KS, Löw R, Liebisch TC, Pfau T, Hofferberth S (2016) Observation of mixed singlet-triplet Rb_2 Rydberg molecules. Phys Rev A 93:032512
123. Ashida Y, Furukawa S, Ueda M (2017) Parity-time-symmetric quantum critical phenomena. Nat Commun 8:15791
124. Ashida Y, Furukawa S, Ueda M (2016) Quantum critical behavior influenced by measurement backaction in ultracold gases. Phys Rev A 94:053615
125. Ashida Y, Ueda M (2018) Full-counting many-particle dynamics: nonlocal and chiral propagation of correlations. Phys Rev Lett 120:185301
126. Ashida Y, Saito K, Ueda M (2018) Thermalization and heating dynamics in open generic many-body systems. Phys Rev Lett 121:170402
127. Ashida Y, Ueda M (2017) Multiparticle quantum dynamics under real-time observation. Phys Rev A 95:022124
128. Ashida Y, Shi T, Bañuls MC, Cirac JI, Demler E (2018) Solving quantum impurity problems in and out of equilibrium with the variational approach. Phys Rev Lett 121:026805
129. Ashida Y, Shi T, Bañuls MC, Cirac JI, Demler E (2018) Variational principle for quantum impurity systems in and out of equilibrium: application to Kondo problems. Phys Rev B 98:024103
130. Kanász-Nagy M, Ashida Y, Shi T, Moca CP, Ikeda TN, Fölling S, Cirac JI, Zaránd G, Demler EA (2018) Exploring the anisotropic Kondo model in and out of equilibrium with alkaline-earth atoms. Phys Rev B 97:155156
131. Ashida Y, Shi T, Schmidt R, Sadeghpour HR, Cirac JI, Demler E (2019) Quantum rydberg central spin model. Phys Rev Lett. 123:183001
132. Ashida Y, Shi T, Schmidt R, Sadeghpour HR, Cirac JI, Demler E (2019) Efficient variational approach to dynamics of a spatially extended bosonic kondo model. Phys Rev A 100:043618
133. Ashida Y, Schmidt R, Tarruell L, Demler E (2018) Many-body interferometry of magnetic polaron dynamics. Phys Rev B 97:060302

Chapter 2
Continuous Observation of Quantum Systems

Abstract We review a general theory to describe the nonunitary evolution of quantum systems under measurement, which is the main subject of the first part of this Thesis. Quantum measurement theory provides us with a theoretical framework to discuss how a quantum system exhibits an unavoidable change due to a measurement process. In particular, theory of continuous measurements gives a unified description to study the nonunitary dynamics of quantum systems subject to weak and frequent repeated measurements. In Sect. 2.1, we formulate a quantum measurement process based on an indirect measurement model and review its mathematical property. In Sect. 2.2, we apply it to formulate a theory of continuous observations, in which measurements are performed continuously in time.

Keywords Quantum measurement theory · Quantum trajectory · Nonunitary dynamics · Dissipation

2.1 General Theory of the Nonunitary Evolution

2.1.1 Indirect Measurement Model

We consider a general quantum measurement process described by a physical interaction between a system S with Hilbert space $\mathcal{H}_S$ and an environment E with Hilbert space $\mathcal{H}_E$ [1–3]. The composite system $S + E$ is assumed to be isolated. We start from a separable initial quantum state of the whole system described by

$$\hat{\rho}_S(0) \otimes \hat{\rho}_E(0), \tag{2.1}$$

where $\hat{\rho}_S(0)$ is the initial state of the system S, and $\hat{\rho}_E(0)$ is the initial state of the environment E (see Fig. 2.1). After the time evolution with a unitary operator $\hat{U}$, the system and the environment interact and get entangled:

$$\hat{U}\left(\hat{\rho}_S(0) \otimes \hat{\rho}_E(0)\right)\hat{U}^\dagger. \tag{2.2}$$

© Springer Nature Singapore Pte Ltd. 2020
Y. Ashida, *Quantum Many-Body Physics in Open Systems: Measurement and Strong Correlations*, Springer Theses,
https://doi.org/10.1007/978-981-15-2580-3_2

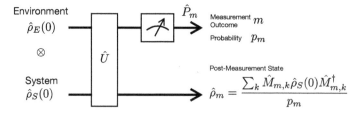

Fig. 2.1 Schematic figure illustrating the model of the indirect measurement. The total system is initially prepared in a separable state of a system S and an environment E. They get entangled after a unitary evolution $\hat{U}$. A projection measurement $\hat{P}_m$ is performed on the environment, generating a measurement outcome m with a probability p_m. Due to the measurement backaction, the system alters its state to the post-measurement state $\hat{\rho}_m$. This process is characterized by a set of measurement operators $\hat{M}_{m,k}$, where k labels the bases in the spectral decompositions of $\hat{\rho}_E(0)$ and $\hat{P}_m$ (see Eq. 2.10)

Then, we perform a measurement on the environment, which is described by the projection operator $\hat{P}_m$ on the subspace of the Hilbert space of the environment E with m being a measurement outcome. A set of projection operators satisfies the completeness condition

$$\sum_m \hat{P}_m = \hat{I}_E, \qquad (2.3)$$

where $\hat{I}_E$ is the identity operator acting on $\mathcal{H}_E$. The probability p_m of obtaining the outcome m is then given by

$$p_m = \mathrm{Tr}_{S+E}\left[\left(\hat{I}_S \otimes \hat{P}_m\right) \hat{U} \left(\hat{\rho}_S(0) \otimes \hat{\rho}_E(0)\right) \hat{U}^\dagger \left(\hat{I}_S \otimes \hat{P}_m\right)\right]. \qquad (2.4)$$

The post-measurement state $\hat{\rho}_m$ of the system S after obtaining the outcome m becomes

$$\hat{\rho}_m = \frac{1}{p_m} \mathrm{Tr}_E\left[\left(\hat{I}_S \otimes \hat{P}_m\right) \hat{U} \left(\hat{\rho}_S(0) \otimes \hat{\rho}_E(0)\right) \hat{U}^\dagger \left(\hat{I}_S \otimes \hat{P}_m\right)\right]. \qquad (2.5)$$

We introduce the spectral decompositions of $\hat{\rho}_E(0)$ and $\hat{P}_m$ as

$$\hat{\rho}_E(0) = \sum_i p_i |\psi_i\rangle_{EE}\langle\psi_i|, \qquad (2.6)$$

$$\hat{P}_m = \sum_j |\phi_{m,j}\rangle_{EE}\langle\phi_{m,j}|. \qquad (2.7)$$

Then, Eq. (2.5) can be written as

$$\hat{\rho}_m = \frac{\sum_k \hat{M}_{m,k} \hat{\rho}_S \hat{M}_{m,k}^\dagger}{p_m} \qquad (2.8)$$

2.1 General Theory of the Nonunitary Evolution

with the probability

$$p_m = \sum_k \text{Tr}_S \left[\hat{\rho}_S \hat{M}_{m,k}^\dagger \hat{M}_{m,k} \right]. \tag{2.9}$$

Here, the operators $\hat{M}_{m,k}$ are called measurement operators, or the Kraus operators [4, 5], and are defined by

$$\hat{M}_{m,k} \equiv \hat{M}_{m,(i,j)} \equiv \sqrt{p_i} \; {}_E\langle \phi_{m,j} | \hat{U} | \psi_i \rangle_E, \tag{2.10}$$

where $k \equiv (i, j)$ labels the bases in the spectral decompositions of $\hat{\rho}_E(0)$ and $\hat{P}_m$. We can show that the measurement operators satisfy the normalization condition:

$$\sum_{m,k} \hat{M}_{m,k}^\dagger \hat{M}_{m,k} = \hat{I}_S, \tag{2.11}$$

where $\hat{I}_S$ is the identity operator acting on the system. If we are interested only in the probability distribution $\{p_m\}$ of the measurement outcomes and not in the post-measurement states $\{\hat{\rho}_m\}$, we can consider a positive operator-valued measure (POVM) $\{\hat{E}_m\}$ which is defined by

$$\hat{E}_m \equiv \sum_k \hat{M}_{m,k}^\dagger \hat{M}_{m,k}. \tag{2.12}$$

The probability of obtaining the outcome m is then given by

$$p_m = \text{Tr}_S \left[\hat{\rho}_S \hat{E}_m \right]. \tag{2.13}$$

We can show the normalization condition $\sum_m p_m = 1$ from the completeness relation $\sum_m \hat{E}_m = \hat{I}_S$.

2.1.2 Mathematical Characterization of Measurement Processes

In the previous subsection, we formulate measurement processes in terms of an indirect measurement model. There, the measurement process associated with the measurement outcome m was characterized as a linear map $\mathcal{E}_m$ acting on the density matrix $\hat{\rho}_S$ (cf. Eq. 2.8):

$$\mathcal{E}_m(\hat{\rho}_S) = \sum_k \hat{M}_{m,k} \hat{\rho}_S \hat{M}_{m,k}^\dagger, \tag{2.14}$$

where operators $\hat{M}_{m,k}$ satisfy the normalization condition $\sum_{m,k} \hat{M}^\dagger_{m,k} \hat{M}_{m,k} = \hat{I}_S$. The probability p_m of an outcome m being obtained and the corresponding post-measurement state $\hat{\rho}_m$ are

$$p_m = \text{Tr}_S\left[\mathcal{E}_m(\hat{\rho}_S)\right], \tag{2.15}$$

$$\hat{\rho}_m = \frac{\mathcal{E}_m(\hat{\rho}_S)}{p_m}. \tag{2.16}$$

In fact, we can characterize a linear map $\mathcal{E} = \sum_m \mathcal{E}_m$ of measurement processes from several physical requirements in a mathematically rigorous manner.

Definition 1.1 (*positivity*) Let $\mathcal{H}$ and $\mathcal{H}'$ be the Hilbert space and $\mathcal{E}$ be a linear mapping from $\mathcal{L}(\mathcal{H})$ to $\mathcal{L}(\mathcal{H}')$. $\mathcal{E}$ is called positive if $\hat{O} \geq 0$ implies $\mathcal{E}(\hat{O}) \geq 0$ for all $\hat{O} \in \mathcal{L}(\mathcal{H})$. Note that $\hat{O} \geq 0$ means that $\langle \psi | \hat{O} | \psi \rangle \geq 0$ for all $|\psi\rangle \in \mathcal{H}$.

Definition 1.2 (*n-positivity*) Let $\mathcal{H}$ and $\mathcal{H}'$ be the Hilbert spaces and $\mathcal{E}$ be a linear mapping from $\mathcal{L}(\mathcal{H})$ to $\mathcal{L}(\mathcal{H}')$. Let $\mathcal{H}_n$ be an n-dimensional Hilbert space. $\mathcal{E}$ is called n-positive if $\mathcal{E} \otimes \mathcal{I}_n : \mathcal{L}(\mathcal{H} \otimes \mathcal{H}_n) \to \mathcal{L}(\mathcal{H}' \otimes \mathcal{H}_n)$ is positive. Here $\mathcal{I}_n$ represents the identity operator on $\mathcal{L}(\mathcal{H}_n)$.

The positivity of a mapping $\mathcal{E}$ ensures that it always maps a physical quantum state to another physical state whose density matrix must be positive. From these definitions, we introduce the notion of complete positivity as follows.

Definition 1.3 (*complete positivity*) Let $\mathcal{H}$ and $\mathcal{H}'$ be the Hilbert space and $\mathcal{E}$ be a linear mapping from $\mathcal{L}(\mathcal{H})$ to $\mathcal{L}(\mathcal{H}')$. $\mathcal{E}$ is called completely positive if and only if $\mathcal{E}$ is n-positive for all $n \in \mathbb{N}$.

We can now state the important theorem that mathematically characterizaes a general measurement process.

Theorem 1.4 (mathematical characterization of measurement processes [4–7]) *Let $\{\mathcal{E}_m\}_{m \in M}$ be a set of linear mappings acting on $\mathcal{L}(\mathcal{H}_S)$. Let us consider the following two conditions:*
(i) $\mathcal{E} \equiv \sum_m \mathcal{E}_m$ *is trace preserving, i.e.,* $\text{Tr}_S[\mathcal{E}(\hat{O})] = \text{Tr}_S[\hat{O}]$ *for all* $\hat{O} \in \mathcal{L}(\mathcal{H}_S)$.
(ii) $\mathcal{E}_m$ *is completely positive for all* $m \in M$.
If and only if $\{\mathcal{E}_m\}_{m \in M}$ satisfies the conditions (i) and (ii), $\{\mathcal{E}_m\}_{m \in M}$ can be represented by

$$\mathcal{E}_m(\hat{\rho}_S) = \text{Tr}_E\left[\left(\hat{I}_S \otimes \hat{P}_m\right) \hat{U} \left(\hat{\rho}_S \otimes \hat{\rho}_E\right) \hat{U}^\dagger \left(\hat{I}_S \otimes \hat{P}_m\right)\right] \tag{2.17}$$

for some set of $\{\mathcal{H}_E, \hat{\rho}_E, \hat{U}, \{\hat{P}_m\}_{m \in M}\}$, where $\mathcal{H}_E$ is a Hilbert space, $\hat{\rho}_E$ is a density operator of $\mathcal{H}_E$, and $\hat{U}$ is a unitary operator acting on $\mathcal{H}_S \otimes \mathcal{H}_E$, and $\{\hat{P}_m\}_{m \in M}$ is a set of projection operators acting on the subspace of $\mathcal{H}_E$ corresponding to each measurement outcome m.

2.1 General Theory of the Nonunitary Evolution

For a given completely positive and trace preserving (CPTP) mapping, Theorem 1.4. guarantees the existence of an environment and a unitary operator which can reproduce the CPTP mapping. However, the constructed environment and the unitary operator are often artificial and rather mathematical objects that cannot necessarily give us a physical insight. Therefore, to deal with a real physical system, it is highly desirable to provide a concrete formulation of a quantum measurement process from a microscopic model governing the physical interaction between the measured system and the environment [8–10]. In the next subsection, we perform such an analysis for continuously observed quantum systems, in which repeated indirect measurements are frequently performed on the system. In particular, we will derive the simple stochastic time-evolution equations governing the noisy quantum dynamics under continuous observations. These will provide the basics of our study on many-body systems under continuous observation, which is the main subject of the first part of this Thesis.

2.2 Continuous Observation of Quantum Systems

Continuous measurement theory, which is also often called as the quantum trajectory approach to open quantum systems, has originally been developed in the field of quantum optics in parallel by several groups having rather different motivations such as quantum measurement [11–14] and laser cooling of atoms [15]. It provides a clear physical picture of the nonunitary and noisy dynamics of systems under continuous observation. The quantum trajectory approach is also important as an efficient numerical method for open quantum systems since it allows one to solve the master equation by taking the ensemble average over stochastic time evolutions of pure quantum states, thus avoiding the complexity of tracking the evolutions of the full density matrix. In this section, we briefly review the quantum trajectory approach from a perspective of quantum measurement.

2.2.1 Repeated Indirect Measurements

We assume that the system repeatedly interacts with the measuring apparatus (meter) with the Hilbert space $\mathcal{H}_M$ during a short interaction time τ, and after each interaction, we perform a projection measurement on the meter and obtain a sequence of measurement outcomes (see Fig. 2.2). The meter is reset to a state $|\psi_0\rangle_M$ after each interaction such that it retains no memory about the system, ensuring that the dynamics is Markovian.

We choose an initial state as the product state between the system and the meter:

$$\hat{\rho}(0) = \hat{\rho}_S(0) \otimes \hat{P}_0, \quad \hat{P}_0 = |\psi_0\rangle_{MM}\langle\psi_0|, \tag{2.18}$$

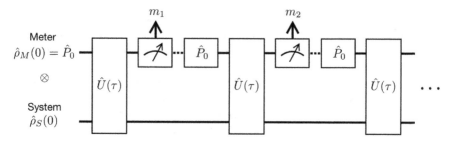

Fig. 2.2 Schematic figure illustrating the model of repeated indirect measurements. The initial state is prepared in the product state between a system and a measuring apparatus (meter). The total system evolves with a unitary operator $\hat{U}(\tau)$ during a time interval τ. A projection measurement is performed on the meter, after which the meter is reset to the initial state $\hat{P}_0 = |\psi_0\rangle_{MM}\langle\psi_0|$. The whole process is repeated and a sequence of measurement outcomes $\{m_1, m_2 \ldots\}$ is obtained

and consider the Hamiltonian

$$\hat{H} = \hat{H}_S + \hat{V}, \quad \hat{V} = \gamma \sum_{m=1}^{M} \hat{A}_m \otimes \hat{B}_m + \text{H.c.}, \qquad (2.19)$$

where $\hat{H}_S$ is a Hamiltonian governing the unitary dynamics of the system, $\gamma \in \mathbb{R}$ characterizes the strength of the coupling between the system and the meter, $\hat{A}_m$ is a linear operator acting on $\mathcal{H}_S$. We assume that $\hat{B}_m$ acts on $\mathcal{H}_M$ and changes a state of the meter into the subspace of $\mathcal{H}_M$ with a measurement outcome m:

$$\hat{P}_{m'}\hat{B}_m = \delta_{m'm}\hat{B}_m \quad (m' = 0, 1, \ldots, M; \ m = 1, 2, \ldots, M), \qquad (2.20)$$

where $\hat{P}_{m'}$ is a projection measurement on the subspace of $\mathcal{H}_M$ providing a measurement outcome m'. A set of $\hat{P}_{m'}$ satisfies the completeness condition $\sum_{m'=0}^{M} \hat{P}_{m'} = \hat{I}$.

For each indirect measurement process, there are two possibilities. The first one is to observe a change of the state of the meter, which corresponds to obtaining an outcome $m = 1, 2, \ldots, M$. The second one is to observe no change in the meter, i.e., it remains in the reset state $|\psi_0\rangle_M$ and thus gives the outcome 0. We first address the first case. The nonunitary mapping $\mathcal{E}_m$ of the system $\hat{\rho}_S$ corresponding to obtaining measurement outcome $m = 1, 2, \ldots, M$ is given by

$$\mathcal{E}_m(\hat{\rho}_S) = \text{Tr}_M\left[\hat{P}_m\hat{U}(\tau)\left(\hat{\rho}_S \otimes \hat{P}_0\right)\hat{U}^\dagger(\tau)\hat{P}_m\right], \qquad (2.21)$$

where we define $\hat{U}(t) = e^{-i\hat{H}t}$ and Tr_M denotes the trace over the meter. To simplify this, it is convenient to transform to the interaction picture:

$$\hat{U}_I(t) = \hat{U}_S^\dagger(t)\hat{U}(t) \qquad (2.22)$$

2.2 Continuous Observation of Quantum Systems

with $\hat{U}_S(t) = e^{-i\hat{H}_S t}$. The unitary operator $\hat{U}_I$ satisfies the time-evolution equation

$$\frac{\partial \hat{U}_I(t)}{\partial t} = -i\hat{V}_I(t)\hat{U}_I(t), \qquad (2.23)$$

where we denote $\hat{V}_I(t) = \hat{U}_S^\dagger(t)\hat{V}\hat{U}_S(t)$. We assume that the interaction time τ is short such that the measurement backaction is weak $\gamma\tau \ll 1$. The unitary operator $\hat{U}_I(\tau)$ can then be expanded as

$$\hat{U}_I(\tau) \simeq \hat{I} - i\tau\hat{V}_I(\tau) - \frac{\tau^2}{2}\hat{V}_I^2(\tau). \qquad (2.24)$$

The mapping $\mathcal{E}_m$ in Eq. (2.21) is approximated by

$$\mathcal{E}_m(\hat{\rho}_S) \simeq \tau^2 \hat{U}_S(\tau)\text{Tr}_M\left[\hat{P}_m V_I(\tau)\left(\hat{\rho}_S \otimes \hat{P}_0\right)\hat{V}_I^\dagger(\tau)\hat{P}_m\right]\hat{U}_S^\dagger(\tau)$$
$$\simeq \tau \hat{L}_m \hat{\rho}_S(\tau)\hat{L}_m^\dagger, \qquad (2.25)$$

where $\hat{\rho}_S(\tau) = \hat{U}_S(\tau)\hat{\rho}_S\hat{U}_S^\dagger(\tau)$ and we introduce operators $\hat{L}_m$ by

$$\hat{L}_m = \sqrt{\gamma^2\tau\langle\hat{B}_m^\dagger\hat{B}_m\rangle_0}\hat{A}_m. \qquad (2.26)$$

We here denote $\langle\cdots\rangle_0$ as an expectation value with respect to $|\psi_0\rangle_M$. The probability p_m of obtaining an outcome $m = 1, 2, \ldots, M$ is given by

$$p_m = \text{Tr}_S[\mathcal{E}_m(\hat{\rho}_S)] = \tau\text{Tr}_S[\hat{L}_m\hat{\rho}_S(\tau)\hat{L}_m^\dagger]. \qquad (2.27)$$

We next consider the second case in which one observes no change of the state of the meter and thus obtains the outcome 0. The nonunitary mapping $\mathcal{E}_0$ in this case is

$$\mathcal{E}_0(\hat{\rho}_S) = \text{Tr}_M\left[\hat{P}_0\hat{U}(\tau)\left(\hat{\rho}_S \otimes \hat{P}_0\right)\hat{U}^\dagger(\tau)\hat{P}_0\right]$$
$$\simeq \hat{U}_S(\tau)\left\{\hat{\rho}_S - \frac{\tau^2}{2}\text{Tr}_M[\hat{P}_0(\hat{V}_I^2(\tau)\hat{\rho}_S + \hat{\rho}_S\hat{V}_I^{\dagger 2}(\tau))\hat{P}_0]\right\}\hat{U}_S^\dagger(\tau)$$
$$= \hat{\rho}_S(\tau) - \frac{\tau}{2}\left\{\sum_{m=1}^M \hat{L}_m^\dagger\hat{L}_m, \hat{\rho}_S(\tau)\right\}, \qquad (2.28)$$

where $\{\hat{O}, \hat{O}'\} \equiv \hat{O}\hat{O}' + \hat{O}'\hat{O}$. The probability p_0 of no change being observed is

$$p_0 = \text{Tr}_S[\mathcal{E}_0(\hat{\rho}_S)] = 1 - \tau\sum_{m=1}^M \text{Tr}_S[\hat{L}_m\hat{\rho}_S(\tau)\hat{L}_m^\dagger]. \qquad (2.29)$$

We note that the probabilities satisfy the normalization condition $\sum_{m'=0}^M p_{m'} = 1$.

2.2.2 Quantum Jump Process

Based on the formalism introduced in the previous subsection, we can derive the simple stochastic time-evolution equation describing the dynamics of quantum systems under continuous observation. To achieve this, we study the evolution during a small time interval $dt = N\tau$ which contains many repetitions of interactions with the meter, i.e., $N \gg 1$. Considering a sufficiently weak system-meter coupling $\gamma\tau \ll 1$, the probability $p_m \sim O(\gamma^2\tau^2)$ ($m = 1, 2, \ldots, M$) of observing a change of the meter during the time interval dt can be assumed to be small. We call such a change as a "jump" process. More specifically, we assume that dt is small enough so that (i) a jump process occurs at most once in the time interval dt, i.e., the probability of more than one jump process being observed is negligible, and (ii) the time interval dt is short compared to the time scale of the dynamics of the system, resulting in the approximation

$$\hat{U}_S(dt) \simeq \hat{I} - i\hat{H}_S dt. \tag{2.30}$$

Then, the nonunitary mapping Φ_m of the system during dt is

$$\Phi_m(\hat{\rho}_S) = \sum_{i=1}^{N} \left(\mathcal{E}_0^{N-i} \circ \mathcal{E}_m \circ \mathcal{E}_0^{i-1} \right)(\hat{\rho}_S)$$

$$= \hat{L}_m \hat{\rho}_S \hat{L}_m^\dagger dt + O(dt^2) \tag{2.31}$$

for a jump process with $m = 1, 2, \ldots, M$ and

$$\Phi_0(\hat{\rho}_S) = \mathcal{E}_0^N(\hat{\rho}_S)$$

$$= \left(\hat{I}_S - i\hat{H}_{\text{eff}} dt \right) \hat{\rho}_S \left(\hat{I}_S + i\hat{H}_{\text{eff}}^\dagger dt \right) + O(dt^2) \tag{2.32}$$

for the no-jump process. Here, we introduce an effective non-Hermitian Hamiltonian defined by

$$\hat{H}_{\text{eff}} = \hat{H}_S - \frac{i}{2} \sum_{m=1}^{M} \hat{L}_m^\dagger \hat{L}_m. \tag{2.33}$$

During the time interval dt, a jump process Φ_m and the no-count process Φ_0 occur with probabilities $\text{Tr}_S[\hat{L}_m \hat{\rho}_S \hat{L}_m^\dagger] dt$ and $1 - \sum_{m=1}^{M} \text{Tr}_S[\hat{L}_m \hat{\rho}_S \hat{L}_m^\dagger] dt$, respectively.

To describe this stochastic time evolution in the form of a stochastic differential equation, it is useful to introduce measurement operators:

$$\hat{M}_0 = 1 - i\hat{H}_{\text{eff}} dt, \tag{2.34}$$

$$\hat{M}_m = \hat{L}_m \sqrt{dt} \quad (m = 1, 2, \ldots, M), \tag{2.35}$$

where $\hat{M}_0$ acts on a quantum state if no jumps are observed during the time interval $[t, t + dt]$ and $\hat{M}_m$ acts on it if a jump process m is observed. Note that the

2.2 Continuous Observation of Quantum Systems

measurement operators satisfy the normalization condition (aside from a negligible contribution of the order of $O(dt^2)$):

$$\sum_{m'=0}^{M} \hat{M}_{m'}^{\dagger} \hat{M}_{m'} = 1. \tag{2.36}$$

For the sake of simplicity, let us assume that the initial state is pure and thus the state $|\psi\rangle_S$ remains so in the course of time evolution. As outlined above, the detection of a jump process is a stochastic process, reflecting the probabilistic nature of quantum measurement. Its probability is characterized by the expectation value of the square of the measurement operator $\hat{M}_m$ with respect to a quantum state $|\psi\rangle_S$ (cf. Eq. (2.9)). In the language of stochastic processes, this is formulated as a discrete random variable dN_m known as a marked point process [16] whose mean value is given by[1]

$$E[dN_m] = \langle \hat{M}_m^{\dagger} \hat{M}_m \rangle_S = \langle \hat{L}_m^{\dagger} \hat{L}_m \rangle_S dt, \tag{2.37}$$

where $E[\cdot]$ represents the ensemble average over the stochastic process and $\langle \cdots \rangle_S$ denotes an expectation value with respect to a quantum state $|\psi\rangle_S$ of the system. These random variables satisfy the following stochastic calculus:

$$dN_m dN_n = \delta_{mn} dN_m. \tag{2.38}$$

Using these notations, the stochastic change of a quantum state $|\psi\rangle$ in the time interval $[t, t+dt]$ can be obtained as

$$|\psi\rangle_S \to |\psi\rangle_S + d|\psi\rangle_S = \left(1 - \sum_{m=1}^{M} E[dN_m]\right) \frac{\hat{M}_0 |\psi\rangle_S}{\sqrt{\langle \hat{M}_0^{\dagger} \hat{M}_0 \rangle_S}} + \sum_{m=1}^{M} dN_m \frac{\hat{M}_m |\psi\rangle_S}{\sqrt{\langle \hat{M}_m^{\dagger} \hat{M}_m \rangle_S}}. \tag{2.39}$$

Physically, the first term on the right-hand side describes the no-count process occurring with the probability $1 - \sum_{m=1}^{M} E[dN_m]$ and the second term describes the detection of a jump process m occurring with a probability $E[dN_m]$. We note that the denominator in each term is introduced to ensure the normalization of the state vector. From Eqs. (2.34) and (2.35), we can rewrite Eq. (2.39) as

$$d|\psi\rangle_S = \left(1 - i\hat{H}_{\text{eff}} + \frac{1}{2} \sum_{m=1}^{M} \langle \hat{L}_m^{\dagger} \hat{L}_m \rangle_S\right) dt |\psi\rangle_S + \sum_{m=1}^{M} \left(\frac{\hat{L}_m |\psi\rangle_S}{\sqrt{\langle \hat{L}_m^{\dagger} \hat{L}_m \rangle_S}} - |\psi\rangle_S\right) dN_m. \tag{2.40}$$

The first term on the right-hand side describes the non-Hermitian time evolution, in which the factor $\sum_{m=1}^{M} \langle \hat{L}_m^{\dagger} \hat{L}_m \rangle_S / 2$ keeps the normalization of the state vector. In the second term, when the jump event m is detected, an operator $\hat{L}_m$ acts on the

---

[1] We remark that dN_m is not a simple Poisson process as its intensity depends on a stochastic vector $|\psi\rangle_S$.

quantum state and causes its discontinuous change ("jump"). A specific realization of this stochastic time evolution is referred to as the quantum trajectory.

Taking the ensemble average over all possible trajectories, one reproduces the Lindblad master equation [17, 18]. To see this explicitly, let us rewrite Eq. (2.40) using the density matrix $\hat{\rho}_S = |\psi\rangle_{SS}\langle\psi|$:

$$d\hat{\rho}_S = -i\left(\hat{H}_{\text{eff}}\hat{\rho}_S - \hat{\rho}_S\hat{H}_{\text{eff}}^\dagger\right)dt + \sum_{m=1}^{M}\langle\hat{L}_m^\dagger\hat{L}_m\rangle_S\hat{\rho}_S dt + \sum_{m=1}^{M}\left(\frac{\hat{L}_m\hat{\rho}_S\hat{L}_m^\dagger}{\langle\hat{L}_m^\dagger\hat{L}_m\rangle_S} - \hat{\rho}_S\right)dN_m, \quad (2.41)$$

where we take the leading order of $O(dt)$ and use the stochastic calculus (2.38). We note that this equation remains valid for a generic density matrix $\hat{\rho}_S$ that is not necessarily pure. Introducing the ensemble-averaged density matrix $E[\hat{\rho}_S]$ and taking the average of Eq. (2.41), one can show that the density matrix obeys the Lindblad master equation [17, 18]:

$$\frac{dE[\hat{\rho}_S]}{dt} = -i\left(\hat{H}_{\text{eff}}E[\hat{\rho}_S] - E[\hat{\rho}_S]\hat{H}_{\text{eff}}^\dagger\right) + \sum_{m=1}^{M}\hat{L}_m E[\hat{\rho}_S]\hat{L}_m^\dagger. \quad (2.42)$$

Distinct subclasses of the quantum trajectory dynamics

Based on the quantum trajectory dynamics formulated above, we introduce several distinct classes of continuously monitored dynamics, which will play crucial roles in the following Chapters.

Single trajectory dynamics realized with the complete information
First, if an observer who can measure the meter has an ability to access the complete information about measurement outcomes, i.e., all the time records and types of quantum jumps, the dynamics is described by a single realization of the quantum trajectory dynamics:

$$\hat{\rho}_S(t; \{t_1, t_2, \ldots, t_n\}, \{m_1, m_2, \ldots, m_n\}) \propto |\psi_{\text{traj}}\rangle_{SS}\langle\psi_{\text{traj}}|,$$

$$|\psi_{\text{traj}}\rangle_S = \prod_{k=1}^{n}\left[\hat{\mathcal{U}}_{\text{eff}}(\Delta t_k)\hat{L}_{m_k}\right]\hat{\mathcal{U}}_{\text{eff}}(t_1)|\psi_0\rangle_S, \quad (2.43)$$

where $0 < t_1 < t_2 < \cdots < t_n < t$ are occurrence times of jumps whose types are $\{m_1, m_2, \ldots, m_n\}$. We denote a time difference as $\Delta t_k = t_{k+1} - t_k$ with $t_{n+1} \equiv t$, an effective non-Hermitian time evolution as $\hat{\mathcal{U}}_{\text{eff}}(t) = e^{-i\hat{H}_{\text{eff}}t}$, and $|\psi_0\rangle_S$ as an initial state of the system. In Chap. 4, we will address thermalization and heating dynamics in such single-trajectory many-body dynamics [19].

Full-counting dynamics realized with partial information
Second, even if an observer cannot have access to the complete information, she/he may still have an ability to access incomplete information about measurement outcomes. One possible example is a capability to measure the total number of quantum jumps occurred during a certain time interval, but not their types and occurrence

2.2 Continuous Observation of Quantum Systems

times. We can then introduce the density matrix conditioned on the number n of quantum jumps that have occurred during the time interval $[0, t]$ as

$$\hat{\rho}^{(n)}(t) \propto \sum_{\alpha \in \mathcal{D}_n} |\psi_\alpha\rangle_{SS}\langle\psi_\alpha|$$

$$= \sum_{\{m_k\}_{k=1}^n} \int_0^t dt_n \cdots \int_0^{t_2} dt_1 \prod_{k=1}^n \left[\hat{\mathcal{U}}_{\text{eff}}(\Delta t_k)\hat{L}_{m_k}\right] \hat{\mathcal{U}}_{\text{eff}}(t_1)\hat{\rho}_S(0)\hat{\mathcal{U}}_{\text{eff}}^\dagger(t_1) \prod_{k=1}^n \left[\hat{L}_{m_k}^\dagger \hat{\mathcal{U}}_{\text{eff}}^\dagger(\Delta t_k)\right],$$

(2.44)

where the ensemble average is taken over the subspace $\mathcal{D}_n$ spanned by all the possible trajectories having n quantum jumps during $[0, t]$. In other words, this can be interpreted as coarse-grained dynamics of pure quantum trajectories (2.43), where the number of jumps is known while the information about times and types of jumps have been lost. We term this class of the dynamics as the full-counting dynamics [20]. In Chap. 4, we will investigate the emergence of distinct out-of-equilibrium many-particle phenomena in the full-counting dynamics.

Non-Hermitian evolution realized with no jump events
The no-jump process $\hat{\rho}^{(0)}(t)$ (which is $n = 0$ case in Eq. (2.44)) is the simplest example of the full-counting dynamics. Its time evolution is described by an effective non-Hermitian Hamiltonian:

$$\hat{\rho}^{(0)}(t) \propto \hat{\mathcal{U}}_{\text{eff}}(t)\hat{\rho}_S(0)\hat{\mathcal{U}}_{\text{eff}}^\dagger(t). \qquad (2.45)$$

While occurrence of quantum jumps will be increasingly likely at long times in general, the non-Hermitian evolution (2.45) still provides physical insights into quasi-equilibrium properties in a short-time regime, where contributions from quantum jumps are not significant or can be eliminated by employing postselections. In Chap. 3, we will discuss unconventional quantum critical behavior in such a non-Hermitian many-body system influenced by measurement backaction from continuous observation [21, 22].

Remark on cases that reduce to classical non-Hermitian dynamics

We remark on two cases in which the continuously monitored non-Hermitian dynamics (2.45) and the dissipative dynamics described by the master equation (2.42) become equivalent. The first case is when a system is subject to a one-body loss (i.e., a jump operator is proportional to the annihilation operator of a particle) and its quantum state is described by a coherent state (i.e., an eigenstate of the annihilation operator). In this case, the jump term, which is the last term in Eq. (2.41), does not alter the quantum state and thus can be neglected upon the normalization of the state. Physically, this case is relevant to a (mean-field) Bose-Einstein condensate subject to a one-body loss process [23–25], and also can be considered as a microscopic justification of the phenomenological non-Hermitian description employed in classical optics [26, 27]. The second case is when a system is subject to a one-body loss and it contains just a single particle. In this case, the jump term becomes trivial since it reduces to the particle-vacuum state and thus can be neglected. As neither quantum entanglement nor particle correlations play roles in these two cases, physical phenomena in such systems can, in principle, be also found in classical systems. In this sense, we denote these systems as "classical" systems throughout this Thesis. Thanks to recent experimental developments especially in classical optics, our understanding of such (one-body) classical non-Hermitian phenomena have been significantly advanced in this decade [26, 27].

Our aim is different from this direction. We will address genuine quantum systems, in which continuously monitored dynamics (2.40) and dissipative dynamics (2.42) are intrinsically distinct. In particular, we will study new aspects of quantum many-body phenomena, where genuine quantum effects such as measurement backaction, many-body correlations and entanglement play important roles. This is the main theme of the first part of this Thesis.

2.2.3 Diffusive Limit

In the previous subsection, we consider a measurement process in which an operator $\hat{L}_m$ induces a discontinuous change of a quantum state. Here, we discuss yet another type of continuous measurement associated with a diffusive stochastic process. To do so, we assume that measurement backaction induced by an operator $\hat{L}_m$ is weak in the sense that it satisfies

$$\hat{L}_m = \sqrt{\Gamma}(\hat{I}_S - l\hat{a}_m), \tag{2.46}$$

where Γ characterizes the detection rate of jump events and $l \ll 1$ is a small dimensionless parameter. We also assume that detections of jump events are frequent so that an expectation value δN_m of the number of jump m observed during a time interval δt is sufficiently large. From the central-limit theorem, it can then be approximated as

$$\begin{aligned}\delta N_m &\simeq \langle \hat{L}_m^\dagger \hat{L}_m \rangle_S \delta t + \sqrt{\langle \hat{L}_m^\dagger \hat{L}_m \rangle_S} \delta W_m \\ &\simeq \Gamma(\hat{I}_S - l\langle \hat{a}_m + \hat{a}_m^\dagger \rangle_S)\delta t + \sqrt{\Gamma}\left(\hat{I}_S - \frac{l}{2}\langle \hat{a}_m + \hat{a}_m^\dagger \rangle_S\right)\delta W_m. \end{aligned} \tag{2.47}$$

Here, $\delta W_m \in \mathcal{N}(0, \delta t)$ is a random variable obeying the normal distribution with the zero mean and the variance δt. We then take the limit of weak $l \to 0$ and frequent $\Gamma \to \infty$ measurement, while Γl^2 is kept constant. This limit is also known as the diffusive measurement limit [1, 3]. The resulting change of a quantum state $\hat{\rho}_S \to \hat{\rho}_S + \delta\hat{\rho}_S$ during the interval δt can be obtained, at the leading order, by (cf. Eq. (2.41))

2.2 Continuous Observation of Quantum Systems

$$\delta\hat{\rho}_S = \left[-i[\hat{H}_S, \hat{\rho}_S] - \frac{\Gamma l^2}{2} \sum_{m=1}^{M} \left(\{\hat{a}_m^\dagger \hat{a}_m, \hat{\rho}_S\} - 2\hat{a}_m \hat{\rho}_S \hat{a}_m^\dagger \right) \right] \delta t$$

$$+ \sqrt{\Gamma l^2} \sum_{m=1}^{M} \left[(\hat{a}_m - \langle \hat{a}_m \rangle_S) \hat{\rho}_S + \hat{\rho}_S (\hat{a}_m^\dagger - \langle \hat{a}_m^\dagger \rangle_S) \right] \delta W_m. \qquad (2.48)$$

Finally, taking the limit $\delta t \to 0$, we obtain the following diffusive stochastic differential equation:

$$d\hat{\rho}_S = \left[-i[\hat{H}_S, \hat{\rho}_S] - \frac{1}{2} \sum_{m=1}^{M} \left(\{\hat{l}_m^\dagger \hat{l}_m, \hat{\rho}_S\} - 2\hat{l}_m \hat{\rho}_S \hat{l}_m^\dagger \right) \right] dt$$

$$+ \sum_{m=1}^{M} \left[\left(\hat{l}_m - \langle \hat{l}_m \rangle_S \right) \hat{\rho}_S + \hat{\rho}_S \left(\hat{l}_m^\dagger - \langle \hat{l}_m^\dagger \rangle_S \right) \right] dW_m, \qquad (2.49)$$

where we introduce operators $\hat{l}_m = \sqrt{\Gamma l^2} \hat{a}_m$ and the Wiener stochastic processes satisfying

$$E[dW_m] = 0, \quad dW_m dW_n = \delta_{mn} dt. \qquad (2.50)$$

As for a pure state, the diffusive stochastic time-evolution equation (2.49) can be also written in terms of the wavefunction $|\psi\rangle_S$ as follows:

$$d|\psi\rangle_S = \left[\hat{I}_S - i\hat{H}_S - \frac{1}{2} \sum_{m=1}^{M} \left(\hat{l}_m^\dagger \hat{l}_m - \hat{l}_m \langle \hat{l}_m + \hat{l}_m^\dagger \rangle_S + \frac{1}{4} \langle \hat{l}_m + \hat{l}_m^\dagger \rangle_S^2 \right) \right] dt |\psi\rangle_S$$

$$+ \sum_{m=1}^{M} \left(\hat{l}_m - \frac{1}{2} \langle \hat{l}_m + \hat{l}_m^\dagger \rangle_S \right) dW_m |\psi\rangle_S. \qquad (2.51)$$

In Chap. 4, we will derive the diffusive stochastic Schrödinger equations appropriate for describing many-particle systems under a minimally destructive spatial observation [28].

2.2.4 Physical Example: Site-Resolved Measurement of Atoms

We discuss a specific physical example for a site-resolved position measurement of atoms via dispersive light scattering [29]. The microscopic light-atom interaction Hamiltonian is given by

$$\hat{V}_{\text{mic}} = \hat{V}_{\text{mic}}^{(-)} + \hat{V}_{\text{mic}}^{(+)},$$
$$\hat{V}_{\text{mic}}^{(-)} = -\int d\mathbf{r}\, \mathbf{d} \cdot \hat{\mathbf{E}}^{(-)}(\mathbf{r}) \hat{\Psi}_g^\dagger(\mathbf{r}) \hat{\Psi}_e(\mathbf{r}), \quad \hat{V}_{\text{mic}}^{(+)} = \hat{V}_{\text{mic}}^{(-)\dagger}, \qquad (2.52)$$

where $\hat{\Psi}_{g,e}(\mathbf{r})$ are the atomic field operators of the excited and ground states, the dipole moment is defined by $\mathbf{d} \equiv \langle e|\hat{\mathbf{d}}|g\rangle$ with $\hat{\mathbf{d}}$ being the dipole moment operator of the transition. The electric field is given by

$$\hat{\mathbf{E}}(\mathbf{r}) = \sum_{\mathbf{k}',\sigma} \sqrt{\frac{\hbar\omega_{k'}}{2\epsilon_0 V}} \mathbf{e}_{\mathbf{k}',\sigma} \left(\hat{a}_{\mathbf{k}',\sigma}^\dagger e^{-i\mathbf{k}'\cdot\mathbf{r}} + \text{H.c.} \right)$$
$$\equiv \hat{\mathbf{E}}^{(-)}(\mathbf{r}) + \hat{\mathbf{E}}^{(+)}(\mathbf{r}), \qquad (2.53)$$

where $\hat{a}_{\mathbf{k},\sigma}^\dagger$ ($\hat{a}_{\mathbf{k},\sigma}$) is the creation (annihilation) operator of a photon with wave vector $\mathbf{k}$ and polarization σ, and we introduce the positive and negative frequency components of the electric field.

We consider dispersive light scattering of atoms by using an off-resonant coherent probe light whose frequency and amplitude are ω_L and $\mathbf{E}_P$. We introduce the rotating coordinate field $\hat{\tilde{\Psi}}_e = e^{i\omega_L t} \hat{\Psi}_e$, resulting in the following Heisenberg equation of motion for the excited-atomic field

$$\dot{\hat{\tilde{\Psi}}}_e = i\Delta \hat{\tilde{\Psi}}_e + \frac{i}{\hbar}\mathbf{d}^* \cdot \mathbf{E}_P \hat{\Psi}_g, \qquad (2.54)$$

where $\Delta = \omega_L - \omega_0$ is the detuning of the probe light from the atomic transition frequency ω_0. Since we assume an off-resonant condition ($|\Delta| \gg |\mathbf{d}^* \cdot \mathbf{E}_P/\hbar|$), we can perform the adiabatic elimination of the excited state $\dot{\hat{\tilde{\Psi}}}_e \simeq 0$:

$$\hat{\tilde{\Psi}}_e \simeq -\frac{\mathbf{d}^* \cdot \mathbf{E}_P}{\hbar\Delta} \hat{\Psi}_g. \qquad (2.55)$$

This procedure is equivalent to taking into account only the leading term of $\mathbf{d}^* \cdot \mathbf{E}_P/(\hbar\Delta)$. We can rewrite the microscopic Hamiltonian (2.52) as

$$\hat{V}_{\text{mic}}^{(-)} \simeq \frac{\mathbf{d}^* \cdot \mathbf{E}_P}{\hbar\Delta} \int d\mathbf{r}\, \mathbf{d} \cdot \hat{\mathbf{E}}^{(-)}(\mathbf{r}) \hat{\Psi}_g^\dagger(\mathbf{r}) \hat{\Psi}_g(\mathbf{r}). \qquad (2.56)$$

For the sake of simplicity, we consider atoms tightly trapped in a one-dimensional lattice as realized in the setup of quantum gas microscopy. Specifically, we adopt the following tight-binding approximation:

$$\hat{\Psi}_g(\mathbf{r}) = \Phi(y,z) \sum_m w(x - md) \hat{b}_m, \qquad (2.57)$$

2.2 Continuous Observation of Quantum Systems

where $\Phi(y,z) = w_y(y)w_z(z)$ is the wave function confined in the transverse direction, $w(x)$ is the Wannier function of the lowest Bloch band centered at $x = 0$, and $\hat{b}_m$ is the annihilation operator of the atom at site m. The resulting effective microscopic Hamiltonian is

$$\hat{V}_{\text{mic}} \simeq \gamma \sum_m \left(\hat{E}_m^{(-)} + \hat{E}_m^{(+)} \right) \hat{n}_m, \tag{2.58}$$

where $\hat{n}_m = \hat{b}_m^\dagger \hat{b}_m$ is an occupation-number operator of atoms at site m and we assume that the probe light is polarized into the direction of the dipole moment, leading to $\gamma = |d|^2 E_P/(\hbar\Delta)$. The operators $\hat{E}_m^{(-)}$ and $\hat{E}_m^{(+)}$ correspond to the annihilation and creation operators of photon modes localized around site m. This Hamiltonian (2.58) can be considered as the system-meter interaction Hamiltonian (2.19) in the indirect measurement model.

We now imagine that photodetectors are prepared at each site. A photodetection at each site corresponds to a destructive measurement of photons scattered by an atom at that site and is described by an operation acting on the photon field

$$|\text{vac}\rangle\langle 1_m| \equiv |\text{vac}\rangle\langle\text{vac}|\hat{E}_m^{(+)}, \tag{2.59}$$

where $|\text{vac}\rangle$ denotes the photon vacuum and m is a detected lattice site. In the language of the indirect measurement model, this process can be interpreted as the combination of a projection measurement and the subsequent reset to the initial state of the meter (i.e., the photon vacuum). Finally, a jump operator acting on the Hilbert space of atoms is given by

$$\hat{L}_m \propto \sqrt{\langle \hat{E}_m^{(+)} \hat{E}_m^{(-)} \rangle_{\text{vac}}} \hat{n}_m, \tag{2.60}$$

which should be compared with Eq. (2.26). While, in a more realistic treatment directly relevant to quantum gas microscopy, we have to take into account effects of diffraction of scattered fields (cf. Ref. [29]), the conclusions are essentially the same as presented here.

References

1. Breuer H-P, Petruccione F (2002) The theory of open quantum systems. Oxford University Press, Oxford, England
2. Nielsen M, Chuang IL (2010) Quantum computation and quantum information. Cambridge University Press, Cambridge, England
3. Wiseman H, Milburn G (2010) Quantum measurement and control. Cambridge University Press, Cambridge, England
4. Hellwig K-E, Kraus K (1970) Operations and measurements II. Comm Math Phys 16:142–147
5. Kraus K (1971) General state changes in quantum theory. Ann Phys 64:311–335
6. Stinespring WF (1955) Positive functions on C*-algebras. Proc Am Math Soc 6:211–216
7. Ozawa M (1984) Quantum measuring processes of continuous observables. J Math Phys 25:79–87

8. Imoto N, Ueda M, Ogawa T (1990) Microscopic theory of the continuous measurement of photon number. Phys Rev A 41:4127–4130
9. Ueda M, Imoto N, Nagaoka H, Ogawa T (1992) Continuous quantum-nondemolition measurement of photon number. Phys Rev A 46:2859–2869
10. Miyashita S, Ezaki H, Hanamura E (1998) Stochastic model of nonclassical light emission from a microcavity. Phys Rev A 57:2046–2055
11. Ueda M (1989) Probability-density-functional description of quantum photodetection processes. Quantum Opt J Eur Opt Soc Part B 1:131
12. Ueda M (1990) Nonequilibrium open-system theory for continuous photodetection processes: a probability-density-functional description. Phys Rev A 41:3875–3890
13. Dum R, Zoller P, Ritsch H (1992) Monte Carlo simulation of the atomic master equation for spontaneous emission. Phys Rev A 45:4879–4887
14. Carmichael H (1993) An open system approach to quantum optics. Springer, Berlin
15. Dalibard J, Castin Y, Mølmer K (1992) Wave-function approach to dissipative processes in quantum optics. Phys Rev Lett 68:580–583
16. Barchielli A, Gregoratti M (2012) Quantum measurements in continuous time, non-Markovian evolutions and feedback. Philos Trans R Soc A 370:5364–5385
17. Gorini V, Kossakowski A, Sudarshan ECG (1976) Completely positive dynamical semigroups of n level systems. J Math Phys 17:821–825
18. Lindblad G (1976) On the generators of quantum dynamical semigroups. Commun Math Phys 48:119–130
19. Ashida Y, Saito K, Ueda M (2018) Thermalization and heating dynamics in open generic many-body systems. Phys Rev Lett 121:170402
20. Ashida Y, Ueda M (2018) Full-counting many-particle dynamics: nonlocal and chiral propagation of correlations. Phys Rev Lett 120:185301
21. Ashida Y, Furukawa S, Ueda M (2017) Parity-time-symmetric quantum critical phenomena. Nat Commun 8:15791
22. Ashida Y, Furukawa S, Ueda M (2016) Quantum critical behavior influenced by measurement backaction in ultracold gases. Phys Rev A 94:053615
23. Würtz P, Langen T, Gericke T, Koglbauer A, Ott H (2009) Experimental demonstration of single-site addressability in a two-dimensional optical lattice. Phys Rev Lett 103:080404
24. Barontini G, Labouvie R, Stubenrauch F, Vogler A, Guarrera V, Ott H (2013) Controlling the dynamics of an open many-body quantum system with localized dissipation. Phys Rev Lett 110:035302
25. Labouvie R, Santra B, Heun S, Ott H (2016) Bistability in a driven-dissipative superfluid. Phys Rev Lett 116:235302
26. El-Ganainy R, Makris KG, Khajavikhan M, Musslimani ZH, Rotter S, Christodoulides DN (2018) Non-Hermitian physics and PT symmetry. Nat Phys 14:11
27. Zhao H, Feng L (2018) Parity-time symmetric photonics. Natl Sci Rev 5:183–199
28. Ashida Y, Ueda M (2017) Multiparticle quantum dynamics under real-time observation. Phys Rev A 95:022124
29. Ashida Y, Ueda M (2015) Diffraction-unlimited position measurement of ultracold atoms in an optical lattice. Phys Rev Lett 115:095301

Chapter 3
Quantum Critical Phenomena

Abstract Quantum critical phenomena originate from collective behavior of strongly correlated particles and lie at the heart of universal low-energy properties in many-body systems. The strong correlation between quantum particles is particularly prominent in a low-dimensional system. In the first part of this Chapter, we identify what types of measurements are relevant to one-dimensional low-energy properties and address how they qualitatively alter the underlying quantum critical behavior. In the second part, we study how the measurement backaction influences on quantum phase transitions in higher dimensions by focusing on the Bose-Hubbard model. For all the theoretical considerations, we discuss possible experimental implementations in ultracold atomic gases.

Keywords Quantum critical phenomena · Non-Hermitian systems · Parity-time symmetry · Tomonaga-Luttinger liquid · Sine-Gordon model

3.1 Introduction

3.1.1 Motivation: Measurement Backaction on Strongly Correlated Systems

Quantum systems under continuous observations inevitably undergo nonunitary evolutions due to measurement backaction, as we have reviewed in Chap. 2. In this Chapter, we explore effects of measurements on quantum critical phenomena (see Fig. 3.1). This is mainly motivated by recent experimental realizations of quantum gas microscopy [1–9], which enable one to measure many-body systems at the single-atom level. With the technique at such an ultimate resolution, the measurement backaction is expected to be significant. A natural question arising here is how the underlying quantum critical behavior will be modified under measurement and whether or not the concept of the universality need be extended. The frontier of ultracold atomic experiments motivates us to reveal influences of measurement backaction

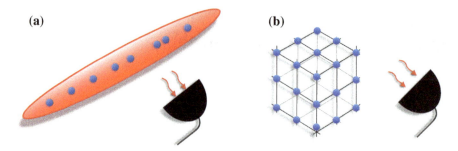

Fig. 3.1 Schematic illustrations of strongly correlated systems under continuous observations. **a** While the universal behavior of isolated one-dimensional quantum many-body systems can be described by the Tomonaga-Luttinger liquid, we show that measurement backaction acted by an external observer can qualitatively alter their low-energy critical behavior and phase diagrams. **b** Ultracold atoms trapped in an optical lattice are described by the Bose-Hubbard model and exhibit a quantum phase transition known as the superfluid-to-Mott insulator transition. We show that measurement backaction can shift the transition point so that the Mott phase is expanded. Reproduced from Fig. 1 of Ref. [10]. Copyright © 2016 by the American Physical Society

on strongly correlated many-body systems beyond the standard framework of quantum many-body theory.

We address this question by studying the simplest quantum trajectory under continuous measurement. With continuous monitoring, there are two possibilities at each moment depending on whether or not one observes a quantum jump (see discussions in the previous Chapter). The occurrences of quantum jumps typically lead to detrimental effects on quantum critical behavior since they induce heating or loss of particles. In the other possibility of no jumps being observed, the underlying low-energy behavior can sustain quantum criticality while there still exist nontrivial effects from continuous monitoring, as they manifest themselves as the non-Hermiticity of the effective Hamiltonian governing the nonunitary evolution [11–13]. We study the influences of measurement backaction on quantum criticality by analyzing this effective non-Hermitian Hamiltonian. While occurrences of quantum jumps will be increasingly likely at long times, the analysis can give physical insights into how a weak continuous observation affects the underlying quantum critical behavior especially in a short-time regime. There, destructive effects from quantum jumps are not significant or can be eliminated by employing postselections [14–17].

Before going into the detailed analyses, let us outline the main idea here. We consider a quantum many-body system whose Hamiltonian $\hat{H}$ exhibits quantum critical behavior, and assume that the system is under continuous observation characterized by a set of jump operators $\{\hat{L}_i\}$. We then analyze the following effective non-Hermitian Hamiltonian:

$$\hat{H}_{\text{eff}} = \hat{H} - \frac{i\hat{\Gamma}}{2}, \tag{3.1}$$

3.1 Introduction

$$\hat{\Gamma} = \gamma \sum_i \hat{L}_i^\dagger \hat{L}_i, \quad (3.2)$$

where $\hat{\Gamma}$ denotes the backaction from continuous observation and γ characterizes its strength. The effective Hamiltonian (3.1) in general has complex eigenvalues. We interpret that their real parts describe effective values of energies while their imaginary parts provide rates at which the corresponding eigenstates decay out of the Hilbert space under consideration. If the original Hamiltonian commutes with all the jump operators, i.e., $[\hat{H}, \hat{L}_i] = 0$ for $\forall i$, the imaginary term in $\hat{H}_{\text{eff}}$ merely shifts the imaginary part of the eigenvalues; the real parts of the eigenspectrum and the corresponding eigenstates do not change. Thus, no qualitative changes in the critical behavior will happen. In contrast, if $\hat{H}$ does not commute with some of the jump operators, i.e., $\exists i$ such that $[\hat{H}, \hat{L}_i] \neq 0$, there will emerge nontrivial effects from measurements, as the measurement now alters the eigenstates and the eigenenergies of $\hat{H}_{\text{eff}}$. Throughout this Chapter, we focus on an effective ground state defined by the state having the lowest real part of the spectrum. We find that the effective ground state is particularly important in the continuously monitored dynamics since it also has the minimal (or almost minimal) decay rate and thus survives longest under the non-Hermitian evolution. We stress that the dynamics of interest here is different from dissipative, unconditional dynamics obeying a master equation, in which the dissipative process tends to monotonically destroy quantum criticality, often ending up in trivial states such as an infinite-temperature or the vacuum state. In particular, it has recently been pointed out that such steady states have physical properties analogous to high-temperature states [18, 19] or classical thermal systems [20–22]. In contrast, by analyzing the dynamics conditioned on measurement outcomes, we here show that measurement backaction can give rise to qualitatively new quantum critical phenomena beyond the conventional paradigm of many-body physics [10, 23].

From a broader perspective, a theoretical approach utilizing the non-Hermitian effective Hamiltonian has proved instrumental in a number of fields such as classical and quantum atomic, molecular and optical (AMO) physics [24–39], nuclear physics [40, 41], open quantum systems [11, 12, 38, 39, 42, 43], condensed matter physics [44–48], mesoscopic systems [49–51], biological network theory [52–54] and quantum chemistry [55]. Studies of phase transitions in non-Hermitian systems can be traced to the pioneering work by Fisher [56] in 1978 on an exotic critical point in the non-Hermitian Ising model, which is now known as the Yang-Lee edge singularity. Also, in a wide class of non-Hermitian Hamiltonians satisfying the parity-time (PT) symmetry, the real-to-complex spectral phase transition has been found [57–59]. The transition is typically accompanied by an exceptional point [60] in the discrete spectrum or the spectral singularity [61] in the continuum spectrum. While these properties also hold for a certain class of antilinear symmetries [62], one advantage of the PT symmetry is that it enables one to implement this symmetry by spatially engineering gain-loss structures, leading to a rich interplay between theory and experiment in classical optics [37, 63–66], superconductors [67], atomic physics [68], and optomechanics [69]. However, the previous works mainly concern the classical (one-body) aspects and the role of strong correlations has yet to be clarified. In this context, our studies fill this gap and create a bridge between the fields of non-Hermitian physics and many-body physics.

In this Chapter, we proceed as follows. In the next two subsections, we review the universal properties of one-dimensional many-body systems and identify what types of non-Hermitian perturbations are relevant to the low-energy behavior. It turns out that there are two possible relevant non-Hermitian terms: the quadratic term and the periodic potential term. In Sect. 3.2, we analyze the former and show that the measurement backaction bifurcates critical exponents into two, making a sharp contrast to the conventional Tomonaga-Luttinger liquid in which only a single parameter governs the critical behavior. We discuss a possible realization in a one-dimensional Bose gas with a two-body loss. In Sect. 3.3, we study the latter relevant non-Hermitian perturbation and show that it induces new types of phase transitions, renormalization group (RG) fixed points and RG flows. We propose a possible experimental realization with a spatially modulated one-body loss. In Sect. 3.4, we investigate effects of measurements on quantum phase transitions in higher dimensions. In particular, we show that the measurement backaction can shift a transition point of the superfluid-to-Mott insulator transition. Our analyses presented in this Chapter can readily be generalized to other many-body systems with different types of measurements. In Sect. 3.5, we propose several concrete experimental systems and discuss experimental parameters to test our theoretical predictions. In Sect. 3.6, we conclude this Chapter with an outlook.

3.1.2 Universal Low-Energy Behavior in One-Dimensional Quantum Systems

Interacting fermions in dimensions higher than one can typically be described by Landau's Fermi liquid theory [70]. There, in the vicinity of the Fermi surface, interactions do not have drastic effects; they are taken into account by simply modifying an effective mass of quasiparticle excitations. In one-dimensional (1D) systems, the situation is completely different. Since particles cannot transport without affecting the motions of all the other ones, interactions inevitably make excitations collective. In 1D systems, interactions thus dramatically enhance the strong correlations between particles, invalidating the Fermi liquid picture. This peculiarity of 1D interacting many-body systems calls for an alternative description beyond the Fermi liquid theory.

The Tomonaga-Luttinger liquid (TLL) theory provides a universal description to study the low-energy behavior of 1D interacting fermions and bosons [71]. The basic idea is to represent underlying collective excitations by bosonic quantum fields and then reformulate the theory in terms of those bosonic degrees of freedom. If there are no relevant contributions in the low-energy limit, the resulting effective

3.1 Introduction

field theory is quadratic in terms of the bosonic operators and thus exactly solvable. This state of matter is known as the TLL. In the presence of relevant perturbations, the most fundamental field theory is the sine-Gordon model [72, 73] that exhibits the quantum phase transition known as the Berezinskii-Kosterlitz-Thouless (BKT) transition [74–76]. In this subsection, we review some basic physical properties of the TLL and the sine-Gordon model.

Brief summary of the Tomonaga-Luttinger liquid

The low-energy universal behavior of 1D many-body systems can be described by the effective field theory known as the TLL [71]:

$$\hat{H}_0 = \frac{\hbar v}{2\pi} \int dx \left[K \left(\partial_x \hat{\theta} \right)^2 + \frac{1}{K} \left(\partial_x \hat{\phi} \right)^2 \right], \tag{3.3}$$

where v is the sound velocity, K is the so-called TLL parameter that characterizes the interaction strength in a microscopic model, and $\hat{\phi}$ and $\hat{\theta}$ are scalar fields satisfying the following commutation relation:

$$[\hat{\phi}(x), \partial_x \hat{\theta}(x')] = -i\pi \delta(x - x'). \tag{3.4}$$

Physically, these fields represent the degrees of freedom of collective excitations in the underlying microscopic model. For instance, in the case of spinful fermions, the low-energy behavior is described by the two independent scalar fields $\hat{\phi}_c$ and $\hat{\phi}_s$, which represent collective excitations of charge (i.e., density fluctuations) and spin, respectively. This is the celebrated spin-charge separation in 1D systems.

The Hamiltonian $\hat{H}_0$ is quadratic and thus exactly solvable. Its energy dispersion is gapless and linear. More specifically, (aside a constant offset) its spectrum is given by

$$\hat{H}_0 = \hbar v \sum_{|k|<\Lambda} |k| \hat{b}_k^\dagger \hat{b}_k, \tag{3.5}$$

where $\hat{b}_k$ ($\hat{b}_k^\dagger$) is an annihilation (creation) operator of a bosonic excitation labeled by a wavevector $k = 2\pi n/L$ with $n = \ldots -2, -1, 1, 2 \ldots$. We introduce L as the system size and Λ as a momentum cutoff. The operators $\hat{b}_k$ and $\hat{b}_k^\dagger$ can be related to the field operators $\hat{\phi}$ and $\hat{\theta}$ via the mode expansions:

$$\hat{\phi}(x) = -\sum_{|k|<\Lambda} i \cdot \text{sgn}(k) \sqrt{\frac{\pi K}{2L|k|}} e^{-a|k|/2 - ikx} (\hat{b}_k^\dagger + \hat{b}_{-k}), \tag{3.6}$$

$$\hat{\theta}(x) = \sum_{|k|<\Lambda} i \sqrt{\frac{\pi}{2KL|k|}} e^{-a|k|/2 - ikx} (\hat{b}_k^\dagger - \hat{b}_{-k}), \tag{3.7}$$

where

$$a = \frac{1}{\Lambda} \tag{3.8}$$

is a short-distance cutoff. The gapless nature of the spectrum indicates the criticality of the ground state. We can analytically show the critical decays of the correlation functions as follows (the derivations will be given later):

$$C_p^\theta(x) = \langle e^{ip\hat{\theta}(x)} e^{-ip\hat{\theta}(0)} \rangle = \left(\frac{a}{|x|}\right)^{p^2/(2K)}, \tag{3.9}$$

$$C_p^\phi(x) = \langle e^{ip\hat{\phi}(x)} e^{-ip\hat{\phi}(0)} \rangle = \left(\frac{a}{|x|}\right)^{p^2 K/2}, \tag{3.10}$$

where $\langle \cdots \rangle$ denotes an expectation value with respect to the ground state of $\hat{H}_0$ and p is a real number. Both functions exhibit critical decays but with different critical exponents $p^2/(2K)$ and $p^2 K/2$, which are characterized by a *single* parameter known as the TLL parameter K.

To gain physical insights, let us consider a specific example of 1D spinless bosons. The bosonic field $\hat{\Psi}(x)$ in the microscopic theory can be represented by the phase field $\hat{\theta}(x)$ and the density operator $\hat{\rho}(x)$ as follows:

$$\hat{\Psi}(x) \simeq \sqrt{\hat{\rho}(x)} e^{i\hat{\theta}(x)}. \tag{3.11}$$

The low-energy collective excitations of the density $\hat{\rho}(x)$ can be related to a scalar field $\hat{\phi}(x)$. Let us derive its representation based on a phenomenological argument [71]. We start from the expression of the classical density

$$\rho(x) = \sum_n \delta(x - x_n), \tag{3.12}$$

where x_n is the position of the n-th particle. Let ρ_0 be the mean density of particles and $d = 1/\rho_0$ be the average distance. We introduce a monotonically increasing continuous function $\chi(x)$ of the position, which takes the value $\chi(x_n) = 2\pi n$. We then rewrite the density as

$$\rho(x) = \sum_n \delta(x - x_n) = \sum_n \partial_x \chi(x) \delta(\chi(x) - 2\pi n) = \frac{\partial_x \chi(x)}{2\pi} \sum_{p=-\infty}^{\infty} e^{ip\chi(x)}, \tag{3.13}$$

where we use the Poisson summation formula to derive the last expression. Since we are interested in the low-energy collective-mode fluctuations, it is useful to introduce a relative field $\phi(x)$ from the crystalline equilibrium particle configuration $\chi_0(x) = 2\pi x/d$ as

$$\chi(x) = \chi_0(x) + 2\phi(x) = 2\pi \rho_0 x + 2\phi(x). \tag{3.14}$$

3.1 Introduction

If we formally interpret the density $\rho(x)$ and the function $\phi(x)$ as quantum fields, we arrive at the correct bosonized expression of $\hat{\rho}(x)$ in terms of a scalar field $\hat{\phi}(x)$ as[1]

$$\hat{\rho}(x) \simeq \left[\rho_0 + \frac{\partial_x \hat{\phi}(x)}{\pi}\right] \sum_{p=-\infty}^{\infty} e^{2ip(\pi\rho_0 x + \hat{\phi}(x))}. \tag{3.15}$$

Using Eqs. (3.11) and (3.15), we can express the original microscopic Hamiltonian (cf. $\gamma = 0$ case in Eq. (3.67) below) of a 1D Bose gas in terms of the scalar fields $\hat{\phi}$ and $\hat{\theta}$. The resulting effective field theory is the TLL in Eq. (3.3), where the parameters in the microscopic model are renormalized into the single TLL parameter K. The correlation function $C^\theta_{p=1}(x)$ can be related to a one-particle correlation function of a 1D Bose gas by

$$\langle \hat{\Psi}^\dagger(x)\hat{\Psi}(0)\rangle \propto \left(\frac{1}{|x|}\right)^{1/(2K)}. \tag{3.16}$$

Similarly, $C^\phi_{p=2}(x)$ characterizes its density-density correlation function as

$$\langle \hat{\rho}(x)\hat{\rho}(0)\rangle - \rho_0^2 = -\frac{K}{2\pi x^2} + \text{const.} \times \frac{\cos(2\pi\rho_0 x)}{|x|^{2K}}. \tag{3.17}$$

We will study how these critical behavior are modified due to measurement backaction in Sect. 3.2.

Brief summary of the sine-Gordon model

Underlying microscopic interactions in 1D systems can generate additional perturbations on top of the TLL. This is the case, for example, in spinful fermions or a 1D Bose gas subject to a spatially periodic potential.[2] To discuss effects of the perturbations to the TLL, let us start consider a Hamiltonian

$$\hat{H} = \hat{H}_0 + \int dx V(\hat{\phi}), \tag{3.18}$$

where $V(\hat{\phi})$ is a potential term of the field $\hat{\phi}$ satisfying the condition[3]

$$V(\hat{\phi}) = V(\hat{\phi} + 2\pi). \tag{3.19}$$

---

[1] A more formal derivation can be found, for example, in Ref. [77].
[2] We will discuss later the latter example in detail to propose a possible experimental test of our theoretical results.
[3] Note that the original bosonic or fermionic field operator $\hat{\Psi}(x) \propto \sqrt{\hat{\rho}(x)}$ is invariant under $\hat{\phi}(x) \to \hat{\phi}(x) + 2\pi$ as inferred from Eq. (3.15). The resulting effective field theory should also satisfy this symmetry.

It can be expanded as

$$V(\hat{\phi}) = \frac{1}{\pi} \sum_{n=1}^{\infty} \left[\alpha_n^c \cos(n\hat{\phi}) + \alpha_n^s \sin(n\hat{\phi}) \right], \tag{3.20}$$

where $\alpha_n^{c,s}$ are real parameters. We can study how the perturbations affect the low-energy behavior on the basis of the renormalization group (RG) analysis.

The basic idea of the RG analysis is as follows. We start from an effective field theory with couplings $\alpha(\Lambda)$ and the TLL parameter $K(\Lambda)$ at a momentum cutoff Λ. The goal of the RG analysis is to find a theory having an equivalent low-energy (i.e., long-distance) behavior but at a lower momentum cutoff $\tilde{\Lambda} < \Lambda$ (i.e., a larger distance cutoff $\tilde{a} > a$) with the renormalized coupling $\alpha(\tilde{\Lambda})$ and the TLL parameter $K(\tilde{\Lambda})$. To achieve this, we integrate out high-momentum degrees of freedom which are irrelevant to the low-energy behavior and examine how couplings $\alpha(\Lambda)$ and $K(\Lambda)$ alter as we renormalize a cutoff scale Λ. A set of differential equations that govern these changes of couplings are known as the RG equations. They define the flows of couplings as functions of a cutoff scale Λ, which are known as the RG flows. Physically, Λ-dependent couplings $\alpha(\Lambda)$ and the TLL parameter $K(\Lambda)$ characterize their renormalized values at an energy scale Λ, or equivalently, at a distance scale $a = 1/\Lambda$. We remark that the aim of RG analysis is not to solve the entire problem of the model, but to determine its low-energy properties by mapping the original model to a (possibly) simpler problem while keeping the low-energy physics unaltered.

Let us consider how the potential (3.20) renormalizes along the RG flows. Analyzing the scaling dimensions of the couplings, we can obtain their RG equations at the leading order by

$$\frac{dg_n^{c,s}}{dl} = \left(2 - \frac{n^2 K}{4}\right) g_n^{c,s}, \tag{3.21}$$

where $l = -\ln(\Lambda/\Lambda_0)$ is the logarithmic RG scale with Λ_0 being an initial cutoff and we rescale the couplings $\alpha_n^{c,s}$ by introducing the dimensionless parameters $g_n^{c,s}$ as

$$g_n^{c,s} = \frac{\alpha_n^{c,s}}{\hbar v \Lambda^2}. \tag{3.22}$$

Equation (3.21) clearly shows that a perturbation is more irrelevant for a higher-order n. It suffices to consider the most relevant contribution that is the lowest-order m for which the coefficients are nonzero, i.e., $g_m^{c,s} \neq 0$ and $g_n^{c,s} = 0$ for $n < m$. We then rescale the field operators and the TLL parameter as (cf. Eqs. (3.6) and (3.7))

$$\hat{\phi} \to (m/2)\hat{\phi}, \quad \hat{\theta} \to (2/m)\hat{\theta}, \quad K \to (m^2/4)K. \tag{3.23}$$

The resulting effective field theory is

3.1 Introduction

$$\hat{H}_0 + \frac{1}{\pi} \int dx \left[\alpha_c \cos(2\hat{\phi}) + \alpha_s \sin(2\hat{\phi}) \right]. \quad (3.24)$$

Finally, employing the translational invariance $\hat{\phi} \to \hat{\phi} - \pi/8$, we arrive at the sine-Gordon model:

$$\hat{H}_{sG} = \int dx \left\{ \frac{\hbar v}{2\pi} \left[K \left(\partial_x \hat{\theta} \right)^2 + \frac{1}{K} \left(\partial_x \hat{\phi} \right)^2 \right] + \frac{\alpha}{\pi} \cos(2\hat{\phi}) \right\}. \quad (3.25)$$

It is well known that the RG equations of K and $g = \alpha/(\hbar v \Lambda^2)$ in the sine-Gordon model are given by (up to the third order in g) [78]

$$\frac{dK}{dl} = -g^2 K^2,$$
$$\frac{dg}{dl} = (2 - K)g + 5g^3. \quad (3.26)$$

Figure 3.2 shows the RG phase diagram of the sine-Gordon model. In the regime $K < 2$, the potential perturbation is relevant and its strength g flows into the strong-coupling limit, suppressing the fluctuations of the field $\hat{\phi}$ by the cosine potential $\cos(2\hat{\phi})$. In this limit, the phase $\hat{\phi}$ is locked and the system exhibits the gapped phase (blue regon). For example, in a 1D Bose gas subject to a periodic potential, the locking indicates the suppression of density fluctuations and thus the gapped phase corresponds to the Mott insulator (MI) phase as we will discuss later. From the equivalence between a 1D quantum system and a 2D classical system, we can also view this locking as the condensation of bounded pairs of the vortex and antivortex

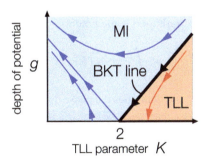

Fig. 3.2 Renormalization group flows of the sine-Gordon model. When the Tomonaga-Luttinger liquid (TLL) parameter K satisfies $K < 2$, the potential perturbation (whose strength is denoted by g) is relevant and the system flows into the gapped phase (blue region). In contrast, for $K > 2$, the perturbation is irrelevant and the low-energy behavior is described by the critical phase known as the Tomonaga-Luttinger liquid (red region). In between these two phases, there exists the Berezinskii-Kosterlitz-Thouless (BKT) transition. As for a one-dimensional bosonic system to be discussed later, the gapped phase is the Mott insulator (MI) phase

that are topological excitations in the phase field ϕ. In the other regime $K \geq 2$, the perturbation is irrelevant or marginal and its strength g flows into zero (red region). The effect of the perturbation appears as a renormalization of the value of the TLL parameter; an addition of g effectively decreases K, ending up a renormalized value K^*. The long-distance critical behavior is thus described by the TLL with this renormalized TLL parameter K^*. Physically, a decrease of K along the RG flows indicates the degradation of superfluid correlations at a large distance (cf. Eq. (3.16)). In between the two phases, the system exhibits the BKT transition. In the classical picture, this transition can be viewed as the proliferation of the vortex-antivortex pairs.

Relevant non-Hermitian perturbations

We are interested in a non-Hermitian perturbation to the TLL Hamiltonian, where the non-Hermiticity originates from measurement backaction due to continuous monitoring (see the introduction in this Chapter). Our aim is to understand how such non-Hermiticity alters the quantum critical behavior of 1D many-body systems. To achieve this aim, we first have to identify what types of non-Hermitian perturbations are relevant in the RG analysis, which can be done by checking the scaling dimensions of the perturbations in the same manner as in the Hermitian case.

There are two types of the possible relevant perturbations. The first one is the quadratic term:

$$\hat{\Gamma}_0 \propto -i \int dx \left(\partial_x \hat{\phi}\right)^2, \tag{3.27}$$

which has the same scaling dimension as in the quadratic terms of the ordinary TLL. This term will effectively make the TLL parameter K complex and thus modify the critical exponents as we detail in the next section. Physically, $\hat{\Gamma}_0$ can arise as the backaction from a density measurement or a two-body loss process of particles.

The second possible perturbation is the potential term

$$\hat{\Gamma}_V \propto -i \int dx \cos\left(2\hat{\phi}(x) + \delta\right), \tag{3.28}$$

where δ is a real parameter and quantifies the phase difference of the imaginary potential compared with a real relevant potential $\cos(2\hat{\phi})$ in the sine-Gordon model (3.25). Such a perturbation physically originates from, for example, a dissipative potential inducing a one-body loss of particles. For $\delta \neq \pm \pi/2$, the imaginary potential term $\hat{\Gamma}_V$ can make K complex along the RG flows. This leads to similar effects as in the quadratic perturbation $\hat{\Gamma}_0$. To identify effects unique to the non-Hermitian potential term $\hat{\Gamma}_V$, we focus on the case of $\delta = \pm \pi/2$ at which K remains to be real along RG flows. From Eqs. (3.25) and (3.28), this condition leads to the following generalized sine-Gordon model:

$$\hat{H} = \hat{H}_{sG} + \hat{\Gamma}_V = \int dx \left\{ \frac{\hbar v}{2\pi} \left[K \left(\partial_x \hat{\theta}\right)^2 + \frac{1}{K} \left(\partial_x \hat{\phi}\right)^2 \right] + \frac{\alpha_r}{\pi} \cos(2\hat{\phi}) - \frac{i\alpha_i}{\pi} \sin(2\hat{\phi}) \right\}. \tag{3.29}$$

3.1 Introduction

where α_r (α_i) is the strength of a real (imaginary) potential perturbation and we choose the sign $\delta = -\pi/2$ without loss of generality. We note that this model satisfies the combination of the parity symmetry ($\hat{P}$) and the time-reversal symmetry ($\hat{T}$):

$$\hat{P}\hat{T}\hat{H}\left(\hat{P}\hat{T}\right)^{-1} = \hat{H}. \tag{3.30}$$

This follows from the fact that $\hat{P}$ reverses the sign of $\hat{\phi}$ and that $\hat{T}$ is an antiunitary operator[4]

$$\hat{P}\hat{\phi}\hat{P}^{-1} = -\hat{\phi}, \quad \hat{T}\hat{\phi}\hat{T}^{-1} = \hat{\phi}, \tag{3.31}$$
$$\hat{T}i\hat{T}^{-1} = -i. \tag{3.32}$$

The parity-time (PT) symmetric non-Hermitian systems can exhibit an exotic real-to-complex spectrum transition that accompanies the exceptional point at which the Hamiltonian is not diagonalizable. Such a transition has no counterpart in Hermitian systems. While it has been recognized that the exceptional point and the spectrum transition can induce interesting one-body phenomena especially in the context of classical optics [58, 59], we will show that the non-Hermiticity can lead to novel types of many-body phenomena such as the emergence of new RG fixed points and the exotic RG flows violating the c-theorem as detailed in Sect. 3.3.

3.2 Backaction on Quantum Criticality I: The Quadratic Term

3.2.1 Model

In this section, we analyze the effect of the non-Hermitian quadratic perturbation

$$\hat{\Gamma}_0 = -i\gamma \int dx \left(\partial_x \hat{\phi}\right)^2 \tag{3.33}$$

to the 1D critical behavior. Physically, such a contribution can appear in systems accompanying a two-body loss of particles or those subject to a density measurement as we will discuss in Sect. 3.2.3. It is useful to rewrite the TLL Hamiltonian as

$$\hat{H}_0 = \frac{\hbar}{2\pi} \int dx \left[v_J \left(\partial_x \hat{\theta}\right)^2 + v_N \left(\partial_x \hat{\phi}\right)^2 \right], \tag{3.34}$$

where the phase stiffness v_J and the density stiffness v_N are defined as

---

[4]The odd parity of $\hat{\phi}$ can be inferred from the fact that $\partial_x \hat{\phi}$ has the same parity as in the density operator $\hat{\rho}(x)$ which is a scalar (see Eq. (3.15)).

$$v_J = vK, \tag{3.35}$$

$$v_N = \frac{v}{K}. \tag{3.36}$$

The role of the quadratic perturbation $\hat{\Gamma}_0$ is to renormalize the density stiffness v_N to a complex value $\tilde{v}_N e^{-i\delta_\gamma}$:

$$\hat{H}_{\text{eff}} = \hat{H}_0 + \hat{\Gamma}_0 = \frac{\hbar}{2\pi} \int dx \left[v_J \left(\partial_x \hat{\theta} \right)^2 + \tilde{v}_N e^{-i\delta_\gamma} \left(\partial_x \hat{\phi} \right)^2 \right], \tag{3.37}$$

where $\tilde{v}_N(\gamma) > 0$ and $0 \leq \delta_\gamma < \pi/2$ are the real parameters depending on γ. Since the effective Hamiltonian (3.37) is quadratic, it is still solvable and we can obtain its exact complex spectrum. To achieve this, we do the mode expansions, reduce the problem to that of a set of complex harmonic potentials [79, 80], and obtain the exact effective ground state $|\Psi_{g,\gamma}\rangle$ having the lowest energy, i.e., the lowest real part of an eigenvalue. We will show that the effective ground state also has the longest lifetime among all the eigenstates, indicating that it survives longest in the non-Hermitian evolution. If we take the limit of vanishing measurement strength $\gamma \to 0$, the effective ground state $|\Psi_{g,\gamma}\rangle$ reproduces the known ground state of the original TLL Hamiltonian. Calculating the correlation functions with respect to the effective ground state $|\Psi_{g,\gamma}\rangle$, we show that it exhibits unusual critical behavior influenced by measurement.

Let us start from the mode expansions of the fields:

$$\hat{\phi}(x) = -\sum_{k \neq 0} i \cdot \text{sgn}(k) \sqrt{\frac{\pi \tilde{K}_\gamma}{2L|k|}} e^{-a|k|/2 - ikx} (\hat{b}_k^\dagger + \hat{b}_{-k}), \tag{3.38}$$

$$\hat{\theta}(x) = \sum_{k \neq 0} i \sqrt{\frac{\pi}{2\tilde{K}_\gamma L|k|}} e^{-a|k|/2 - ikx} (\hat{b}_k^\dagger - \hat{b}_{-k}), \tag{3.39}$$

where we introduce L as the size of a system and $\tilde{K}_\gamma$ as the following modified TLL parameter

$$\tilde{K}_\gamma \equiv \sqrt{\frac{v_J}{\tilde{v}_N}}, \tag{3.40}$$

$\hat{b}_k$ ($\hat{b}_k^\dagger$) is an annihilation (creation) operator associated with a mode having momentum $k = 2\pi m/L$ ($m = \pm 1, \pm 2, \ldots$), and $a \to +0$ is a short-distance cutoff. The effective Hamiltonian (3.37) is then (aside an irrelevant constant) rewritten as

$$\hat{H}_{\text{eff}} = \hbar \tilde{v} \sum_{k>0} k \left[\frac{e^{-i\delta_\gamma} + 1}{2} (\hat{b}_k^\dagger \hat{b}_k + \hat{b}_{-k}^\dagger \hat{b}_{-k}) + \frac{e^{-i\delta_\gamma} - 1}{2} (\hat{b}_k \hat{b}_{-k} + \hat{b}_k^\dagger \hat{b}_{-k}^\dagger) \right], \tag{3.41}$$

3.2 Backaction on Quantum Criticality I: The Quadratic Term

where we introduce $\tilde{v} \equiv \sqrt{v_J \tilde{v}_N}$. Introducing the position and momentum operators $\hat{x}_k$ and $\hat{p}_k$ as

$$\hat{b}_k = \frac{\hat{x}_k + i\hat{p}_k}{\sqrt{2}}, \quad \hat{b}_k^\dagger = \frac{\hat{x}_k - i\hat{p}_k}{\sqrt{2}}, \tag{3.42}$$

we can express Eq. (3.41) as

$$\hat{H}_{\text{eff}} = \hbar\tilde{v}\sum_{k>0} k\left[\frac{e^{-i\delta_\gamma}+1}{4}(\hat{x}_k^2 + \hat{p}_k^2 + \hat{x}_{-k}^2 + \hat{p}_{-k}^2) + \frac{e^{-i\delta_\gamma}-1}{2}(\hat{x}_k\hat{x}_{-k} - \hat{p}_k\hat{p}_{-k})\right]. \tag{3.43}$$

Furthermore, we use the center-of-mass $\hat{x}_{k,+}$ ($\hat{p}_{k,+}$) and relative $\hat{x}_{k,-}$ ($\hat{p}_{k,-}$) coordinates (momenta) with modes $\pm k$:

$$\hat{x}_k = \frac{\hat{x}_{k,+} + \hat{x}_{k,-}}{\sqrt{2}}, \quad \hat{x}_{-k} = \frac{\hat{x}_{k,+} - \hat{x}_{k,-}}{\sqrt{2}}, \tag{3.44}$$

$$\hat{p}_k = \frac{\hat{p}_{k,+} + \hat{p}_{k,-}}{\sqrt{2}}, \quad \hat{p}_{-k} = \frac{\hat{p}_{k,+} - \hat{p}_{k,-}}{\sqrt{2}}, \tag{3.45}$$

where we choose k to be a positive discrete momentum, i.e., $k = 2\pi m/L$ with $m = 1, 2, \ldots$. Using Eqs. (3.44) and (3.45) to rewrite Eq. (3.43), we obtain

$$\hat{H}_{\text{eff}} = \hbar\tilde{v}\sum_{k>0} k\left(\frac{e^{-i\delta_\gamma}}{2}\hat{x}_{k,+}^2 + \frac{1}{2}\hat{p}_{k,+}^2 + \frac{1}{2}\hat{x}_{k,-}^2 + \frac{e^{-i\delta_\gamma}}{2}\hat{p}_{k,-}^2\right). \tag{3.46}$$

This Hamiltonian is nothing but a set of non-Hermitian harmonic oscillators [79, 80]. To diagonalize it, we choose the coherent-state basis $|\{x_{k,+}, p_{k,-}\}\rangle$, where $\hat{x}_{k,+}$ and $\hat{p}_{k,-}$ are diagonalized such that $\hat{\phi}(x)$ becomes a c-number (cf. Eq. (3.51) below). We can obtain the effective ground-state wavefunction $|\Psi_{g,\gamma}\rangle$ as

$$\langle\{x_{k,+}, p_{k,-}\}|\Psi_{g,\gamma}\rangle \propto \exp\left[-\frac{e^{-i\delta_\gamma/2}}{2}\sum_{k>0}(x_{k,+}^2 + p_{k,-}^2)\right]. \tag{3.47}$$

This state is an extension of the original ground-state wavefunction [81] to the non-Hermitian TLL. The full complex spectrum is given by

$$\hbar\tilde{v}e^{-i\delta_\gamma/2}\sum_{k>0} k(n_{k,+} + n_{k,-} + 1), \tag{3.48}$$

where $n_{k,\pm}$ are nonnegative integers representing quantum numbers in the modes $(k, \pm)$. From the condition $0 \leq \delta_\gamma < \pi/2$, the energies are bounded from below and the ground-state wavefunction can be normalized.[5] We remark that the effective ground state ($n_{k,\pm} = 0$) with the lowest energy also has the minimal imaginary part, i.e., it survives the longest in the non-Hermitian evolution.

Effective parity-time symmetry in the non-Hermitian TLL

We note that there exists a "hidden" parity-time symmetry in the Hamiltonian (3.37). To show this, we rewrite the Hamiltonian as

$$\hat{H}_{\text{eff}} = e^{-i\delta_\gamma/2} \hat{H}'_{\text{eff}}, \quad \hat{H}'_{\text{eff}} = \frac{\hbar}{2\pi} \int dx \left[v_J e^{i\delta_\gamma/2} \left(\partial_x \hat{\theta}\right)^2 + \tilde{v}_N e^{-i\delta_\gamma/2} \left(\partial_x \hat{\phi}\right)^2 \right]. \quad (3.49)$$

where $\hat{H}'_{\text{eff}}$ is invariant under the combination of the time-reversal operation and the following parity transformation[6]:

$$\hat{\theta} \to \sqrt{\frac{\tilde{v}_N}{v_J}} \hat{\phi}, \quad \hat{\phi} \to \sqrt{\frac{v_J}{\tilde{v}_N}} \hat{\theta}. \quad (3.50)$$

Thus, the original Hamiltonian (3.37) satisfies the parity-time symmetry aside a constant complex factor $e^{-i\delta_\gamma/2}$. In fact, the spectrum of the parity-time-symmetric part $\hat{H}'_{\text{eff}}$ is entirely real (i.e., the parity-time symmetry is unbroken) for any δ_γ (cf. Eq. (3.48)).

3.2.2 Correlation Functions: Bifurcating Critical Exponents

We next analyze the critical behavior of the effective ground state that is influenced by the measurement backaction.

Calculation of the correlation function $C^\phi(x)$

Let us calculate the correlation function $C^\phi_{p=2}(x - y) = \langle e^{2i\hat{\phi}(x)} e^{-2i\hat{\phi}(y)} \rangle$, which can be related to the density-density fluctuations for the case of a 1D Bose gas (see Eq. (3.17)). Here, we denote $\langle \cdots \rangle$ as the expectation value with respect to the effective ground state $|\Psi_{g,\gamma}\rangle$. In terms of the variables $\hat{x}_{k,+}$ and $\hat{p}_{k,-}$, we can rewrite $\hat{\phi}(x)$ as

$$\hat{\phi}(x) = i \sum_{k>0} \sqrt{\frac{\pi \tilde{K}_\gamma}{2Lk}} \left[(\hat{x}_{k,+} + i\hat{p}_{k,-}) e^{ikx} - (\hat{x}_{k,+} - i\hat{p}_{k,-}) e^{-ikx} \right]. \quad (3.51)$$

---

[5]Although the wavefunction (3.47) remains normalizable for $\pi/2 \leq \delta_\gamma < \pi$, the full spectrum including the zero-mode contributions is no longer bounded from below in this regime. We also remark that if $\delta_\gamma \geq \pi$, the spectrum is not bounded from below and there are no eigenstates having discrete eigenvalues since the wavefunction cannot be normalized.

[6]Here by a parity transformation, we mean that its square reduces to the identity operator $\hat{P}^2 = \hat{I}$. A spatial parity transformation against a certain plane is just its special example.

3.2 Backaction on Quantum Criticality I: The Quadratic Term

Let $|\{\phi_k\}\rangle$ be an eigenstate of $\hat{\phi}(x)$ ($0 \leq x < L$):

$$\hat{\phi}(x)|\{\phi_k\}\rangle = \phi(x)|\{\phi_k\}\rangle, \qquad (3.52)$$

$$\phi(x) = \sqrt{\frac{\pi}{L}} \sum_{k>0} \left(\phi_k e^{ikx} + \phi_k^* e^{-ikx}\right), \qquad (3.53)$$

where ϕ_k satisfies

$$i(x_{k,+} + ip_{k,-}) = \sqrt{\frac{2k}{\tilde{K}_\gamma}} \phi_k. \qquad (3.54)$$

The wavefunction of the ground state (3.47) can be rewritten as

$$\langle\{\phi_k\}|\Psi_{g,\gamma}\rangle = \frac{1}{\sqrt{\mathcal{N}}} \exp\left(-\frac{e^{-i\delta_\gamma/2}}{\tilde{K}_\gamma} \sum_{k>0} k|\phi_k|^2\right), \qquad (3.55)$$

where $\mathcal{N}$ is a normalization constant. We can now express the correlation function $C^\phi_{p=2}(x - y)$ by

$$\langle e^{2i\hat{\phi}(x)} e^{-2i\hat{\phi}(y)}\rangle = \frac{1}{\mathcal{N}} \int \mathcal{D}\phi \mathcal{D}\phi^*$$
$$\times \exp\left\{\sum_{k>0}\left[-\frac{2k\cos(\delta_\gamma/2)}{\tilde{K}_\gamma}|\phi_k|^2 + 2i\sqrt{\frac{\pi}{L}}\phi_k(e^{ikx} - e^{iky}) + 2i\sqrt{\frac{\pi}{L}}\phi_k^*(e^{-ikx} - e^{-iky})\right]\right\}$$
$$= \exp\left[-\frac{\tilde{K}_\gamma}{\cos(\delta_\gamma/2)} \sum_{k>0} \frac{k_1}{k}(2 - e^{ikr} - e^{-ikr})\right], \qquad (3.56)$$

where we do the Gaussian integrations to obtain the last expression, and define $k_1 \equiv 2\pi/L$ and $r \equiv x - y$. The sum over $k > 0$ can be taken with a regularization trick:

$$\sum_{k>0} \frac{k_1 e^{-ak}}{k}(2 - e^{ikr} - e^{-ikr}) \rightarrow -2\ln\left(\frac{ak_1}{2\sin(k_1 r/2)}\right), \qquad (3.57)$$

where we use $ak_1 \ll 1$. The resulting critical behavior is

$$\langle e^{2i\hat{\phi}(x)} e^{-2i\hat{\phi}(y)}\rangle = \left(\frac{a}{(L/\pi)\sin(\pi r/L)}\right)^{\frac{2\tilde{K}_\gamma}{\cos(\delta_\gamma/2)}} \rightarrow \left(\frac{a}{r}\right)^{2K_\phi}, \qquad (3.58)$$

where we use $L \gg r$ and introduce the critical exponent K_ϕ by

$$K_\phi(\gamma) = \frac{\tilde{K}_\gamma}{\cos\left(\frac{\delta_\gamma}{2}\right)}. \qquad (3.59)$$

Calculation of the correlation function $C^\theta(x)$

We next calculate the correlation function $C^\theta_{p=1}(x-y) = \langle e^{i\hat\theta(x)} e^{-i\hat\theta(y)}\rangle$, which can be related to the one-particle correlation for the case of a 1D Bose gas (see Eq. (3.16)). We expand the operator $\hat\theta$ as

$$\hat\theta(x) = -i\sum_{k>0}\sqrt{\frac{\pi}{2\tilde{K}_\gamma Lk}}\left[i\hat p_{k,+}(e^{ikx}+e^{-ikx}) + \hat x_{k,-}(e^{ikx}-e^{-ikx})\right]. \quad (3.60)$$

This acts as a shift operator on an eigenstate of $\hat x_{k,+}$ and $\hat p_{k,-}$:

$$e^{i\hat\theta(x)}|\{x_{k,+},p_{k,-}\}\rangle = \left|\left\{x_{k,+}-i\sqrt{\frac{\pi}{2\tilde{K}_\gamma Lk}}(e^{ikx}+e^{-ikx}),\, p_{k,-}+\sqrt{\frac{\pi}{2\tilde{K}_\gamma Lk}}(e^{ikx}-e^{-ikx})\right\}\right\rangle. \quad (3.61)$$

From Eq. (3.54), we obtain

$$e^{i\hat\theta(x)}|\{\phi_k\}\rangle = \left|\left\{\phi_k - \frac{i}{k}\sqrt{\frac{\pi}{L}}e^{-ikx}\right\}\right\rangle. \quad (3.62)$$

The correlation function can now be calculated as

$$\langle e^{i\hat\theta(x)}e^{-i\hat\theta(y)}\rangle = \frac{1}{N}\int \mathcal{D}\phi\mathcal{D}\phi^* \exp\left[-\frac{1}{\tilde{K}_\gamma}\sum_{k>0}k\left(e^{i\delta_\gamma}\left|\phi_k - \frac{i}{k}\sqrt{\frac{\pi}{L}}e^{-ikx}\right|^2 + e^{-i\delta_\gamma}\left|\phi_k - \frac{i}{k}\sqrt{\frac{\pi}{L}}e^{-iky}\right|^2\right)\right]$$

$$= \exp\left[-\frac{\pi}{2\tilde{K}_\gamma L\cos(\delta_\gamma/2)}\sum_{k>0}\frac{1}{k}(2 - e^{ikr} - e^{-ikr})\right] = \left(\frac{a}{(L/\pi)\sin(\pi r/L)}\right)^{\frac{1}{2\tilde{K}_\gamma \cos(\delta_\gamma/2)}}$$

$$\to \left(\frac{a}{r}\right)^{1/(2K_\theta)}, \quad (3.63)$$

where we use $L \gg r$ and define the critical exponent K_θ as

$$K_\theta(\gamma) = \tilde{K}_\gamma \cos\left(\frac{\delta_\gamma}{2}\right). \quad (3.64)$$

Physical consequences: shifts and bifurcation of critical exponents

We now discuss physical consequences of the measurement backaction for the case of a 1D Bose gas. As shown above, the backaction leads to the two TLL parameters $K_{\theta,\phi}$ in Eqs. (3.64) and (3.59), which describe the critical behavior of the one-particle correlation and the density-density correlation, respectively:

$$\langle\hat\Psi^\dagger(r)\hat\Psi(0)\rangle \propto \left(\frac{1}{r}\right)^{1/(2K_\theta)}, \quad (3.65)$$

$$\langle\hat\rho(r)\hat\rho(0)\rangle - \rho_0^2 = -\frac{K_\phi}{2\pi^2 r^2} + \text{const.} \times \frac{\cos(2\pi\rho_0 r)}{r^{2K_\phi}}. \quad (3.66)$$

3.2 Backaction on Quantum Criticality I: The Quadratic Term

The emergence of the two characteristic parameters K_θ and K_ϕ signifies the critical behavior beyond the conventional TLL class, in which the single parameter K characterizes the critical behavior (see Eqs. (3.16) and (3.17)). Thus, the non-Hermiticity bifurcates the single critical exponent into the two characterized by K_θ and K_ϕ. In 1D ultracold bosons, the critical exponent K_θ can be found by utilizing interferometric techniques [82, 83], while K_ϕ can be measured by in-situ observations of density fluctuations [84].

3.2.3 Realization in a 1D Ultracold Bose Gas with a Two-Body Loss

As a concrete microscopic physical realization of the non-Hermitian TLL, we here discuss a 1D interacting Bose gas subject to a two-body loss process. The effective non-Hermitian Hamiltonian is the following generalized Lieb-Liniger Hamiltonian:

$$\hat{H} = \int dx \left[-\frac{\hbar^2}{2m} \hat{\Psi}^\dagger(x) \frac{\partial^2}{\partial x^2} \hat{\Psi}(x) + \frac{g - i\gamma}{2} \hat{\Psi}^\dagger(x) \hat{\Psi}^\dagger(x) \hat{\Psi}(x) \hat{\Psi}(x) \right], \quad (3.67)$$

where m is the atomic mass and $\hat{\Psi}(x)$ is the bosonic field operator. The strength $g > 0$ of a repulsive contact interaction is given by $g = 2\hbar\omega a_r$, where ω is the trap frequency in the transverse direction, and a_r is the elastic scattering length between atoms. The backaction from a two-body loss process modifies the interaction parameter as $g \to g - i\gamma$ in the Lieb-Liniger Hamiltonian with $\gamma = 2\hbar\omega a_i$ being the measurement strength determined from the inelastic two-body scattering length a_i [85, 86]. The low-energy critical behavior of Eq. (3.67) can be described by the effective non-Hermitian TLL Hamiltonian in Eq. (3.37). Physically, the field $\hat{\theta}$ in the TLL Hamiltonian is related to the phase of the bosonic field operator as in Eq. (3.11), while $\partial_x \hat{\phi}$ is related to the density fluctuations as in Eq. (3.15).

In the original TLL, correlation functions decay algebraically with exponents determined by the single TLL parameter $K = \sqrt{v_J/v_N}$. We can relate the parameters v_J and v_N in the effective field theory to the microscopic parameters in the original Lieb-Liniger Hamiltonian. Firstly, the Galilean invariance ensures that v_J can be obtained as [87]

$$v_J = \frac{\hbar \pi \rho_0}{m}. \quad (3.68)$$

Secondly, v_N takes the following asymptotic forms in the weakly and strongly interacting regimes [88, 89]:

$$v_N = \begin{cases} \frac{v_J u}{\pi^2} \left(1 - \frac{\sqrt{u}}{2\pi}\right) & \text{for } u \ll 1; \\ v_J \left(1 - \frac{8}{u} + O\left(\frac{1}{u^2}\right)\right) & \text{for } u \gg 1, \end{cases} \quad (3.69)$$

where we introduce the normalized (dimensionless) strength of the interaction

$$u = \frac{mg}{\hbar^2 \rho_0}. \qquad (3.70)$$

In contrast, in the presence of the two-body loss process, the backaction leads to the change $g \to g - i\gamma$ and, accordingly, we analytically continue Eq. (3.69) by replacing u with a complex value $u(1 - ig/\gamma) = u(1 - ia_r/a_i)$. This changes v_N to $\tilde{v}_N(\gamma)e^{-i\delta_\gamma}$, where $\tilde{v}_N$ and δ_γ are the real parameters determined from γ. These relations fix the correspondence between the microscopic parameters m, ρ_0, g and γ in the non-Hermitian Lieb-Liniger Hamiltonian (3.67) and the effective parameters v_J and $\tilde{v}_N e^{-i\delta_\gamma}$ in the non-Hermitian TLL in Eq. (3.37). Figure 3.3 plots $K_{\phi,\theta}$ against the dimensionless measurement strength $\gamma/g = a_i/a_r$. The decrease in $K_{\phi,\theta}$ can be interpreted as a manifestation of the continuous quantum Zeno effect [90, 91], which effectively enhances the repulsive interactions [86]. The split between the two characteristic parameters $K_{\phi,\theta}$ is caused by the additional degree of freedom δ_γ in the parameter space of the effective Hamiltonian (3.37). We remark that these features should also appear in other 1D critical systems than the 1D Bose gas discussed here in view of the universality of the effective field theory.

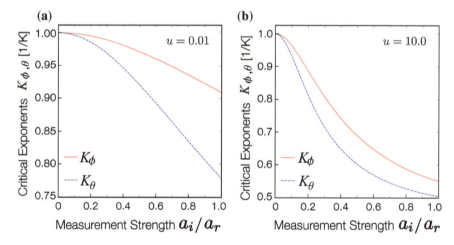

Fig. 3.3 Shifts and bifurcation of the effective TLL parameters K_ϕ (red solid curve) and K_θ (blue dashed curve). We set the interaction strength to be **a** $u = 0.01$ and **b** $u = 10.0$. Reproduced from Fig. 4 of Ref. [10]. Copyright © 2016 by the American Physical Society

3.3 Backaction on Quantum Criticality II: The Potential Term

3.3.1 Model and Its Symmetry

We next consider the other relevant non-Hermitian perturbation, which is the following imaginary potential term

$$\hat{\Gamma}_V = -\frac{i\alpha_i}{\pi} \int dx \, \sin(2\hat{\phi}). \tag{3.71}$$

Physically, this contribution appears, for example, when the system is subject to a spatially modulated dissipative potential inducing a one-body loss as we will detai in Sect. 3.3.5. The resulting Hamiltonian considered in this section is

$$\hat{H} = \int dx \left\{ \frac{\hbar v}{2\pi} \left[K(\partial_x \hat{\theta})^2 + \frac{1}{K}(\partial_x \hat{\phi})^2 \right] + \frac{\alpha_r}{\pi} \cos(2\hat{\phi}) - \frac{i\alpha_i}{\pi} \sin(2\hat{\phi}) \right\}. \tag{3.72}$$

Without the imaginary potential, this Hamiltonian reduces to the sine-Gordon model (3.25), which exhibits the celebrated BKT transition between the TLL and a gapped phase (see the RG phase diagram in Fig. 3.2). For example, as for a 1D Bose gas, this transition corresponds to a superfluid-to-Mott-insulator transition [92]. We aim to extend this paradigm by studying the generalized sine-Gordon Hamiltonian (3.72) that includes both the real and imaginary potentials. When the real potential is relevant, it suppresses the fluctuations of $\hat{\phi}$, stabilizing a non-critical, gapped phase. In contrast, we find that the imaginary potential has an opposite effect; when it becomes relevant, it facilitates the fluctuations of $\hat{\phi}$ and enhances correlations in the conjugate field $\hat{\theta}$. It is this competition between the real and imaginary potentials that makes the present effective field theory particularly rich and nontrivial.[7]

The field theory (3.72) satisfies the parity-time (PT) symmetry (cf. Eq. (3.30)). The PT symmetry is said to be unbroken if every eigenstate of the Hamiltonian satisfies the PT symmetry; then, the entire spectrum is real even though the Hamiltonian is not Hermitian. The PT symmetry is said to be spontaneously broken if some eigenstates of the Hamiltonian are not the eigenstates of the PT operator; then, some pairs of eigenvalues become complex conjugate to each other. This real-to-complex spectral transition is often called the PT transition. The transition is typically accompanied by the coalescence of eigenstates and that of the corresponding eigenvalues at an exceptional point [60] in the discrete spectrum or the spectral singularity [61] in the continuum spectrum.

---

[7]It is worthwhile to recall that the phase difference between the real and imaginary potentials in Eq. (3.72) is chosen so that the TLL parameter K remains to be real along the RG flows (see discussions below Eq. (3.28)), as we have already clarified the physical consequences of the complexification of K in the previous subsection.

In our many-body Hamiltonian (3.72), if the depth of the real potential exceeds that of the imaginary potential ($\alpha_r > \alpha_i$), $\hat{H}$ has an entirely real spectrum, i.e, the PT symmetry is unbroken. This can be proved by the theorem [93] which states that the spectrum is real if and only if there exists an operator $\hat{O}$ satisfying

$$\hat{O}^{-1}\hat{H}\hat{O} = \hat{H}', \tag{3.73}$$

where $\hat{H}'$ is a Hermitian operator. We can explicitly construct such an operator for $\alpha_r > \alpha_i$ as

$$\hat{O} = e^{-\eta\hat{\theta}_0/2}, \quad \eta = \operatorname{arctanh}(\alpha_i/\alpha_r), \tag{3.74}$$

where $\hat{\theta}_0$ is a constant part of the field $\hat{\theta}$, which shifts its conjugate field as $\hat{\phi} \to \hat{\phi} - i\eta$. Then, the potential term in $\hat{H}$ is transformed to

$$\hat{O}^{-1}\left[\frac{\alpha_r}{\pi}\cos(2\hat{\phi}) - \frac{i\alpha_i}{\pi}\sin(2\hat{\phi})\right]\hat{O} = \frac{\sqrt{\alpha_r^2 - \alpha_i^2}}{\pi}\int dx\, \cos(2\hat{\phi}) \tag{3.75}$$

and thus the field theory reduces to the ordinary sine-Gordon Hamiltonian. The divergence of η at $\alpha_r = \alpha_i$ signals the spontaneous breaking of the PT symmetry [94]. When $\alpha_i > \alpha_r$, some of excited eigenstates of the Hamiltonian are no longer eigenstates of the PT operator, generating complex conjugate pairs of eigenvalues.

The main findings we present in this section are the following. First, at the threshold of the PT transition, we find that the spectral singularity and quantum criticality conspire to yield an unconventional RG fixed point, which has no counterpart in Hermitian systems. Second, when an imaginary potential becomes relevant, we find that a local gain-loss structure triggers an enhancement of superfluid correlation that is facilitated by anomalous RG flows violating the c-theorem [95]. This makes a sharp contrast to the suppression of superfluid correlation due to the real potential as expected in the conventional BKT transition. In the following subsections, we show these results based on the RG analyses and numerical calculations using the exact diagonalization and the infinite time-evolving block decimation (iTEBD) algorithm.

3.3.2 Perturbative Renormalization Group Analysis

To reveal the universal behavior of the effective Hamiltonian $\hat{H}$, we first perform a perturbative RG analysis to obtain the following set of flow equations which are valid up to the third order in the couplings:

$$\frac{dK}{dl} = -\left(g_r^2 - g_i^2\right)K^2, \tag{3.76}$$

3.3 Backaction on Quantum Criticality II: The Potential Term

$$\frac{dg_r}{dl} = (2-K)g_r + 5g_r^3 - 5g_i^2 g_r, \tag{3.77}$$

$$\frac{dg_i}{dl} = (2-K)g_i - 5g_i^3 + 5g_r^2 g_i, \tag{3.78}$$

where $g_{r,i}$ are the dimensionless couplings defined as

$$g_{r,i} = \frac{\alpha_{r,i}}{\hbar v \Lambda^2}. \tag{3.79}$$

The velocity v stays constant to all orders in $g_{r,i}$ because of the Lorentz invariance of the theory. The RG equations can be obtained by deriving the RG equations for the Hermitian potential $(\alpha_r/\pi)\cos(2\hat{\phi}) + (\alpha_i/\pi)\sin(2\hat{\phi})$ and then analytically continuing them by replacing $\alpha_i \to -i\alpha_i$. The first step can be done by following the standard procedure [78] of the perturbative RG analysis for the ordinary sine-Gordon model (cf. Eq. (3.26)). The vanishing of the terms proportional to g_i, g_i^3 and $g_i g_r^2$ in the RG equation (3.77) of g_r can be understood from the different parity of $\cos(2\hat{\phi})$ and $\sin(2\hat{\phi})$ potentials (the similar arguments also hold in the RG equation (3.78) of g_i).

The RG equations can be solved analytically for small $g_{r,i}$ and $\delta \equiv K - 2$ by noting the following two quantities

$$C_1 = \frac{g_i}{g_r}, \tag{3.80}$$

$$C_2 = \delta^2 - 4(g_r^2 - g_i^2) - 10\delta(g_r^2 - g_i^2) + \delta^3, \tag{3.81}$$

which are conserved up to the third order of $g_{r,i}$ and δ. The resulting three-dimensional RG phase diagram is shown in Fig. 3.4a. Its bottom plain at $g_i = 0$ gives the two-dimensional phase diagram of the conventional sine-Gordon model (see Fig. 3.2). Let us explain each regime in the phase diagram.

(i) $C_1 < 1$

When PT symmetry is unbroken, i.e., $g_i < g_r$ (or equivalently, $C_1 < 1$), the spectrum is equivalent to that of the original sine-Gordon model with a modified parameter (cf. Eq. (3.75)). Thus, the conventional RG flow diagram is reproduced (compare Fig. 3.4b with Fig. 3.2). The BKT boundary corresponds to the curved surface defined by $C_2 = 0$ with $\delta > 0$. We note that the critical behavior in the PT unbroken regime is also the same as in the original sine-Gordon model because the operator $\hat{O}$ only changes the zero modes of $\hat{\phi}$, which do not contribute to the ground-state critical properties. As the non-Hermiticity signifies the measurement backaction, the transition across the surface $C_2 = 0$ with $\delta > 0$ induced by increasing g_i may be regarded as measurement-induced.

(ii) $C_1 = 1$

A new type of transition emerges on the PT transition plane, which is defined by $g_i = g_r$ in the strongly correlated regime $K < 2$ (or equivalently, on the plane $C_1 = 1$ with $\delta < 0$). The BKT and PT phase boundaries merge on the line defined by $K = 2$

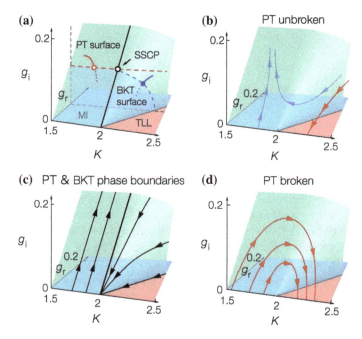

Fig. 3.4 Renormalization group (RG) phase diagram and flows in the generalized sine-Gordon model. **a** The three-dimensional phase diagram in (K, g_r, g_i), where K is the TLL parameter and g_r (g_i) describes the depth of the real (imaginary) potential. The Tomonaga-Luttinger liquid (TLL) and the Mott insulator (MI) phases are separated by the curved surface of the Berezinskii-Kosterlitz-Thouless (BKT) transition in $K > 2$ and the plain surface of the PT transition (red open circle) in $K < 2$. An exotic renormalization group (RG) fixed point lies on the critical line with $K = 2$ and $g_r = g_i$, which we term as a spectral singular critical point (SSCP). **b** RG flows in a PT-unbroken region ($g_i < g_r$) (compare it with Fig. 3.2). **c** RG flows on the two phase boundaries. **d** Unconventional semicircular RG flows in a PT-broken region ($g_i > g_r$), which violate the c-theorem [95]. The increase of K along the RG flows indicates the anomalous enhancement of the superfluid correlation and significant changes in critical exponents in a large-distance region. Reproduced from Fig. 1 of Ref. [23] by the author licensed under a Creative Commons Attribution 4.0 International License

and $g_i = g_r$ (i.e., the line $C_2 = 0$ and $C_1 = 1$), which is constituted from a set of RG fixed points (black thick line in Fig. 3.4c). On the PT threshold plane ($C_1 = 1$), the spectral singularity [61] arises where two eigenvalues as well as their eigenstates coalesce in the continuum spectrum. Our finding thus indicates that in many-body systems the coexistence of the spectral singularity and the quantum criticality can result in exotic RG fixed points unique to non-Hermitian systems. We remark that the quantum field theory on this scale invariant line $K = 2$ and $g_i = g_r$ is known as (a special type of) quantum Liouville theory [96], which has recently attracted much attention in high-energy physics. In this context, it is remarkable that Ref. [97] has demonstrated that the scale-invariant theory at $K = 2$ and $g_i = g_r$ is indeed a

3.3 Backaction on Quantum Criticality II: The Potential Term

new universality class unique to non-Hermitian systems by explicitly calculating its three-point correlation functions.

(iii) $C_1 > 1$

Unconventional RG flows emerge when the PT symmetry is broken ($g_i > g_r$ or equivalently $C_1 > 1$). They start from the $K < 2$ regime and initially lead to increases of $g_{r,i}$ and K. After entering the $K > 2$ regime, the flow winds and converges to the fixed line with $g_{r,i} = 0$ (see Fig. 3.4d), which corresponds to the TLL phase with $K > 2$. Physically, this significant increase in the TLL parameter K along the RG flows indicates that the superfluid correlation decays more slowly at a larger distance (see Eq. (3.16)). This enhancement of superfluid correlation should be considered as anomalous because, in the conventional BKT paradigm, a real potential has a completely opposite effect of destroying the criticality and stabilizing the gapped MI phase for $K < 2$ (see Fig. 3.2). It is particularly notable that the semicircular RG flows permit a substantial increase of the TLL parameter K even if the potential strength g_i is initially very small. This indicates that even a very weak measurement backaction can significantly alter the underlying critical behavior in a large-distance regime.

We remark that these RG flows violate the c-theorem, which states that in (Hermitian) conformal field theories the central charge c must monotonically decrease along RG flows when a relevant perturbation is added [95]. In the anomalous RG flows found here, the system remains in the TLL phase and thus the central charge remains $c = 1$ even though the relevant perturbation is added. It is worthwhile to mention that, while the c-theorem has recently been generalized to non-Hermitian systems [98], the unbroken PT symmetry is required in order to ensure the presence of a monotonically decreasing positive-definite function along the RG flows. Our finding of the violation of the c-theorem only in the PT broken regime is consistent with this result. We note that similar anomalous RG flows have recently been reported also in the non-Hermitian Kondo models [99, 100], where the authors found the violation of the g-theorem [101] (i.e., the monotonic decrease of a ground-state degeneracy g along the RG flows).

3.3.3 Numerical Demonstration in a Non-Hermitian Spin-Chain Model

To numerically test the RG predictions discussed above, we introduce a lattice Hamiltonian

$$\hat{H}_\mathrm{L} = \sum_{m=1}^{N} \left[-(J + (-1)^m i\gamma) \left(\hat{S}_m^x \hat{S}_{m+1}^x + \hat{S}_m^y \hat{S}_{m+1}^y \right) + \Delta \hat{S}_m^z \hat{S}_{m+1}^z + (-1)^m h_\mathrm{s} \hat{S}_m^z \right], \quad (3.82)$$

whose low-energy behavior is described by the generalized sine-Gordon model $\hat{H}$ in Eq. (3.72). Here $\hat{S}_m^{x,y,z}$ are the spin-1/2 operators at site m and the parameters

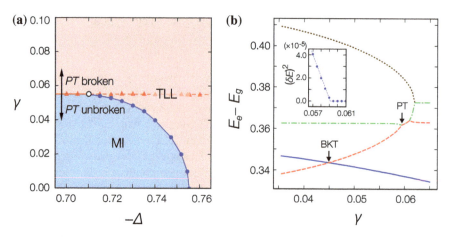

Fig. 3.5 **a** Phase diagram and **b** typical finite-size spectrum of the non-Hermitian spin-chain model whose low-energy effective field theory coincides with the generalized sine-Gordon model. In panel (**a**), the Tomonaga-Luttinger liquid (TLL) and Mott insulator (MI) phases are separated by the PT-symmetry breaking (red line with filled triangles) and the Berezinskii-Kosterlitz-Thouless (BKT) transition (blue curve with filled circles). The merging point (open circle) of the two phase boundaries corresponds to the spectral singular critical point. In panel (**b**), a typical low-energy excitation spectrum of the non-Hermitian spin-chain model is plotted. We show the three lowest levels in the $S^z = 0$ sector (red, green, and yellow curves from the lowest), and the lowest excitation energy in the $S^z = \pm 4$ sector (blue curve), where we denote $S^z = \sum_{m=1}^{N} \hat{S}_m^z$ as a total magnetization. The energy difference δE between the two coalescing levels (e.g., red and green) obeys the square-root scaling (inset) and closes at the PT-symmetry breaking point. We determine the BKT transition point from a crossing point of appropriate energy levels (red and blue). We use the staggered field $h_s = 0.1$ for both panels (**a**) and (**b**). In panel (**a**), we determine the phase boundaries by performing the finite-size scaling analysis (see panels (**d**) and (**e**) in Fig. 3.6), while the data shown in panel (**b**) are the values for $N = 16$ and $-\Delta = 0.735$. Note that we use the unit of $J = 1$. Reproduced from Fig. 2 of Ref. [23] by the author licensed under a Creative Commons Attribution 4.0 International License

$(-\Delta, h_s, \gamma)$ correspond to the ones (K, g_r, g_i) in the effective field theory (3.72). We have used this model to numerically test the RG phase diagram in Fig. 3.4a and the anomalous RG flows in Fig. 3.4d. The obtained results are consistent with the RG predictions as shown in the determined phase diagram in Fig. 3.5a and in the observed enhancement of the TLL parameter in a large-distance regime in Fig. 3.7b. Our numerical results thus demonstrate that the RG analysis is indeed instrumental to study critical properties of a non-Hermitian many-body system. Below we discuss how we obtain these results in detail.

(i) Identifying the BKT transition point

To determine the BKT transition point, we calculate the exact finite-size spectrum and find a crossing of low-energy levels having appropriate quantum numbers. This can be done by employing the so-called level spectroscopy method [102, 103]. The key idea of this method is to relate the low-energy spectrum to the running coupling constants that appear in the RG equations. Under the periodic boundary condition, the lattice

3.3 Backaction on Quantum Criticality II: The Potential Term

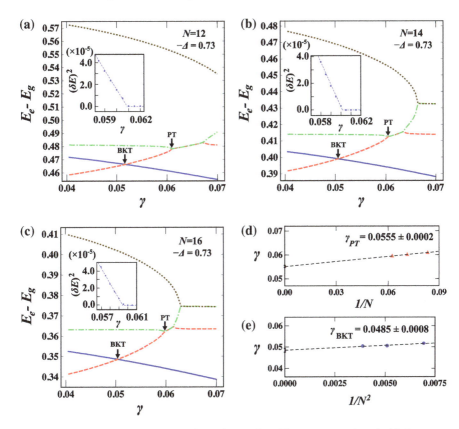

Fig. 3.6 Exact finite-size spectra for different system sizes. The spectra are plotted with the parameters $-\Delta = 0.73$ and $h_s = 0.1$ for different system sizes: **a** $N = 12$, **b** $N = 14$, and **c** $N = 16$. Here the three lowest excited levels in the $(S^z = 0, q = 0, P = T = 1)$ sector (red, green, and yellow curves from the lowest), and the lowest excitation energy in the $(S^z = \pm 4, q = 0, P = 1)$ sector (blue curve) are plotted. The BKT transition point corresponds to the crossing point of the two energy levels in $(S^z = 0, q = 0, P = T = 1)$ (red) and $(S^z = \pm 4, q = 0, P = 1)$ (blue). The PT transition point corresponds to the first coalescence point of two low-energy levels (e.g., red and green), which is confirmed to be an exceptional point of the spectrum by testing the square-root scaling of the energy difference δE between the two coalescing levels (inset). **d** The PT threshold (γ_{PT}) and **e** the BKT transition point (γ_{BKT}) are determined by extrapolating finite-size data to the thermodynamic limit. Reproduced from Supplementary Fig. 4 of Ref. [23] by the author licensed under a Creative Commons Attribution 4.0 International License

Hamiltonian is invariant with respect to spin rotation about the z axis, translation by two sites, space inversion, and spin reversal. The corresponding conserved quantum numbers are the total magnetization $S^z \equiv \sum_{i=1}^{N} S_i^z$, the wavenumber $q = 2\pi k/L$ ($k \in \mathbb{Z}$, $L \equiv N/2$), the parity $P = \pm 1$, and the spin reversal $T = \pm 1$. The ground state with energy $E_g(L)$ resides in the sector $(S^z = 0, q = 0, P = T = 1)$. Following Ref. [103], we denote the second lowest eigenenergy in this sector by E_0 and the lowest eigenenergy in the sector $(S^z = \pm 4, q = 0, P = 1)$ by E_3. Near the BKT transition line, these excitation energies satisfy [102]

$$E_0(L) - E_g(L) = \frac{2\pi v}{L}\left(2 + \frac{1}{3}\delta(l) - \frac{8}{3}g'(l)\right), \qquad (3.83)$$

$$E_3(L) - E_g(L) = \frac{2\pi v}{L}(2 - \delta(l)), \qquad (3.84)$$

where $\delta \equiv K - 2$, $g' \equiv \sqrt{g_r^2 - g_i^2}$, and the logarithmic RG scale l is related to the system size L via $e^l = L/\pi$. At the lowest order of the RG flow Eq. (3.26), the boundary of the BKT transition corresponds to $\delta = 2g'$. Since $E_0 = E_3$ is equivalent to this condition, the BKT transition point is determined from the crossing point of these two energy levels. In our model, this corresponds to the crossing of the levels shown as the red dashed line and the blue solid line in Fig. 3.5b. In numerical calculations, we obtain the excitation energy of the level ($S^z = \pm 4, q = 0, P = 1$) by multiplying that of the level ($S^z = \pm 1, q = 0, P = 1$) by a factor of 16 to minimize possible finite-size effects due to an increase in the total magnetization S^z. We note that, even though we consider a non-Hermitian model here, the level spectroscopy method is still applicable because the BKT phase boundary entirely lies within the PT-unbroken region in which the low-energy spectrum is equivalent to that of the sine-Gordon model as we have proved by Eq. (3.75). We calculate the transition point for different system sizes (see Fig. 3.6a–c), and extrapolate it to the thermodynamic limit to determine the BKT transition point (see Fig. 3.6d). Since $-\Delta = \cos(\pi/2K)$ for $h_s = \gamma = 0$ and the BKT transition occurs near $K = 2$, our analysis focuses on a region around $-\Delta = \cos(\pi/4) = 1/\sqrt{2}$.

(ii) Identifying the PT transition point

We identify the PT transition point as the first coalescence point in the low-energy spectrum with increasing the non-Hermitian term γ. To confirm that the identified point indeed represents an exceptional point of the spectrum, we plot the square of the energy difference $(\delta E)^2$ and test the square-root scaling of δE which is a signature of the coalescence of two eigenstates [60, 104] (see the inset figures in Figs. 3.5b and 3.6a–c). We then perform a linear fit to the $(\delta E)^2$-γ plot and determine the PT threshold γ_{PT} for different system sizes. Finally, we extrapolate it to the thermodynamic limit and determine the PT symmetry breaking point (see Fig. 3.6d).

(iii) Testing the anomalous RG flows

To numerically test the anomalous semicircular RG flows predicted in the PT-broken regime, we have calculated a large-distance behavior of a correlation function by employing the infinite time-evolving block decimation (iTEBD) algorithm [105]. The iTEBD is still applicable to investigate the properties of the effective ground state of interest here. The reason is that it allows us to obtain the imaginary-time evolution $\exp(-\hat{H}\tau)|\Psi_0\rangle/\|\exp(-\hat{H}\tau)|\Psi_0\rangle\|$ for an infinite system size; using a sufficiently long imaginary time τ, we can reach the effective ground state, i.e., the state having the lowest real part of eigenvalues in the entire spectrum. We remark that the imaginary part of the eigenvalue does not contribute to the calculation as

3.3 Backaction on Quantum Criticality II: The Potential Term

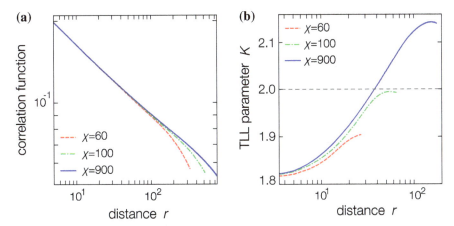

Fig. 3.7 Anomalous enhancements of superfluid correlation and the Tomonaga-Luttinger liquid parameter K in the PT-broken quantum critical phase. In panel **a**, we plot the spin-spin correlation function $\mathrm{Re}[\langle \hat{S}_r^+ \hat{S}_0^- \rangle]$, which exhibits the critical decay. In panel **b**, we plot K against the distance r. The TLL parameter is extracted by performing the linear fitting of the correlation function in the log-log plot around the distance r to determine the critical exponent, which can be related to the TLL parameter via $\langle \hat{S}_r^+ \hat{S}_0^- \rangle \propto (1/r)^{1/(2K)}$. We use $-\Delta = 0.61$, $h_s = 0.1$, and $\gamma = 0.08$, and χ denotes the bond dimension of the matrix-product state used in the iTEBD calculations. Reproduced from Fig. 3 of Ref. [23] by the author licensed under a Creative Commons Attribution 4.0 International License

it merely modifies the phase of the wavefunction. The TLL parameter K can then be extracted from the calculated correlation function via $\langle \hat{S}_r^+ \hat{S}_0^- \rangle \propto (1/r)^{1/(2K)}$ (cf. Eq. (3.16)).

Figure 3.7 shows the critical decay with a varying critical exponent; remarkably, the corresponding TLL parameter significantly increases and eventually surpasses $K = 2$ in a long-distance regime. We can interpret a physical origin of this enhancement of the superfluid correlation in the following way. In analogy with the dynamical system [106], we may imagine that a local gain-loss structure introduced by the imaginary term causes locally equilibrated flows of particles in the ground state. This indicates the enhancement of fluctuations in the density, or equivalently, the suppression of fluctuations in the conjugate phase, leading to the increase of the superfluid correlation.

3.3.4 Nonperturbative Renormalization Group Analysis

We have so far analyzed the perturbative regime in which both $g_{r,i}$ and $\delta = K - 2$ are small. The performed perturbative RG analysis should remain valid only in this restricted region. Yet, the anomalous RG flows found in the PT broken regime (see Fig. 3.4d) indicate that the strength of g_i can be significantly large in the intermediate region of the flows if we start from a large value of $-\delta = 2 - K > 0$. In such a nonperturbative regime, the perturbative analysis is expected to be no longer valid. It is natural to ask the fate of the anomalous RG flows in a large coupling regime.

To address this question, we employ the nonperturbative analysis based on the functional renormalization group (FRG). In general, a FRG equation is defined by [107, 108]

$$\Lambda \partial_\Lambda \Gamma_\Lambda[\phi] = \frac{1}{2} \text{Tr} \left[\frac{\Lambda \partial_\Lambda R_\Lambda}{\Gamma_\Lambda^{(2)}[\phi] + R_\Lambda} \right], \quad (3.85)$$

where Γ_Λ is the effective action for the field ϕ at a momentum scale Λ, $\Gamma_\Lambda^{(2)}$ is the second functional derivative of the effective action with respect to ϕ, Tr stands for the integration over all momenta, and R_Λ is the regulator function which we choose to be

$$R_\Lambda(p) = p^2 r(y), \quad y = p^2 / \Lambda^2, \quad (3.86)$$

where we use the power-law regulator function $r(y) = 1/y$. Using the truncated derivative expansion with the action being expanded in powers of the derivative of the field, we get

$$\Gamma_\Lambda[\phi] = \int d^2x \left[\frac{1}{2} z_\Lambda(\phi)(\partial_\mu \phi)^2 + V_\Lambda(\phi) + \cdots \right], \quad (3.87)$$

where $z_\Lambda(\phi)$ is the wavefunction renormalization factor and $V_\Lambda(\phi)$ is the momentum-dependent potential term. The sine-Gordon model can be analyzed by solving the FRG equation over the functional subspace spanned by the following ansatz:

$$\Gamma_\Lambda[\phi] = \int d^2x \left[\frac{1}{2} z_\Lambda(\partial_\mu \phi)^2 + g_{r,\Lambda} \cos(2\phi) \right], \quad (3.88)$$

where $g_{r,\Lambda}$ is the momentum-dependent depth of the real potential, z_Λ is the momentum-dependent (but field-independent) wavefunction renormalization that can be related to the TLL parameter K by

$$z = \frac{1}{4\pi K}. \quad (3.89)$$

For the power-law regulator (3.86), the momentum integrals can be analytically performed. The resulting RG equations are [109, 110]:

$$\frac{dg_r}{dl} = -\frac{1}{2\pi z g_r} \left[1 - \sqrt{1 - g_r^2} \right] + 2g_r, \quad (3.90)$$

3.3 Backaction on Quantum Criticality II: The Potential Term

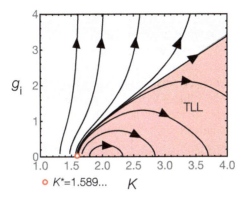

Fig. 3.8 RG flows of the generalized sine-Gordon model in the nonperturbative regime of the plane $g_r = 0$ (PT broken phase). In the vicinity of $K \simeq 2$, the RG flows reproduce the semicircular flows found in the perturbative analysis (cf. Fig. 3.4d). Below $K < K^*$, the flows no longer go back to the TLL fixed line and g_i flows into the strong-coupling limit

$$\frac{dz}{dl} = \frac{1}{24\pi} \frac{g_r^2}{\left(1 - g_r^2\right)^{3/2}}, \quad (3.91)$$

where $l = -\ln \Lambda$ is the logarithmic RG scale. These equations reproduce the well-known RG phase diagram of the sine-Gordon model [110] (cf. Fig. 3.2). We note that for small coupling g_r the equations are consistent with the perturbative results in Eq. (3.26).[8]

$g_r = 0$ and $g_i \neq 0$ case: isine-Gordon model

We now move onto the analysis of the generalized sine-Gordon model. For the sake of simplicity, we first focus on the model with only an imaginary potential, namely, the isine-Gordon model:

$$\Gamma_\Lambda = \int d^2x \left[\frac{1}{2} z_\Lambda (\partial_\mu \phi)^2 - i g_{i,\Lambda} \cos(\phi) \right], \quad (3.92)$$

where g_i is the depth of the imaginary potential. Analytically continuing Eqs. (3.90) and (3.91) by replacing $g_r \to -ig_i$, we obtain the following RG equations:

$$\frac{dg_i}{dl} = \frac{1}{2\pi z g_i} \left[1 - \sqrt{1 + g_i^2} \right] + 2g_i, \quad (3.93)$$

$$\frac{dz}{dl} = -\frac{1}{24\pi} \frac{g_i^2}{\left(1 + g_i^2\right)^{3/2}}. \quad (3.94)$$

---

[8] Note that the definitions of g_r in the perturbative RG analysis and the FRG analysis coincide aside from a constant factor.

The corresponding RG phase diagram is shown in Fig. 3.8. In the vicinity of the perturbative regime around $K = 2$, we reproduce the semicircular RG flows directing into the TLL fixed line in $K > 2$, which are consistent with the perturbative analysis (see Fig. 3.4d). These anomalous flows end up with a large TLL parameter $K > 2$ and persist even in the nonperturbative regime as long as the initial TLL parameter K is larger than a certain value $K^* \simeq 1.589$. As we approach K^* from above, the TLL parameter diverges. At the threshold value $K = K^*$, we find a new type of phase transition; the separatrix starting from this point (red open circle in Fig. 3.8) defines the phase boundary. We note that the curve separating the two phases is asymptotically given by $g_i \simeq K$ for large g_i and K. Below the threshold $K < K^*$, the depth g_i of the imaginary potential grows up to the strong-coupling limit. It remains an important open question to address in detail the nature of this phase; in analogy to the continuous quantum Zeno effect, it is expected that the diverging g_i (i.e., the strong measurement strength) will make the field ϕ localized and thus this phase can be a gapped, noncritical phase.

General case in the generalized sine-Gordon model

For the sake of completeness, we here provide the RG equations for a general case of the PT-symmetric sine-Gordon model in Eq. (3.72). To do so, let us begin with the Hermitian model:

$$\Gamma_\Lambda = \int d^2x \left[\frac{1}{2} z_\Lambda (\partial_\mu \phi)^2 + g_{r,\Lambda} \cos(\phi) + g_{i,\Lambda} \sin(\phi) \right]$$

$$= \int d^2x \left[\frac{1}{2} z_\Lambda (\partial_\mu \phi)^2 + \sqrt{g_{r,\Lambda}^2 + g_{i,\Lambda}^2} \cos(\phi - \theta_0) \right], \quad (3.95)$$

where $\theta_0 = \arctan[g_i/g_r]$. Because of the translational invariance $\phi \to \phi + \theta_0$, this model is equivalent to the sine-Gordon model in Eq. (3.88). We thus obtain the RG equation by replacing g_r with $\sqrt{g_r^2 + g_s^2}$ in Eqs. (3.90) and (3.91):

$$g_r \frac{dg_r}{dl} + g_i \frac{dg_i}{dl} = -\frac{1}{2\pi z} \left[1 - \sqrt{1 - (g_r^2 + g_i^2)} \right] + 2(g_r^2 + g_i^2), \quad (3.96)$$

$$\frac{dz}{dl} = \frac{1}{24\pi} \frac{g_r^2 + g_i^2}{\left(1 - (g_r^2 + g_i^2)\right)^{3/2}}. \quad (3.97)$$

Since the constant term θ_0 should be invariant under the RG transformation, we obtain:

$$\frac{g_i}{g_r} = \text{const.} \iff \frac{d}{dl}\left(\frac{g_i}{g_r}\right) = 0 \iff \frac{dg_i/dl}{dg_r/dl} = \frac{g_i}{g_r}. \quad (3.98)$$

From Eqs. (3.96), (3.97), and (3.98), we obtain the RG equations for g_r, g_i, and z as follows:

3.3 Backaction on Quantum Criticality II: The Potential Term

$$\frac{dg_r}{dl} = -\frac{g_r}{2\pi z(g_r^2 + g_i^2)}\left[1 - \sqrt{1 - (g_r^2 + g_i^2)}\right] + 2g_r, \quad (3.99)$$

$$\frac{dg_i}{dl} = -\frac{g_i}{2\pi z(g_r^2 + g_i^2)}\left[1 - \sqrt{1 - (g_r^2 + g_i^2)}\right] + 2g_i, \quad (3.100)$$

$$\frac{dz}{dl} = \frac{1}{24\pi}\frac{g_r^2 + g_i^2}{(1 - (g_r^2 + g_i^2))^{3/2}}. \quad (3.101)$$

We now consider the PT-symmetric non-Hermitian sine-Gordon model:

$$\Gamma_\Lambda = \int d^2x \left[\frac{1}{2}z_\Lambda(\partial_\mu\phi)^2 + g_{r,\Lambda}\cos(\phi) - ig_{i,\Lambda}\sin(\phi)\right]. \quad (3.102)$$

To obtain the RG equations of this model, we analytically continue Eqs. (3.99), (3.100), and (3.101) by replacing $g_i \to -ig_i$:

$$\frac{dg_r}{dl} = -\frac{g_r}{2\pi z(g_r^2 - g_i^2)}\left[1 - \sqrt{1 - (g_r^2 - g_i^2)}\right] + 2g_r, \quad (3.103)$$

$$\frac{dg_i}{dl} = -\frac{g_i}{2\pi z(g_r^2 - g_i^2)}\left[1 - \sqrt{1 - (g_r^2 - g_i^2)}\right] + 2g_i, \quad (3.104)$$

$$\frac{dz}{dl} = \frac{1}{24\pi}\frac{g_r^2 - g_i^2}{(1 - (g_r^2 - g_i^2))^{3/2}}. \quad (3.105)$$

Several remarks are in order. Firstly, on the PT transition boundary ($g_i = g_r$) with $K < 2$, the unconventional RG flows directing into the strong-coupling limit (cf. Fig. 3.4c) still exist in the nonperturbative regime. Secondly, the fixed points on the line $K = 2$ and $g_r = g_s$ found in the perturbative analysis remain to be fixed points even in the nonperturbative regime. Finally, as we found in the previous subsection for the $g_r = 0$ case, a new type of phase transition between the TLL phase and a possible noncritical phase with the strong imaginary potential will appear. We note that this transition has not been found in the perturbative analysis. The 2D phase boundary for this transition can be asymptotically given by the surface $g_i^2 - g_r^2 \simeq K^2$ for large g_i.

3.3.5 Realization in Ultracold Gases with a One-Body Loss

We here discuss a possible physical realization of the generalized sine-Gordon model (3.72) using ultracold atoms (see Fig. 3.9a). We start from a 1D Bose gas subject to an off-resonant optical lattice. This system is described by the Lieb-Liniger model [111] with the periodic potential:

$$\hat{\mathcal{H}} = \int dx \left\{\hat{\psi}^\dagger(x)\left[-\frac{\hbar^2\nabla^2}{2m} + V_r\cos\left(\frac{2\pi x}{d}\right)\right]\hat{\psi}(x) + \frac{g}{2}\hat{\psi}^\dagger(x)\hat{\psi}^\dagger(x)\hat{\psi}(x)\hat{\psi}(x)\right\}, \quad (3.106)$$

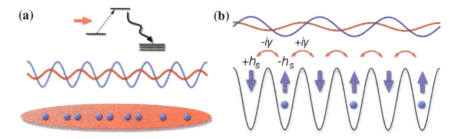

Fig. 3.9 Possible experimental realization of the generalized sine-Gordon model in a one-dimensional dissipative ultracold gases. **a** One-dimensional ultracold bosonic atoms are subject to a complex potential whose real (blue curve) and imaginary (red curve) parts are generated by far-detuned and weak near-resonant waves. The latter induces a dissipative effect corresponding to a one-body loss. **b** A non-Hermitian spin-chain model whose low-energy universal behavior reproduces that of the generalized sine-Gordon model. An additional deep optical lattice, which does not affect the universal behavior, is introduced to tightly localize atoms. The real and the imaginary parts of the complex potential lead to the staggered potentials $\pm h_s$ and imaginary hoppings $\pm i\gamma$ in the lattice model. We represent a lattice site occupied (not occupied) by a hard-core boson as the up (down) spin. Reproduced from Fig. 4 of Ref. [23] by the author licensed under a Creative Commons Attribution 4.0 International License

where $\hat{\Psi}(x)$ denotes the bosonic field operator, m is the mass of an atom, V_r is a lattice depth, d is a lattice constant, and g is an interaction strength between atoms. We then introduce a dissipative optical lattice potential causing a one-body loss of atoms, which can be created by a weak near-resonant light. We can interpret an atomic loss as a one-body loss if the total decay rate Γ of the excited state $|e\rangle$ is faster than the spontaneous emission rate from $|e\rangle$ to the ground state $|g\rangle$ and the Rabi frequency Ω [32, 112, 113] (see Fig. 3.10). This scheme can be possible by, e.g., employing appropriate energy levels [114] or recoil energies due to light-induced transitions [1]. From the second-order perturbation theory [60] with respect to the Rabi coupling, we can adiabatically eliminate the excited state, resulting in the following effective time-evolution equation of the ground-state atoms (see Appendix A for the detailed derivation):

$$\frac{d\hat{\rho}}{dt} = -\frac{i}{\hbar}\left(\hat{\mathcal{H}}_{\text{eff}}\hat{\rho} - \hat{\rho}\hat{\mathcal{H}}_{\text{eff}}^{\dagger}\right) + \int dx \frac{|\Omega(x)|^2}{\Gamma}\hat{\Psi}(x)\hat{\rho}\hat{\Psi}^{\dagger}(x), \quad (3.107)$$

$$\hat{\mathcal{H}}_{\text{eff}} \equiv \hat{\mathcal{H}} - i\hbar \int dx \frac{|\Omega(x)|^2}{2\Gamma}\hat{\Psi}^{\dagger}(x)\hat{\Psi}(x). \quad (3.108)$$

In general, the dissipative (unconditional) dynamics described by a master equation tend to destroy subtle correlations underlying quantum critical phenomena. In contrast, we are here interested in the conditional non-Hermitian evolution, where the dynamics is free from the destructive jump processes while nontrivial effects due to measurement backaction still appear via non-Hermiticity of the effective Hamiltonian (3.108). Such conditional dynamics can be studied by employing postselections

3.3 Backaction on Quantum Criticality II: The Potential Term

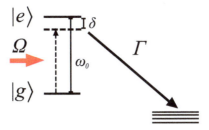

Fig. 3.10 Energy-level diagram of an atom. The excited state $|e\rangle$ has the frequency ω_0 relative to the ground state $|g\rangle$ and fast decay modes with the total decay rate Γ. A weak near-resonant light with the Rabi frequency Ω and detuning δ creates an effective imaginary potential for the ground-state atom, provided that Γ is much larger than the spontaneous decay rate from $|e\rangle$ to $|g\rangle$. Reproduced from Supplementary Fig. 1 of Ref. [23] by the author licensed under a Creative Commons Attribution 4.0 International License

[12, 13]; the achieved experimental fidelity has already been high enough to allow experimenters to implement various types of postselections [15–17] (see also Sect. 3.5).

In the effective Hamiltonian (3.108), the spatially modulated Rabi frequency $\Omega(x)$ is proportional to the electric field creating a dissipative optical lattice. To realize the model (3.72), we choose $|\Omega(x)|^2 = \Omega^2(1 + \sin(2\pi x/d))/2$ such that the effective non-Hermitian Hamiltonian becomes

$$\hat{\mathcal{H}}_{\text{eff}} = \int dx \left\{ \hat{\Psi}^\dagger(x) \left[-\frac{\hbar^2 \nabla^2}{2m} + V(x) \right] \hat{\Psi}(x) + \frac{g}{2} \hat{\Psi}^\dagger(x)\hat{\Psi}^\dagger(x)\hat{\Psi}(x)\hat{\Psi}(x) \right\}, \quad (3.109)$$

where we introduce the complex potential

$$V(x) = V_{\text{r}} \cos\left(\frac{2\pi x}{d}\right) - i V_{\text{i}} \sin\left(\frac{2\pi x}{d}\right), \quad (3.110)$$

$$V_{\text{i}} = \frac{\hbar \Omega^2}{4\Gamma}. \quad (3.111)$$

Here, we ignore the constant term $-iV_{\text{i}}N$ proportional to the total number N of atoms. This constant term is irrelevant in the non-Hermitian dynamics because it is cancelled upon the normalization of the quantum state. Note that the effective Hamiltonian (3.109) satisfies the PT symmetry because the potential satisfies the condition $V(x) = V^*(-x)$.

The low-energy effective field theory of $\hat{\mathcal{H}}_{\text{eff}}$ in Eq. (3.109) can be obtained by following the standard procedure [77]. Firstly, as we have discussed in the previous sections, an interacting 1D Bose gas without the potential $V(x)$ is described at low energies by the TLL Hamiltonian (3.3), where the fields $\hat{\phi}$ and $\hat{\theta}$ are related to the bosonic field $\hat{\Psi}(x)$ via Eqs. (3.11) and (3.15). Secondly, we then discuss the perturbative role of the potential term

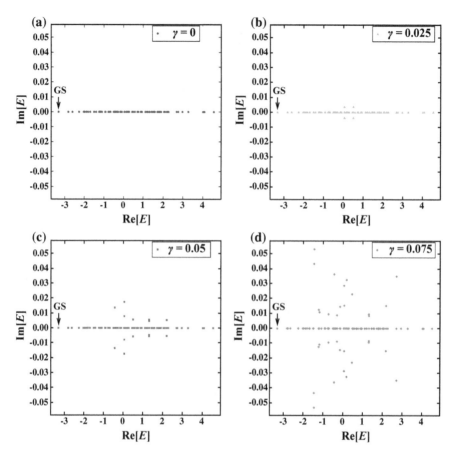

Fig. 3.11 Spectra of the lattice spin-chain model (3.82) plotted with the parameters $-\Delta = 0.5$, $h_s = 0.0$, $N = 12$ for different strengths of the imaginary hopping **a** $\gamma = 0$, **b** $\gamma = 0.025$, **c** $\gamma = 0.05$, and **d** $\gamma = 0.075$. When the PT symmetry is broken, some of the excited states have pairs of complex eigenvalues which are conjugate to each other, while the ground state remains to have a real eigenvalue. The plotted energy levels reside in the sector ($S^z = 0, q = 0, P = T = 1$). The ground state (GS) is indicated by the black arrow. Reproduced from Supplementary Fig. 2 of Ref. [23] by the author licensed under a Creative Commons Attribution 4.0 International License

$$\hat{H}_V = \int dx\, V(x) \hat{\Psi}^\dagger(x)\hat{\Psi}(x) = \int dx\, V(x)\hat{\rho}(x). \quad (3.112)$$

Since we are interested in the commensurate phase transition, we assume the unit filling $\rho_0 d = 1$, i.e., one atom per site. By substituting Eqs. (3.110) and (3.15) into Eq. (3.112) and ignoring fast oscillating terms, we obtain [87]

$$\hat{H}_V = \rho_0 V_\mathrm{r} \int dx \cos\left[2\hat{\phi}(x)\right] - i\rho_0 V_\mathrm{i} \int dx \sin\left[2\hat{\phi}(x)\right]. \quad (3.113)$$

3.3 Backaction on Quantum Criticality II: The Potential Term

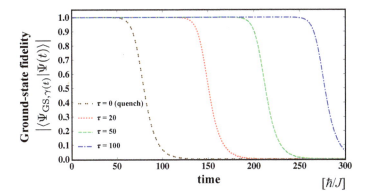

Fig. 3.12 Ground-state fidelity in the PT-broken regime. The time evolution of the ground-state fidelity of the system in the PT-broken regime is plotted for several different values of τ. The imaginary hopping term γ is ramped up with different timescales $\tau = 0, 20, 50, 100$. The ground state $|\Psi_{GS,\gamma(t)}\rangle$ is calculated from the exact diagonalization of the Hamiltonian at each time step. The parameters are set to $-\Delta = 0.5$, $h_s = 0$, $\gamma = 0.05$, and $N = 12$. Reproduced from Supplementary Fig. 3 of Ref. [23] by the author licensed under a Creative Commons Attribution 4.0 International License

Defining $\alpha_{r,i} \equiv \pi\rho_0 V_{r,i}$, we arrive at the PT-symmetric potential term in the generalized sine-Godron model (3.72). It is worthwhile to mention that the lattice Hamiltonian (3.82) can also be realized by introducing a deep lattice, which has a period $d/2$ and does not affect the universal behavior, in addition to the complex potential (see Fig. 3.7b).

Remark on the dynamics in the PT-broken regime

We make a remark on the ground state of the PT-symmetric spin-chain model (3.82) in the PT-broken regime. When the PT symmetry is broken, some excited eigenstates turn out to have complex pairs of eigenvalues while the ground state remains to have a real eigenvalue (see Fig. 3.11a–d for typical spectra). In particular, there exist high-lying unstable modes having positive imaginary parts of eigenvalues. As a result, if the system is significantly perturbed and highly excited, the amplitudes of these modes can grow in time and eventually govern the physical properties of the system. This is reminiscent of the phenomenon known as parametric instability or self-pulsing [115] in exciton-polariton systems, which can destroy the off-diagonal quasi-long-range order in 1D Bose systems [116–118].

In contrast, our main focus here is on the effective ground state that sustains the quantum critical behavior. We stress that this state can indeed be relevant in our setup, where the system is initially prepared in the zero-temperature state of the Hermitian Hamiltonian and then the imaginary part of the potential is adiabatically ramped up. We numerically demonstrate in Fig. 3.12 that the system remains in the ground state with almost unit fidelity for a long-time interval. Here we consider the spin-chain model in Eq. (3.82) and adiabatically ramp up the imaginary term with the time dependence $\gamma(t) = \gamma \times \left(1 - 2/\left(e^{(t/\tau)^2} + 1\right)\right)$, where τ characterizes the timescale of the operation. The initial state $|\Psi(0)\rangle$ is chosen to be the ground state of the Hamiltonian with $\gamma(0) = 0$, and the time evolution $|\Psi(t)\rangle$ is calculated by diagonalizing the Hamiltonian at each time step. Figure 3.12 shows the ground-state fidelity $|\langle \Psi_{\text{GS},\gamma(t)}|\Psi(t)\rangle|$ of the instantaneous Hamiltonian with $\gamma(t)$, indicating that the system remains in the ground state for a time much longer than the ramping time τ. Using a typical experimental time scale $\hbar/J = 3.6/(2\pi)$ ms [119], the lifetime of the ground state can reach ~ 150 ms, which is sufficiently long compared with a typical operation time of ultracold atom experiments [119].[9]

3.3.6 Short Summary

It is useful to make a short summary of this subsection.

- Generalizing the sine-Gordon model to the PT-symmetric non-Hermitian case, we find that the two types of phase transitions emerge depending on the TLL parameter K. One belongs to the celebrated BKT transition that appears in the weakly interacting regime ($K > 2$). The other is the PT transition found in the strongly interacting regime ($K < 2$). The latter accompanies the spectral singularity, i.e., the nondiagonalizability of the Hamiltonian, and thus is unique to a non-Hermitian system.
- Using the perturbative and functional RG analyses, we determine the phase diagram of the generalized sine-Gordon model and find that the BKT and PT phase transition surfaces are continuously connected. The connecting line consists of nontrivial RG fixed points. On this fixed line, the theory is scale invariant and constitutes the universality class that has no counterpart in Hermitian systems.
- The RG analyses predict that, in the PT-broken phase, there appear the unconventional semicircular RG flows that violate the c-theorem. In this regime, the TLL parameter K can increase to $K > 2$ even if the flow starts from the region $K < 2$. Physically, this indicates that the competition between the local gain-loss structure in the PT-symmetric system and strong correlations can result in the significant enhancement of the superfluidity.
- Using the exact diagonalizations and the iTEBD algorithm, we numerically confirm the above RG predictions. The analyses clearly demonstrate that the RG analysis is indeed useful to study critical properties of a non-Hermitian many-body system.
- In the nonperturbative regime of the PT-broken phase, the RG flows starting from the strongly interacting regime ($K < K^* \simeq 1.589$) flow into the strong coupling

---

[9]While the system size is rather small here, we note that the first signature of the enhancement of superfluid correlation can appear from a relatively small size of, say, ~ 10 sites (see Fig. 3.7b).

limit of the imaginary potential. While we speculate that this phase is noncritical, its nature merits further study.
• Our findings can be experimentally tested by a 1D Bose gas subject to a spatially modulated one-body loss (see Sect. 3.5 for experimental situations).

3.4 Backaction on Quantum Phase Transitions

We have so far studied the influence of measurement backaction relevant to 1D quantum critical phenomena. In this section, we discuss its influence on quantum phase transitions in higher dimensions by taking the Bose-Hubbard model as a concrete physical example. This model exhibits the superfluid-to-Mott insulator transition [92] and has been realized in ultracold atoms trapped in an optical lattice [120, 121]. We are interested in how the backaction from an external observer affects the quantum critical point.

3.4.1 Model

The Bose-Hubbard Hamiltonian [92] is defined by

$$\hat{H}_{\text{BH}} = \hat{H}_0 + \hat{V}, \tag{3.114}$$

$$\hat{H}_0 = \frac{U}{2} \sum_i \hat{n}_i (\hat{n}_i - 1) - \mu \sum_i \hat{n}_i, \tag{3.115}$$

$$\hat{V} = -J \sum_{\langle i,j \rangle} (\hat{b}_i^\dagger \hat{b}_j + \text{H.c.}). \tag{3.116}$$

Here, $\hat{b}_i^\dagger$ ($\hat{b}_i$) is a creation (annihilation) operator of bosons at site i, $\hat{n}_i \equiv \hat{b}_i^\dagger \hat{b}_i$, U represents the strength of the on-site interaction, J characterizes the hopping amplitude, and μ is the chemical potential. If the interaction energy is dominant (i.e., $J \ll U$), the ground state is the gapped Mott insulator phase. With an increase in the kinetic energy J/U, the phase transition to the superfluid phase occurs [122] at the critical point $(J/U)_c$, where the energy gap closes. In the $\mu - J$ phase diagram, a tip of each Mott lobe corresponds to a critical value $(J/U)_c$ at an integer filling ρ (as illustrated in Fig. 3.14d). Under continuous monitoring, we consider an effective non-Hermitian Hamiltonian $\hat{H}_{\text{eff}}$

$$\hat{H}_{\text{eff}} = \hat{H}_{\text{BH}} - \frac{i\gamma}{2} \sum_i \hat{L}_i^\dagger \hat{L}_i, \tag{3.117}$$

where a jump operator $\hat{L}_i$ characterizes a measurement process performed and γ is the measurement strength. Unless the second term commutes with $\hat{H}_{\text{BH}}$, the measurement backaction shifts the energies of the effective Hamiltonian. Such a measurement-induced shift can manifest itself as a shift in the quantum critical point. A specific form of a jump operator $\hat{L}_i$ is determined by an underlying measurement process; for example, if the process accompanies a two-body loss [85], we take $\hat{L}_i = \hat{b}_i^2$. Another example is a site-resolved measurement of atoms [1] corresponding to $\hat{L}_i = \hat{n}_i$.

3.4.2 Measurement-Induced Shift of the Quantum Critical Point

Mean-field analysis

We first study the shift of the quantum phase transition point by a mean-field analysis [92]. To be concrete, we consider the unit filling $\rho = 1$ in this subsection. We denote the annihilation operator as $\hat{b}_i = \beta + \delta\hat{b}_i$ with a mean field β, substitute it in the kinetic energy $\hat{V}$, and neglect the second-order terms of $\delta\hat{b}_i$. The resulting mean-field Hamiltonian is given by

$$\hat{H}_{\text{eff}}^{\text{MF}} = \hat{V}^{\text{MF}} + \hat{H}_0 - \frac{i\gamma}{2}\sum_i \hat{L}_i^\dagger \hat{L}_i, \tag{3.118}$$

$$\hat{V}^{\text{MF}} = -Jz\sum_i (\beta^*\hat{b}_i + \beta\hat{b}_i^\dagger - |\beta|^2), \tag{3.119}$$

where z is the number of neighboring sites. The lowest real part of the eigenvalues of $\hat{H}_{\text{eff}}^{\text{MF}}$ provides the effective ground-state energy $E_{\beta,\gamma}$ of the system. In the vicinity of the transition point, we can expand it as $E_{\beta,\gamma} = a_0 + a_2(\gamma)|\beta|^2 + a_4(\gamma)|\beta|^4 + \cdots$. We determine the coefficient $a_2(\gamma)$ from the second-order perturbation with respect to $\hat{V}^{\text{MF}}$. Then, the condition $a_2(\gamma) = 0$ determines the phase boundary $(J/U)_\gamma$. The tip of the Mott lobe (i.e., $\partial(J/U)_\gamma/\partial\mu = 0$ under $\partial^2(J/U)_\gamma/\partial\mu^2 < 0$) corresponds to the critical point $(J/U)_{c,\gamma}$. We define the relative shift of the transition point by

$$\Delta_{c,\gamma} \equiv \frac{(J/U)_{c,\gamma} - (J/U)_{c,0}}{(J/U)_{c,0}}, \tag{3.120}$$

which can be obtained in the mean-field analysis as

$$\Delta_{c,\gamma} = \frac{2+\sqrt{2}}{2}\left(\sqrt{2}c_0^2 + c_1^2\right)\left(\frac{\gamma}{U}\right)^2 + \frac{5}{16}(4 + 3\sqrt{2})\left(2c_0^2 - c_1^2\right)^2\left(\frac{\gamma}{U}\right)^4 + o\left(\left(\frac{\gamma}{U}\right)^6\right). \tag{3.121}$$

We define the coefficient by $c_\rho \equiv (\langle\rho+1|\hat{L}^\dagger\hat{L}|\rho+1\rangle - \langle\rho|\hat{L}^\dagger\hat{L}|\rho\rangle)/2$, in which we omit the site label i. For instance, it takes $c_\rho = \rho$ for the two-body loss process $\hat{L} = \hat{b}^2$. From Eq. (3.121), we can conclude that a general measurement process

3.4 Backaction on Quantum Phase Transitions

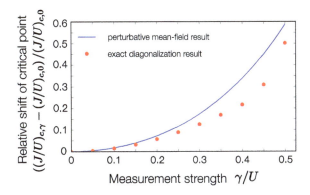

Fig. 3.13 Relative amount of the shift $\Delta_{c,\gamma}$ of the quantum critical point due to the measurement backaction plotted against the measurement strength γ/U. To be concrete, we choose $\hat{L}_i = \hat{b}^2$, in which a quantum jump process leads to a two-body loss of atoms. We compare the perturbative mean-field result given by Eq. (3.121) (blue solid curve) with the numerical results (red dots) calculated by the exact diagonalization (see the text for details). Reproduced from Fig. 2 of Ref. [10]. Copyright © 2016 by the American Physical Society

shifts the quantum critical point in favor of the Mott phase. In the case of two-body loss or site-resolved measurement, this shift is a reminiscent of the suppression of hopping due to the continuous quantum Zeno effect [85, 90, 91, 123–125]. We note that the strong-coupling-expansion analysis presented below indicates that the effective ground state has the lowest imaginary part of the eigenvalue (as well as its real part) and thus survives longest in the non-Hermitian dynamics. We also obtain the transition point by exactly diagonalizing the non-Hermitian Hamiltonian. To be consistent with the mean-field analysis, we choose an infinite-range hopping with the amplitude J/N. Figure 3.13 shows that the perturbative formula (3.121) is indeed valid for small γ/U. In numerics, we determine the transition point by locating the point corresponding to the minimal gap.

Strong-coupling-expansion analysis

In the $\mu - J$ diagram shown in Fig. 3.14d, atom-number fluctuations at each site are enhanced in the regimes close to integer values of μ/U (see Fig. 3.14d). In such regimes, we thus have to take into account a statistical mixture of Mott states with different fillings. This can be done by performing the strong-coupling-expansion analysis [126, 127], which provides asymptotically exact results in the limit of $J/U \to 0$. We perform a degenerate perturbation theory with respect to the hopping term $\hat{V}$, and calculate the gap Δ of the eigenenergies between the ground state and the first excited state. We determine the phase boundary as the point at which the energy gap closes $\Delta = 0$. As a consequence, we can obtain an analytical expression for the shifted phase boundary at near integer values of μ/U, showing the characteristic expansion of the Mott lobe due to measurement backaction.

Let us now describe details about the analysis. As mentioned above, we consider the Mott lobes with integer filling $\rho = 1, 2, \ldots$ on a d-dimensional hypercubic lattice.

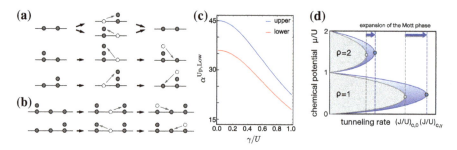

Fig. 3.14 Second-order processes in the strong-coupling-expansion analysis. Panels (**a**) and (**b**) show the processes contributing to the diagonal and off-diagonal matrix elements, respectively. The occupied lattice sites are represented by the filled circles, while holes are indicated by the dashed circles. **c** Coefficients $\alpha_{d,\rho,\gamma}^{\text{Up,Low}}$ for the upper (blue) and lower (red) boundaries of the Mott lobe (see the main text for the definitions of $\alpha_{d,\rho,\gamma}^{\text{Up,Low}}$). We consider $\hat{L}_i = \hat{b}_i^2$ with $d = 3$ and $\rho = 1$. **d** Schematic illustration of the modified μ-J diagram of the non-Hermitian Bose-Hubbard model. The measurement backaction triggers the expansion of the Mott lobe, resulting in the shift of the quantum critical point. Reproduced from Fig. 3 of Ref. [10]. Copyright © 2016 by the American Physical Society

The unperturbed Hamiltonian is

$$\hat{H}_0 = \frac{U}{2}\sum_i \hat{n}_i(\hat{n}_i - 1) - \mu \sum_i \hat{n}_i - \frac{i\gamma}{2}\sum_i \hat{L}_i^\dagger \hat{L}_i. \tag{3.122}$$

Its ground state is still in the Mott-insulator state. The first excited states in the regime close to the upper phase boundary are degenerate states with only a single site being occupied by $(\rho + 1)$ bosons while the other sites being occupied by ρ bosons. In contrast, close to the lower boundary, the first excited states are degenerate with only a single site being occupied by $(\rho - 1)$ bosons while the other sites being occupied by ρ bosons. Using a degenerate perturbation theory [127] with respect to the hopping term $\hat{V} = -J\sum_{\langle i,j \rangle}\left(\hat{a}_i^\dagger \hat{a}_j + \text{H.c.}\right)$ up to the second order, we obtain complex eigenvalues of the ground and first excited states (see Fig. 3.14a, b for the relevant hopping processes). We define the energy gaps by the differences of the real parts of the eigenvalues between the ground and first excited states. They are obtained as

$$\Delta_{d,\rho}^{\text{Up}} = -2d(\rho+1)J + \rho U - \mu - \frac{2d\rho(\rho+1)(2d-3)J^2}{U+(\gamma^2/U)(c_{M,\rho-1}-c_{M,\rho})^2} - \frac{d\rho(\rho+2)J^2}{U+(\gamma^2/4U)(c_{M,\rho-1}-c_{M,\rho+1})^2}, \tag{3.123}$$

$$\Delta_{d,\rho}^{\text{Low}} = -2d\rho J - (\rho-1)U + \mu - \frac{2d\rho(\rho+1)(2d-3)J^2}{U+(\gamma^2/U)(c_{M,\rho-1}-c_{M,\rho})^2} - \frac{d(\rho-1)(\rho+1)J^2}{U+(\gamma^2/4U)(c_{M,\rho-1}-c_{M,\rho+1})^2}, \tag{3.124}$$

where Δ^{Up} and Δ^{Low} represent the gaps in the vicinity of the upper and lower boundaries of the Mott lobe. The phase boundary can be obtained from the gap closing points:

3.4 Backaction on Quantum Phase Transitions

$$\left(\frac{\mu}{U}\right)^{\text{Up}}_{c,\gamma} = \rho - 2d(\rho+1)\frac{J}{U} - \alpha^{\text{Up}}_{d,\rho,\gamma}\left(\frac{J}{U}\right)^2, \tag{3.125}$$

$$\left(\frac{\mu}{U}\right)^{\text{Low}}_{c,\gamma} = \rho - 1 + 2d\rho\frac{J}{U} + \alpha^{\text{Low}}_{d,\rho,\gamma}\left(\frac{J}{U}\right)^2, \tag{3.126}$$

where the coefficients $\alpha^{\text{Up,Low}}_{d,\rho,\gamma}$ are given by

$$\alpha^{\text{Up}}_{d,\rho,\gamma} = \frac{2d\rho(\rho+1)(2d-3)}{1+(\gamma/U)^2(c_{\rho-1}-c_\rho)^2} + \frac{d\rho(\rho+2)}{1+(\gamma/2U)^2(c_{\rho-1}-c_{\rho+1})^2}, \tag{3.127}$$

$$\alpha^{\text{Low}}_{d,\rho,\gamma} = \frac{2d\rho(\rho+1)(2d-3)}{1+(\gamma/U)^2(c_{\rho-1}-c_\rho)^2} + \frac{d(\rho-1)(\rho+1)}{1+(\gamma/2U)^2(c_{\rho-1}-c_{\rho+1})^2}. \tag{3.128}$$

The measurement backaction $\gamma > 0$ decreases these coefficients α as shown in Fig. 3.14c, resulting in effective expansions of the Mott lobes as we can infer from Eqs. (3.125) and (3.126) (see Fig. 3.14d for an illustration). We remark that the results (3.125) and (3.126) reproduce the known results for the (Hermitian) Bose-Hubbard model [126] in the limit $\gamma \to 0$.

We remark that the effective ground state discussed here has the minimal decay rate (i.e., the longest lifetime) in the non-Hermitian evolution. To be specific, we consider the case of $\rho = 1$ and $d = 3$. The decay rates Γ_e and Γ_g of the first excited state and the ground state are given by the imaginary parts of the corresponding eigenvalues. The strong-coupling-expansion analysis provides

$$\Gamma_e^{\text{Up}} - \Gamma_g^{\text{Up}} = \frac{36\gamma J^2}{U^2+\gamma^2(c_{\rho-1}-c_\rho)^2} + \frac{9\gamma J^2}{U^2+(\gamma^2/4)(c_{\rho-1}-c_{\rho+1})^2} > 0, \tag{3.129}$$

$$\Gamma_e^{\text{Low}} - \Gamma_g^{\text{Low}} = \frac{36\gamma J^2}{U^2+\gamma^2(c_{\rho-1}-c_\rho)^2} > 0, \tag{3.130}$$

where Γ^{Up} and Γ^{Low} represent the decay rates of states close to the upper and lower boundaries of the Mott lobe. These relations indicate that the excited state decays faster than the ground state. It is straightforward to extend the analyses to the higher energy states and to show that the decay rate of the ground state is minimal (i.e., it has the longest lifetime).

3.4.3 Realization with Ultracold Gases in an Optical Lattice

We here propose two possible experimental realizations to test the theoretical results presented in this subsection. Firstly, we propose to use an inelastic scattering process to implement atomic systems accompanying a two-body loss of atoms. This process can occur, for example, in the metastable state or by using the light-assisted inelastic collisions as we will detail in the next section. The system evolves in time under

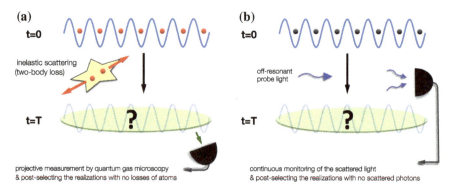

Fig. 3.15 Possible experiments of many-body systems under continuous observation. We propose to use **a** an inelastic two-body loss and **b** an off-resonant light scattering. Reproduced from Fig. 5 of Ref. [10]. Copyright © 2016 by the American Physical Society

effects of an inelastic scattering process by which a pair of atoms is lost from a trap potential (see Fig. 3.15a). This two-body loss process can be characterized by $\hat{L}_i = \hat{b}_i^2$. The total number of particles after a short-time duration can be measured at the single-atom level by quantum gas microscopy [128]. Comparing it with the total number of particles in the initial state, one can postselect the realizations in which the two numbers agree (see Fig. 3.15a). In this way, one can ideally simulate the non-Hermitian dynamics governed by the effective Hamiltonian (3.117). In fact, our simple analysis based on the effective non-Hermitian Hamiltonian correctly captures the experimentally observed shift of the critical point [129], which is measured without any postselections. It merits further study to elucidate to what extent the non-Hermitian description can be applied to strongly correlated dissipative systems accompanying a nonzero number of quantum jumps.

Secondly, we consider a measurement process using an off-resonant probe light (see Fig. 3.15b), where the dominant process becomes an elastic photon scattering process [130]. In this case, the jump operator becomes $\hat{L}_i = \hat{n}_i$ (see e.g., Ref. [131] for a microscopic derivation). One can continuously monitor the scattered light by quantum gas microscopy and may postselect the realizations in which no photons are detected. Yet, only a portion of the scattered photons can be collected in practice and the predicted shift of the critical point can be measured if a possible heating due to undetected photons is not significant, as we detail in the next section.

3.5 Experimental Realizations in Ultracold Gases

We discuss practical experimental situations of the theoretical proposals discussed in Sects. 3.2.3, 3.3.5 and 3.4.3.

3.5 Experimental Realizations in Ultracold Gases

Possible realization of the generalized sine-Gordon model

We discuss a possible experimental realization of the PT-symmetric system proposed in Sect. 3.3.5. To create an imaginary optical potential, we need to realize atomic levels illustrated in Fig. 3.10. The fast decay modes can be realized by (i) choosing appropriate spontaneous emission processes or (ii) employing light-induced transitions.

In scheme (i), one can use the $F = 3$ to $F' = 3$ transition ($5S_{1/2}$ to $5P_{3/2}$) of ^{85}Rb atoms to create an imaginary potential [113], where the excited $F' = 3$ state has a decay channel to the $F = 2$ state. Implementations of complex potentials have also been demonstrated by using other metastable atomic states [32, 112, 114]. The postselection can be implemented by, e.g., applying the state-selective imaging technique [16, 132]. Measuring only the number of atoms residing in the ground state, one can select the realizations in which this number is unchanged between the initial and final states. We note that the experimental fidelity of measuring the atom number with such site-resolved imaging has reached almost unit fidelity (99.5% according to Ref. [2]). Various types of postselections have already been implemented owing to the high experimental fidelity [15–17].

In scheme (ii), one can use a fluorescent transition between the excited state and a state other than the original ground state. In this setup, when the ground-state atom is excited, it is quickly lost from an optical potential due to heating [133]. If the resulting loss rate Γ is much larger than both the spontaneous emission rate and the Rabi frequency Ω, one can adiabatically eliminate the excited state and implement an effective imaginary potential [112]. Other candidate systems are 1D trapped ions or Rydberg atoms with spontaneous decay [38, 134], and the atom-cavity system [135] in which the photon leakage permits continuous observation.

While substantial atomic losses usually lead to experimental difficulties, we remark that our theoretical predictions are accessible by using a very weak imaginary potential with which the atomic loss rate can, in principle, be made arbitrarily small. This is because the key parameter driving the phase transitions is the ratio between g_i and g_r, which is equivalent to the ratio between the amplitudes of the imaginary and real potentials (see Eq. (3.113)). Thus, the imaginary potential required to induce the transition can be made very weak if the depth of the real part of the optical potential is chosen to be sufficiently small. Indeed, such a weak imaginary potential should be created in our proposal since the atomic loss rate is suppressed by a factor of Ω/Γ in the limit of large Γ (see Eq. (3.108)). Because the depth of the real potential can be made small, and the condition on the Rabi frequency $\Delta_{\text{off}} > \Omega$ can easily be met owing to the smallness of the optical depth, the only requirement for the detuning Δ_{off} of the off-resonant light is $\Delta_{\text{off}} > \Gamma$. This point validates our assumption that the real and imaginary potentials have the same periodicity. For example, for the spontaneous emission process in ^{85}Rb or the light-induced transition in the $D2$ transition of ^{87}Rb, Γ is of the order of tens of MHz [113, 133]. Thus, if we set the detuning at $\Delta_{\text{off}} = 100$ GHz, the off-resonant condition is well satisfied, while such a detuning causes a less than 0.1% shift in the optical wavelength.

We finally discuss experimental signatures in our theoretical proposal. First, the measurement-induced BKT transition corresponds to a 1D superfluid-to-Mott-insulator transition for ultracold atoms. This is associated with a power-law divergence in the momentum distribution at zero momentum [87], which can be detected by applying standard techniques such as time-of-flight imaging [121]. Second, the PT symmetry breaking can be probed by detecting the single-mode lasing dynamics of the system. In the PT-broken region, the system has an excited state whose eigenvalue possesses a positive imaginary contribution; such an excited state can acquire an exponentially growing amplitude in the time evolution. Thus, after exciting the system through, e.g., shaking of an optical lattice [136], the system eventually approaches the state having the largest imaginary part of the eigenvalue. Such a single-mode lasing dynamics entails a significant decrease in the entropy of the system, which can be probed from shot-to-shot fluctuations in in-situ imaging of atomic gases [15–17, 136]. Third, the anomalous variation of the critical exponent (see Fig. 3.7) can be investigated through the analysis of the shot-to-shot noise correlations in density fluctuations of a 1D Bose gas, as demonstrated in Ref. [83].

Possible realization of ultracold atoms with a two-body loss or light scattering

We discuss practical experimental situations of our proposals in Sects. 3.2.3 and 3.4.3, where a two-body loss of atoms and off-resonant light scattering are considered as possible measurement processes. Let us first discuss the former. A system accompanying a controlled two-body loss can be realized with implementing inelastic collisions between atoms [85]. The strength of measurement, i.e., the loss rate, can be controlled by changing the intensity of an external light that induces inelastic collisions. One may also use the metastable state of atoms to implement a two-body loss [85, 137]; for example, the $^3P_2$ state of ^{174}Yb atoms has the inelastic scattering length $a_i = 2.8$ nm in addition to the elastic one $a_r = 5.8$ nm [137]. Another promising candidate is ultracold molecules [124], which also have inelastic scattering channels leading to two-body losses.

We next consider experimental situations in the proposal using light scattering to test the shift of the transition point. There, recoil energies due to (scattered but) uncollected photons inevitably lead to heating of the system. We can estimate the expected heating energy per atom as $\delta E = (1-\eta)\gamma\tau \times \hbar^2 k^2/(2m)$, where η is the collection efficiency of photons, $k = 2\pi/\lambda$ represents the wavenumber, γ is the scattering rate, τ represents the duration of the time evolution, and m is the atomic mass. To be concrete, let us consider ^{87}Rb atoms and the wavelength $\lambda = 1064$ nm and set $\gamma/U = 0.2$ corresponding to $\sim 10\%$ shift of the transition point (see Fig. 3.13). If one takes $\tau \sim 1/\gamma$, the temperature that induces the same amount of the shift can be estimated as $T_{\text{th}}/J \simeq 3$ at $U/J \simeq 25$ [138], where we set $k_B = 1$ and $J/\hbar = 3$ms. From the condition $\delta E < T_{\text{th}}/J$, which means that the predicted shift is not masked by the finite-temperature effect, one can obtain a constraint $\eta > 0.08$. This detection fidelity can be met by, for example, quantum gas microscopy in which an impressively high fidelity ($\eta = 0.1 \sim 0.2$) has been achieved [1, 2, 4–8, 128].

Finally, we consider a possible heating effect in the proposal to test the predicted shifts of the critical exponents in a 1D Bose gas under light scattering.

3.5 Experimental Realizations in Ultracold Gases

A finite-temperature effect is characterized by a finite thermal correlation length $\xi_T \equiv \hbar^2 \rho_0 \pi/(mT)$, where ρ_0 is the number density of atoms. If the heating is not so significant that a length scale of interest r in the correlation functions satisfies $r < \xi_T$, then the critical decay of the correlations can still be measured and the predicted changes of the critical exponents can be tested. We also have to take into account the validity of the TLL low-energy description. In a weakly (strongly) interacting 1D Bose gas, the TLL description is applicable when $2\pi/(\rho_0 \xi_T u^2) \lesssim 10^2$ (10^{-2}) [139], where we recall that u is the dimensionless interaction parameter defined in Sect. 3.2. If one considers the density $\rho_0 = 55$ μm^{-1} and a length scale $r = 10$ μm, then one can obtain a constraint $\eta > 0.13$ in the weakly interacting regime from the condition on the TLL description, which can be realized in, e.g., quantum gas microscopy. The constraint is much less stringent for the strongly interacting regime because the TLL description is more robust due to a large value of u.

3.6 Conclusions and Outlook

We have investigated how the measurement backaction from continuous monitoring influences the quantum critical phenomena and quantum phase transitions. Identifying the relevant non-Hermitian perturbations to the Tomonaga-Luttinger liquid, we have revealed unconventional 1D quantum critical phenomena beyond the realm of the standard universality class. We found that the quadratic non-Hermitian perturbation leads to the bifurcating critical exponents that depend on the strength of the measurement. In the presence of the non-Hermitian potential perturbation, a combination of spectral singularity and quantum critical point results in a nontrivial renormalization group fixed point that has no counterpart in Hermitian systems. Moreover, we found anomalous renormalization group flows violating the c-theorem, which lead to enhancements of superfluid correlations in stark contrast to the Berezinskii-Kosterlitz-Thouless paradigm. Our field-theoretic arguments provide a universal model-independent perspective to one-dimensional quantum critical phenomena under measurement. Analyzing the Bose-Hubbard model as a concrete model, we have also studied the influence of the measurement backaction on quantum phase transitions in higher dimensions. We found that the superfluid-to-Mott-insulator transition point is shifted in favor of the Mott lobes. We have proposed experiments using quantum gas microscope with an estimation of possible parameters.

There are several important future directions. Firstly, while we have focused on the simplest quantum trajectory, which is the non-Hermitian dynamics without quantum jumps, it would be desirable to develop a theoretical framework to address all the possible quantum trajectories with a nonzero number of jump events. In the next Chapter, we will introduce the notion of full-counting dynamics to achieve this aim. There, we demonstrate that essential features found in the non-Hermitian dynamics without quantum jumps can persist even in the presence of quantum jumps. From this perspective, we expect that the signature of unique features found in this Chapter can sustain as long as detrimental effects from jump events are not significant; yet, quantitative understanding of effects from quantum jumps remains an important open question. This can be done by applying the formulation developed in the next Chapter to interacting many-body systems.

Secondly, it merits further study to analyze the generalized sine-Gordon model in more detail. For instance, it is interesting to elucidate the relation between the universality found in this work and the non-unitary conformal field theories (CFT), which appear in high-energy physics [96] and statistical mechanics [140]. In this respect, it is worthwhile to note that a certain integrable spin model with PT-symmetric non-Hermitian boundary fields is believed to be described by the non-unitary CFT at the critical exceptional point [141]. It remains as an open question how the spectral singularity alters the behavior of the entanglement entropy, which can experimentally be measured in ultracold atoms [17]. Also, it is a promising direction to further explore unconventional quantum critical phenomena in other nonconservative many-body systems beyond the generalized sine-Gordon model.

Finally, it is intriguing to explore further interesting physical phenomena unique to non-Hermitian systems. Historically, the non-Hermitian approach has proven useful to describe nuclear resonances [25, 26], chaotic scattering [27–29], microwave cavities [30, 31], single atoms [32, 33] and molecules [34, 35]. Recently, this description has also found applications to a wide variety of fields including classical optics [58, 59], condensed matter physics [44–48], biological network [52–54], and chemistry [55]. In view of these developments, it is particularly interesting to extend the conventional notion of topological phenomena in the presence of the non-Hermiticity [142]. There, how the unique aspects in non-Hermitian systems such as topological structures around exceptional points [24] and spectral singularity can lead to a truly novel phenomenon has yet to be clarified.

Appendix A: Adiabatic Elimination of Excited Atomic States

We provide the detailed derivation of the master equation (3.107) for the dissipative dynamics of atoms subject to a one-body loss, which can be obtained after adiabatically eliminating excited atomic states. We consider a situation in which atoms in the system have an energy level diagram shown in Fig. 3.10. Here the excited state $|e\rangle$ has the frequency ω_0 relative to the ground state $|g\rangle$ and fast decay channels to other states with the total decay rate Γ much larger than the spontaneous emission rate from $|e\rangle$ to $|g\rangle$. The system is subject to a weak near-resonant light whose electric filed is given by $\mathbf{E}(\mathbf{x}, t) = 2\mathbf{E}_0(\mathbf{x}) \cos(\omega_L t)$. The dynamics of atoms in the levels $\{|g\rangle, |e\rangle\}$ is then described by the many-body Lindblad equation:

$$\frac{d\hat{\rho}}{dt} = -\frac{i}{\hbar}[\hat{H}, \hat{\rho}] - \frac{\Gamma}{2} \int \left[\hat{\Psi}_e^\dagger(\mathbf{x})\hat{\Psi}_e(\mathbf{x})\hat{\rho} + \hat{\rho}\hat{\Psi}_e^\dagger(\mathbf{x})\hat{\Psi}_e(\mathbf{x}) - 2\hat{\Psi}_e(\mathbf{x})\hat{\rho}\hat{\Psi}_e^\dagger(\mathbf{x}) \right] d\mathbf{x}, \quad (A.1)$$

3.6 Conclusions and Outlook

where $\hat{\Psi}_e$ denotes the field operator of an excited atom and the terms involving Γ describe a loss of atoms in the state $|e\rangle$. Here $\hat{H}$ is the Hamiltonian of the interacting two-level atoms:

$$\hat{H} = \hat{\mathcal{H}}_g + \hat{\mathcal{H}}_e + \hat{V}. \tag{A.2}$$

Going onto the rotating frame and making the rotating-wave approximation, the Hamiltonians $\hat{\mathcal{H}}_g$ and $\hat{\mathcal{H}}_e$ of ground- and excited-state atoms and the interaction Hamiltonian $\hat{V}$ describing the Rabi coupling between the two atomic levels are given by

$$\hat{\mathcal{H}}_g = \int d\mathbf{x} \left[\hat{\Psi}_g^\dagger(\mathbf{x}) \left(-\frac{\hbar^2 \nabla^2}{2m} + U_g(\mathbf{x}) \right) \hat{\Psi}_g(\mathbf{x}) + \frac{g}{2} \hat{\Psi}_g^\dagger(\mathbf{x}) \hat{\Psi}_g^\dagger(\mathbf{x}) \hat{\Psi}_g(\mathbf{x}) \hat{\Psi}_g(\mathbf{x}) \right], \tag{A.3}$$

$$\hat{\mathcal{H}}_e = \int d\mathbf{x}\, \hat{\Psi}_e^\dagger(\mathbf{x}) \left(-\frac{\hbar^2 \nabla^2}{2m} + U_e(\mathbf{x}) + \hbar\delta \right) \hat{\Psi}_e(\mathbf{x}), \tag{A.4}$$

$$\hat{V} = -\frac{\hbar}{2} \int d\mathbf{x} \left(\Omega(\mathbf{x}) \hat{\Psi}_g^\dagger(\mathbf{x}) \hat{\Psi}_e(\mathbf{x}) + \text{H.c.} \right) \equiv \hat{V}_- + \hat{V}_+, \tag{A.5}$$

where $U_{g,e}(\mathbf{x})$'s are optical trapping potentials of the ground- and excited-state atoms created by a far-detuned light, g is the strength of the contact interaction between the ground-state atoms, $\delta = \omega_L - \omega_0$ is the detuning, $\Omega(\mathbf{x}) = 2\mathbf{d} \cdot \mathbf{E}_0(\mathbf{x})/\hbar$ is the Rabi frequency with $\mathbf{d} = \langle e|\hat{\mathbf{d}}|g\rangle$ being the dipole moment, and $\hat{V}_{+(-)}$ are the coupling terms that cause excitation (deexcitation) of the atoms. Let us introduce the non-Hermitian Hamiltonian $\hat{\mathcal{H}}_{e,\text{eff}}$ of the excited-state atoms by

$$\hat{\mathcal{H}}_{e,\text{eff}} = \hat{\mathcal{H}}_e - \frac{i\hbar\Gamma}{2} \int d\mathbf{x}\, \hat{\Psi}_e^\dagger(\mathbf{x}) \hat{\Psi}_e(\mathbf{x}). \tag{A.6}$$

Then, the time-evolution equation (A.1) is written as follows:

$$\frac{d\hat{\rho}}{dt} = -\frac{i}{\hbar} \left[\left(\hat{\mathcal{H}}_g + \hat{\mathcal{H}}_{e,\text{eff}} + \hat{V} \right) \hat{\rho} - \hat{\rho} \left(\hat{\mathcal{H}}_g + \hat{\mathcal{H}}_{e,\text{eff}}^\dagger + \hat{V} \right) \right] + \Gamma \int d\mathbf{x}\, \hat{\Psi}_e(\mathbf{x}) \hat{\rho} \hat{\Psi}_e^\dagger(\mathbf{x}). \tag{A.7}$$

In the limit of a rapid decay $\Gamma \gg \delta, \Omega$, we can adiabatically eliminate the rapidly evolving excited states and obtain the effective dynamics of the ground-state atoms. We achieve this by solving Eq. (A.7) using the second-order perturbation theory with respect to the weak coupling $\hat{V}$ [60]. As shown below, the resulting time-evolution equation for the ground-state atoms is given by Eq. (A.28), and it reduces to the effective non-Hermitian dynamics (A.33) with the effective Hamiltonian (A.29) when we consider the simplest continuously monitored dynamics without quantum jumps.

To perform the perturbative analysis, we work in the interaction picture, where the density matrix is given by

$$\hat{\rho}_I(t) = e^{i\left(\hat{\mathcal{H}}_g + \hat{\mathcal{H}}_{e,\text{eff}}\right)t/\hbar} \hat{\rho}(t) e^{-i\left(\hat{\mathcal{H}}_g + \hat{\mathcal{H}}_{e,\text{eff}}^\dagger\right)t/\hbar}, \tag{A.8}$$

and a general operator $\hat{O}$ is represented by

$$\hat{O}_{\mathrm{I}}(t) = e^{i(\hat{\mathcal{H}}_{\mathrm{g}}+\hat{\mathcal{H}}_{\mathrm{e,eff}})t/\hbar}\hat{O}e^{-i(\hat{\mathcal{H}}_{\mathrm{g}}+\hat{\mathcal{H}}_{\mathrm{e,eff}})t/\hbar}. \tag{A.9}$$

We note that $\hat{\tilde{\rho}}_{\mathrm{I}}$ in Eq. (A.8) is not normalized to unity in general. The time-evolution equation (A.7) is then simplified to

$$\dot{\hat{\tilde{\rho}}}_{\mathrm{I}} = -\frac{i}{\hbar}\left[\hat{\mathcal{V}}_{\mathrm{I}}\hat{\tilde{\rho}}_{\mathrm{I}} - \hat{\tilde{\rho}}_{\mathrm{I}}\hat{\mathcal{V}}_{\mathrm{I}}^{\dagger}\right] + \Gamma\int d\mathbf{x}\hat{\Psi}_{\mathrm{I,e}}(\mathbf{x})\hat{\tilde{\rho}}_{\mathrm{I}}\hat{\Psi}_{\mathrm{I,e}}^{\dagger}(\mathbf{x}). \tag{A.10}$$

We assume that all the atoms reside in the ground state at $t = 0$. Then, we decompose the evolving state $\hat{\tilde{\rho}}_{\mathrm{I}}(t)$ into a perturbation series with respect to the weak coupling $\hat{\mathcal{V}}_{\mathrm{I}}$:

$$\hat{\tilde{\rho}}_{\mathrm{I}}(t) = \hat{\tilde{\rho}}_{\mathrm{I}}^{(0)}(t) + \hat{\tilde{\rho}}_{\mathrm{I}}^{(1)}(t) + \hat{\tilde{\rho}}_{\mathrm{I}}^{(2)}(t) + \cdots, \quad \left|\hat{\tilde{\rho}}_{\mathrm{I}}^{(n)}(t)\right| \propto \left(\frac{|\Omega|}{\Gamma}\right)^{n}\left|\hat{\tilde{\rho}}_{\mathrm{I}}^{(0)}(t)\right|, \tag{A.11}$$

where $|\cdots|$ denotes the trace norm. The recursive equations of the first three terms in the expansion (A.11) are given by

$$\dot{\hat{\tilde{\rho}}}_{\mathrm{I}}^{(0)} = 0, \tag{A.12}$$

$$\dot{\hat{\tilde{\rho}}}_{\mathrm{I}}^{(1)} = -\frac{i}{\hbar}\left[\hat{\mathcal{V}}_{\mathrm{I}}\hat{\tilde{\rho}}_{\mathrm{I}}^{(0)} - \hat{\tilde{\rho}}_{\mathrm{I}}^{(0)}\hat{\mathcal{V}}_{\mathrm{I}}^{\dagger}\right], \tag{A.13}$$

$$\dot{\hat{\tilde{\rho}}}_{\mathrm{I}}^{(2)} = -\frac{i}{\hbar}\left[\hat{\mathcal{V}}_{\mathrm{I}}\hat{\tilde{\rho}}_{\mathrm{I}}^{(1)} - \hat{\tilde{\rho}}_{\mathrm{I}}^{(1)}\hat{\mathcal{V}}_{\mathrm{I}}^{\dagger}\right] + \Gamma\int d\mathbf{x}\hat{\Psi}_{\mathrm{I,e}}(\mathbf{x})\hat{\tilde{\rho}}_{\mathrm{I}}^{(2)}\hat{\Psi}_{\mathrm{I,e}}^{\dagger}(\mathbf{x}). \tag{A.14}$$

From Eq. (A.12), we can take $\hat{\tilde{\rho}}_{\mathrm{I}}^{(0)}$ as a time-independent operator. Equation (A.13) can formally be integrated to give

$$\hat{\tilde{\rho}}_{\mathrm{I}}^{(1)}(t) = -\frac{i}{\hbar}\int_{0}^{t}dt'\left[\hat{\mathcal{V}}_{\mathrm{I}}(t')\hat{\tilde{\rho}}_{\mathrm{I}}^{(0)} - \hat{\tilde{\rho}}_{\mathrm{I}}^{(0)}\hat{\mathcal{V}}_{\mathrm{I}}^{\dagger}(t')\right]. \tag{A.15}$$

To integrate out the excited states and obtain the effective dynamics of the ground-state atoms, we decompose $\hat{\tilde{\rho}}_{\mathrm{I}}^{(2)}$ into the subspaces of the ground- and excited-state atoms. To do so, we introduce the projection $\hat{\mathcal{P}}_{\mathrm{g}}$ onto the ground-state manifold by $\hat{\mathcal{P}}_{\mathrm{g}} = \sum_{N}\hat{\mathcal{P}}_{\mathrm{g}}^{N}$, where $\hat{\mathcal{P}}_{\mathrm{g}}^{N}$ denotes the projection onto the subspace spanned by quantum states containing N ground-state atoms only. We also introduce the projection $\hat{\mathcal{Q}}_{\mathrm{e}}^{1}$ onto quantum states having a single excited-state atom (and an arbitrary number of ground-state atoms). Then, Eq. (A.14) can be decomposed as

$$\hat{\mathcal{P}}_{\mathrm{g}}\dot{\hat{\tilde{\rho}}}_{\mathrm{I}}^{(2)}\hat{\mathcal{P}}_{\mathrm{g}} = -\frac{i}{\hbar}\hat{\mathcal{P}}_{\mathrm{g}}\left[\hat{\mathcal{V}}_{\mathrm{I}}\hat{\tilde{\rho}}_{\mathrm{I}}^{(1)} - \hat{\tilde{\rho}}_{\mathrm{I}}^{(1)}\hat{\mathcal{V}}_{\mathrm{I}}^{\dagger}\right]\hat{\mathcal{P}}_{\mathrm{g}} + \Gamma\hat{\mathcal{P}}_{\mathrm{g}}\int d\mathbf{x}\hat{\Psi}_{\mathrm{I,e}}\hat{\mathcal{Q}}_{\mathrm{e}}^{1}\hat{\tilde{\rho}}_{\mathrm{I}}^{(2)}\hat{\mathcal{Q}}_{\mathrm{e}}^{1}\hat{\Psi}_{\mathrm{I,e}}^{\dagger}\hat{\mathcal{P}}_{\mathrm{g}}, \tag{A.16}$$

$$\hat{\mathcal{Q}}_{\mathrm{e}}^{1}\dot{\hat{\tilde{\rho}}}_{\mathrm{I}}^{(2)}\hat{\mathcal{Q}}_{\mathrm{e}}^{1} = -\frac{i}{\hbar}\hat{\mathcal{Q}}_{\mathrm{e}}^{1}\left[\hat{\mathcal{V}}_{\mathrm{I}}\hat{\tilde{\rho}}_{\mathrm{I}}^{(1)} - \hat{\tilde{\rho}}_{\mathrm{I}}^{(1)}\hat{\mathcal{V}}_{\mathrm{I}}^{\dagger}\right]\hat{\mathcal{Q}}_{\mathrm{e}}^{1}, \tag{A.17}$$

3.6 Conclusions and Outlook

where Eq. (A.17) follows from the fact that $\hat{\rho}_I^{(2)}$ contains, at most, one excited-state atom. We adiabatically eliminate the excited states by integrating out Eq. (A.17):

$$\hat{Q}_e^1 \hat{\rho}_I^{(2)}(t) \hat{Q}_e^1 = -\frac{i}{\hbar} \hat{Q}_e^1 \int_0^t dt' \left[\hat{V}_I(t') \hat{\rho}_I^{(1)}(t') - \hat{\rho}_I^{(1)}(t') \hat{V}_I^\dagger(t') \right] \hat{Q}_e^1. \tag{A.18}$$

Substituting Eqs. (A.15) and (A.18) into (A.16), we obtain

$$\hat{\mathcal{P}}_g \dot{\hat{\rho}}_I^{(2)} \hat{\mathcal{P}}_g = -\frac{1}{\hbar^2} \hat{\mathcal{P}}_g \left[\hat{V}_I(t) \int_0^t dt' \hat{V}_I(t') \hat{\rho}_I^{(0)} + \text{H.c.} \right] \hat{\mathcal{P}}_g$$
$$+ \frac{\Gamma}{\hbar^2} \hat{\mathcal{P}}_g \int d\mathbf{x} \hat{\Psi}_{I,e} \hat{Q}_e^1 \int_0^t dt' \int_0^{t'} dt'' \left[\hat{V}_I(t') \hat{\rho}_I^{(0)} \hat{V}_I^\dagger(t'') + \text{H.c.} \right] \hat{Q}_e^1 \hat{\Psi}_{I,e}^\dagger \hat{\mathcal{P}}_g. \tag{A.19}$$

Here, in the second line in Eq. (A.19), the terms proportional to $\hat{V}_I \hat{V}_I \hat{\rho}_I^{(0)}$ or $\hat{\rho}_I^{(0)} \hat{V}_I^\dagger \hat{V}_I^\dagger$ vanish because of the projection $\hat{Q}_e^1$. Then, since we assume that the time scale of the strong dissipation is fast compared with other time scales appearing in the system, we approximate the leading contributions by

$$e^{-i(\hat{\mathcal{H}}_g + \hat{\mathcal{H}}_{e,\text{eff}})t/\hbar} \hat{\mathcal{P}}_g \simeq \hat{\mathcal{P}}_g, \quad e^{-i(\hat{\mathcal{H}}_g + \hat{\mathcal{H}}_{e,\text{eff}})t/\hbar} \hat{Q}_e^1 \simeq e^{-\Gamma t/2} \hat{Q}_e^1. \tag{A.20}$$

From these equations, it follows that

$$\hat{\mathcal{P}}_g \hat{V}_I(t) = \hat{\mathcal{P}}_g e^{i(\hat{\mathcal{H}}_g + \hat{\mathcal{H}}_{e,\text{eff}})t/\hbar} (\hat{V}_+ + \hat{V}_-) e^{-i(\hat{\mathcal{H}}_g + \hat{\mathcal{H}}_{e,\text{eff}})t/\hbar}$$
$$\simeq \hat{\mathcal{P}}_g \hat{V}_- \hat{Q}_e^1 e^{-i(\hat{\mathcal{H}}_g + \hat{\mathcal{H}}_{e,\text{eff}})t/\hbar}$$
$$\simeq e^{-\Gamma t/2} \hat{\mathcal{P}}_g \hat{V}_- \hat{Q}_e^1. \tag{A.21}$$

Similarly, we obtain

$$\hat{Q}_e^1 \hat{V}_I(t) \hat{\mathcal{P}}_g = \hat{Q}_e^1 e^{i(\hat{\mathcal{H}}_g + \hat{\mathcal{H}}_{e,\text{eff}})t/\hbar} (\hat{V}_+ + \hat{V}_-) e^{-i(\hat{\mathcal{H}}_g + \hat{\mathcal{H}}_{e,\text{eff}})t/\hbar} \hat{\mathcal{P}}_g$$
$$\simeq e^{\Gamma t/2} \hat{Q}_e^1 \hat{V}_+ \hat{\mathcal{P}}_g. \tag{A.22}$$

We then perform the integration in the first line on the right-hand side of Eq. (A.19) and obtain

$$-\frac{1}{\hbar^2} \hat{\mathcal{P}}_g \left[\hat{V}_I(t) \int_0^t dt' \hat{V}_I(t') \hat{\rho}_I^{(0)} + \text{H.c.} \right] \hat{\mathcal{P}}_g \simeq -\frac{1}{\hbar^2} \left[\hat{\mathcal{P}}_g \hat{V}_- \hat{Q}_e^1 e^{-\Gamma t/2} \int_0^t dt' e^{\Gamma t'/2} \hat{Q}_e^1 \hat{V}_+ \hat{\mathcal{P}}_g \hat{\rho}_I^{(0)} + \text{H.c.} \right]$$
$$\simeq -\frac{2}{\hbar^2 \Gamma} \left(\hat{\mathcal{P}}_g \hat{V}_- \hat{V}_+ \hat{\mathcal{P}}_g \hat{\rho}_I^{(0)} + \hat{\rho}_I^{(0)} \hat{\mathcal{P}}_g \hat{V}_- \hat{V}_+ \hat{\mathcal{P}}_g \right)$$
$$= -\left\{ \int d\mathbf{x} \frac{|\Omega(\mathbf{x})|^2}{2\Gamma} \hat{\Psi}_g^\dagger(\mathbf{x}) \hat{\Psi}_g(\mathbf{x}), \hat{\rho}_I^{(0)} \right\}, \tag{A.23}$$

where we use Eqs. (A.21), (A.22), and the relations $\hat{\mathcal{P}}_g\hat{\tilde{\rho}}_I^{(0)}\hat{\mathcal{P}}_g = \hat{\tilde{\rho}}_I^{(0)}$ and $(\hat{Q}_e^1)^2 = \hat{Q}_e^1$ in the first line, and use Eq. (A.5) to derive the last line. To calculate the last line in Eq. (A.19), we approximate

$$\hat{Q}_e^1\int_0^t dt' \int_0^{t'} dt'' \left[\hat{V}_I(t')\hat{\tilde{\rho}}_I^{(0)}\hat{V}_I^\dagger(t'') + \text{H.c.}\right]\hat{Q}_e^1 \simeq 2\int_0^t dt' \int_0^{t'} dt'' e^{\Gamma(t'+t'')/2}\hat{Q}_e^1\hat{V}_+\hat{\tilde{\rho}}_I^{(0)}\hat{V}_-\hat{Q}_e^1$$
$$\simeq \frac{4e^{\Gamma t}}{\Gamma^2}\hat{V}_+\hat{\tilde{\rho}}_I^{(0)}\hat{V}_-, \qquad (\text{A}.24)$$

and

$$\hat{\mathcal{P}}_g\hat{\Psi}_{I,e}\hat{Q}_e^1 \simeq e^{-\Gamma t/2}\hat{\mathcal{P}}_g\hat{\Psi}_e\hat{Q}_e^1, \quad \hat{Q}_e^1\hat{\Psi}_{I,e}^\dagger\hat{\mathcal{P}}_g \simeq e^{-\Gamma t/2}\hat{Q}_e^1\hat{\Psi}_e^\dagger\hat{\mathcal{P}}_g. \qquad (\text{A}.25)$$

The last line in Eq. (A.19) can then be calculated as

$$\frac{\Gamma}{\hbar^2}\hat{\mathcal{P}}_g \int d\mathbf{x}\hat{\Psi}_{I,e}\hat{Q}_e^1 \int_0^t dt' \int_0^{t'} dt'' \left[\hat{V}_I(t')\hat{\tilde{\rho}}_I^{(0)}\hat{V}_I^\dagger(t'') + \text{H.c.}\right]\hat{Q}_e^1\hat{\Psi}_{I,e}^\dagger\hat{\mathcal{P}}_g$$
$$\simeq \frac{4}{\hbar^2\Gamma}\int d\mathbf{x}\hat{\mathcal{P}}_g\hat{\Psi}_e(\mathbf{x})\hat{V}_+\hat{\mathcal{P}}_g\hat{\tilde{\rho}}_I^{(0)}\hat{\mathcal{P}}_g\hat{V}_-\hat{\Psi}_e^\dagger(\mathbf{x})\hat{\mathcal{P}}_g$$
$$\simeq \hat{\mathcal{P}}_g\int d\mathbf{x}\frac{|\Omega(\mathbf{x})|^2}{\Gamma}\hat{\Psi}_g(\mathbf{x})\hat{\tilde{\rho}}_I^{(0)}\hat{\Psi}_g^\dagger(\mathbf{x})\hat{\mathcal{P}}_g, \qquad (\text{A}.26)$$

where we use Eqs. (A.24) and (A.25) and $\hat{\mathcal{P}}_g\hat{\tilde{\rho}}_I^{(0)}\hat{\mathcal{P}}_g = \hat{\tilde{\rho}}_I^{(0)}$ in the second line. To derive the last line, we use the following relation

$$\hat{\mathcal{P}}_g\hat{\Psi}_e(\mathbf{x})\hat{V}_+\hat{\mathcal{P}}_g = -\frac{\hbar\Omega^*(\mathbf{x})}{2}\hat{\mathcal{P}}_g\hat{\Psi}_g(\mathbf{x})\hat{\mathcal{P}}_g. \qquad (\text{A}.27)$$

From Eqs. (A.19), (A.23), (A.26) and $\hat{\mathcal{P}}_g\dot{\hat{\tilde{\rho}}}_I^{(0)}\hat{\mathcal{P}}_g = \hat{\mathcal{P}}_g\dot{\hat{\tilde{\rho}}}_I^{(1)}\hat{\mathcal{P}}_g = 0$, the effective time-evolution equation of the ground-state atoms is obtained as

$$\frac{d\hat{\rho}_g}{dt} = -\frac{i}{\hbar}\left(\hat{\mathcal{H}}_{g,\text{eff}}\hat{\rho}_g - \hat{\rho}_g\hat{\mathcal{H}}_{g,\text{eff}}^\dagger\right) + \int d\mathbf{x}\frac{|\Omega(\mathbf{x})|^2}{\Gamma}\hat{\Psi}_g(\mathbf{x})\hat{\rho}_g\hat{\Psi}_g^\dagger(\mathbf{x}), \qquad (\text{A}.28)$$

$$\hat{\mathcal{H}}_{g,\text{eff}} \equiv \hat{\mathcal{H}}_g - i\hbar\int d\mathbf{x}\frac{|\Omega(\mathbf{x})|^2}{2\Gamma}\hat{\Psi}_g^\dagger(\mathbf{x})\hat{\Psi}_g(\mathbf{x}), \qquad (\text{A}.29)$$

where we go back to the Schrödinger picture and introduce the density matrix $\hat{\rho}_g$ projected onto the ground-state manifold by

$$\hat{\rho}_g(t) = \hat{\mathcal{P}}_g\hat{\rho}(t)\hat{\mathcal{P}}_g \simeq \hat{\mathcal{P}}_g\left(\hat{\rho}^{(0)}(t) + \hat{\rho}^{(1)}(t) + \hat{\rho}^{(2)}(t)\right)\hat{\mathcal{P}}_g. \qquad (\text{A}.30)$$

The non-Hermitian Hamiltonian (A.29) describes the continuously monitored dynamics of the system conditioned on the realization with no quantum jumps being observed, i.e., no atoms escape from the ground state [12, 13, 143]. To clarify this point, let us assume that N ground-state atoms are initially prepared,

3.6 Conclusions and Outlook

i.e., $\hat{\mathcal{P}}_g^N \hat{\rho}(0) \hat{\mathcal{P}}_g^N = \hat{\rho}(0)$. This initial condition implies

$$\hat{\mathcal{P}}_g^{N+l} \hat{\rho}_g(0) \hat{\mathcal{P}}_g^{N+l} = 0 \quad (l = 1, 2, \ldots), \tag{A.31}$$

where we use $\left[\hat{\mathcal{P}}_g^N, \hat{\mathcal{P}}_g\right] = 0$. From Eqs. (A.28) and (A.31), we can in particular show that, during the course of the time evolution,

$$\hat{\mathcal{P}}_g^{N+1} \hat{\rho}_g(t) \hat{\mathcal{P}}_g^{N+1} = 0. \tag{A.32}$$

Let us now consider the dynamics without quantum jumps $\hat{\tilde{\rho}}_g^N(t) \equiv \hat{\mathcal{P}}_g^N \hat{\rho}(t) \hat{\mathcal{P}}_g^N = \hat{\mathcal{P}}_g^N \hat{\rho}_g(t) \hat{\mathcal{P}}_g^N$, where the dynamics is conditioned such that no atoms are lost from the initial state. Using Eqs. (A.28) and (A.32), we can show that this dynamics $\hat{\tilde{\rho}}_g^N$ is governed by the non-Hermitian Hamiltonian (A.29):

$$\frac{d\hat{\tilde{\rho}}_g^N}{dt} = -\frac{i}{\hbar} \left(\hat{\mathcal{H}}_{g,\text{eff}} \hat{\tilde{\rho}}_g^N - \hat{\tilde{\rho}}_g^N \hat{\mathcal{H}}_{g,\text{eff}}^\dagger \right). \tag{A.33}$$

Some remarks are in order here. First, an imaginary potential $-i|\Omega(\mathbf{x})|^2/(2\Gamma)$ in Eq. (A.29) arises from the second-order process of a virtual excitation and deexcitation of the ground-state atoms (see Eq. (A.23)). Since no atoms are lost in this process, the non-Hermitian contribution exists even when we do not observe actual losses of atoms. Physically, such a contribution originates from the measurement backaction associated with continuous monitoring of the population of atoms in the excited state [12]. Second, we note that the expression of the imaginary potential indicates that the loss rate of atoms from the ground state is suppressed by a factor of Ω/Γ for a large Γ. In particular, in the limit of $\Gamma \to \infty$, the dynamics reduces to the Hermitian evolution governed by $\hat{\mathcal{H}}_g$. This limit can be interpreted as the quantum Zeno dynamics [144], where the strong measurement confines the dynamics into the decay-free subspace and the time-evolution obeys the effective "Zeno" Hamiltonian. In our model, such a Hamiltonian is given by $\hat{\mathcal{H}}_g = \hat{\mathcal{P}}_g \hat{H} \hat{\mathcal{P}}_g$, where the total Hamiltonian $\hat{H}$ is projected onto the decay-free, ground-state manifold. In a general case of a strong but finite Γ, we need to perform careful perturbative analyses [85, 86, 124, 145] to obtain the correction terms beyond the quantum Zeno dynamics, as we have conducted above.

References

1. Bakr WS, Gillen JI, Peng A, Fölling S, Greiner M (2009) A quantum gas microscope for detecting single atoms in a Hubbard-regime optical lattice. Nature 462:74–77
2. Sherson JF, Weitenberg C, Endres M, Cheneau M, Bloch I, Kuhr S (2010) Single-atom-resolved fluorescence imaging of an atomic Mott insulator. Nature 467:68–72
3. Miranda M, Inoue R, Okuyama Y, Nakamoto A, Kozuma M (2015) Site-resolved imaging of ytterbium atoms in a two-dimensional optical lattice. Phys Rev A 91:063414
4. Cheuk LW, Nichols MA, Okan M, Gersdorf T, Ramasesh VV, Bakr WS, Lompe T, Zwierlein MW (2015) Quantum-gas microscope for fermionic atoms. Phys Rev Lett 114:193001
5. Parsons MF, Huber F, Mazurenko A, Chiu CS, Setiawan W, Wooley-Brown K, Blatt S, Greiner M (2015) Site-resolved imaging of fermionic ^{6}Li in an optical lattice. Phys Rev Lett 114:213002
6. Haller E, Hudson J, Kelly A, Cotta DA, Bruno P, Bruce GD, Kuhr S (2015) Single-atom imaging of fermions in a quantum-gas microscope. Nat Phys 11:738–742
7. Omran A, Boll M, Hilker TA, Kleinlein K, Salomon G, Bloch I, Gross C (2015) Microscopic observation of Pauli blocking in degenerate fermionic lattice gases. Phys Rev Lett 115:263001
8. Edge GJA, Anderson R, Jervis D, McKay DC, Day R, Trotzky S, Thywissen JH (2015) Imaging and addressing of individual fermionic atoms in an optical lattice. Phys Rev A 92:063406
9. Yamamoto R, Kobayashi J, Kuno T, Kato K, Takahashi Y (2016) An ytterbium quantum gas microscope with narrow-line laser cooling. New J Phys 18:023016
10. Ashida Y, Furukawa S, Ueda M (2016) Quantum critical behavior influenced by measurement backaction in ultracold gases. Phys Rev A 94:053615
11. Dalibard J, Castin Y, Mølmer K (1992) Wave-function approach to dissipative processes in quantum optics. Phys Rev Lett 68:580–583
12. Carmichael H (1993) An open system approach to quantum optics. Springer, Berlin
13. Daley AJ (2014) Quantum trajectories and open many-body quantum systems. Adv Phys 63:77–149
14. Kok P, Munro WJ, Nemoto K, Ralph TC, Dowling JP, Milburn GJ (2007) Linear optical quantum computing with photonic qubits. Rev Mod Phys 79:135–174
15. Endres M, Cheneau M, Fukuhara T, Weitenberg C, Schauß P, Gross C, Mazza L, Bañuls MC, Pollet L, Bloch I, Kuhr S (2011) Observation of correlated particle-hole pairs and string order in low-dimensional Mott insulators. Science 334:200–203
16. Fukuhara T, Hild S, Zeiher J, Schauß P, Bloch I, Endres M, Gross C (2015) Spatially resolved detection of a spin-entanglement wave in a Bose-Hubbard chain. Phys Rev Lett 115:035302
17. Islam R, Ma R, Preiss PM, Tai ME, Lukin A, Rispoli M, Greiner M (2015) Measuring entanglement entropy in a quantum many-body system. Nature 528:77–83
18. Yanay Y, Mueller EJ (2014) Heating from continuous number density measurements in optical lattices. Phys Rev A 90:023611
19. Schachenmayer J, Pollet L, Troyer M, Daley AJ (2014) Spontaneous emission and thermalization of cold bosons in optical lattices. Phys Rev A 89:011601
20. Mitra A, Takei S, Kim YB, Millis AJ (2006) Nonequilibrium quantum criticality in open electronic systems. Phys Rev Lett 97:236808
21. Sieberer LM, Huber SD, Altman E, Diehl S (2013) Dynamical critical phenomena in driven-dissipative systems. Phys Rev Lett 110:195301
22. Täuber UC, Diehl S (2014) Perturbative field-theoretical renormalization group approach to driven-dissipative Bose-Einstein criticality. Phys Rev X 4:021010
23. Ashida Y, Furukawa S, Ueda M (2017) Parity-time-symmetric quantum critical phenomena. Nat Commun 8:15791
24. Rotter I (2009) A non-Hermitian Hamilton operator and the physics of open quantum systems. J Phys A 42:153001
25. Mahaux C, Weidenmüller HA (1969) Shell model approach to nuclear reactions. North Holland, Amsterdam

26. Binsch G (1969) Unified theory of exchange effects on nuclear magnetic resonance line shapes. J Am Chem Soc 91:1304–1309
27. Lewenkopf CH, Weidenmüller HA (1991) Stochastic versus semiclassical approach to quantum chaotic scattering. Ann Phys 212:53–83
28. Stein J, Stöckmann H-J, Stoffregen U (1995) Microwave studies of billiard green functions and propagators. Phys Rev Lett 75:53–56
29. Stöckmann, Quantum chaos-an introduction (Cambridge University Press, Cambridge, 1999)
30. Dembowski C, Gräf H-D, Harney HL, Heine A, Heiss WD, Rehfeld H, Richter A (2001) Experimental observation of the topological structure of exceptional points. Phys Rev Lett 86:787–790
31. Choi Y, Kang S, Lim S, Kim W, Kim J-R, Lee J-H, An K (2010) Quasieigenstate coalescence in an atom-cavity quantum composite. Phys Rev Lett 104:153601
32. Stützle R, Göbel MC, Hörner T, Kierig E, Mourachko I, Oberthaler MK, Efremov MA, Fedorov MV, Yakovlev VP, van Leeuwen KAH, Schleich WP (2005) Observation of non-spreading wave packets in an imaginary potential. Phys Rev Lett 95:110405
33. Cartarius H, Main J, Wunner G (2007) Exceptional points in atomic spectra. Phys Rev Lett 99:173003
34. Brunel C, Lounis B, Tamarat P, Orrit M (1998) Rabi resonances of a single molecule driven by RF and laser fields. Phys Rev Lett 81:2679–2682
35. Rozhkov I, Barkai E (2005) Photon emission from a driven single-molecule source: A renormalization group approach. J Chem Phys 123:074703
36. Gao T, Estrecho E, Bliokh KY, Liew TCH, Fraser MD, Brodbeck S, Kamp M, Schneider C, Hofling S, Yamamoto Y, Nori F, Kivshar YS, Truscott AG, Dall RG, Ostrovskaya EA (2015) Observation of non-Hermitian degeneracies in a chaotic exciton-polariton billiard. Nature 526:554–558
37. Zhen B, Hsu CW, Igarashi Y, Lu L, Kaminer I, Pick A, Chua S-L, Joannopoulos JD, Soljacic M (2015) Spawning rings of exceptional points out of Dirac cones. Nature 525:354–358
38. Lee TE, Chan C-K (2014) Heralded magnetism in non-Hermitian atomic systems. Phys Rev X 4:041001
39. Kozlowski W, Caballero-Benitez SF, Mekhov IB (2016) Non-Hermitian dynamics in the quantum Zeno limit. Phys Rev A 94:012123
40. Gamow G (1928) Zur quantentheorie des atomkernes. Z Für Phys 51:204–212
41. Feshbach H, Porter CE, Weisskopf VF (1954) Model for nuclear reactions with neutrons. Phys Rev 96:448–464
42. Breuer H-P, Petruccione F (2002) The theory of open quantum systems. Oxford University Press, Oxford, England
43. Rudner MS, Levitov LS (2009) Topological transition in a non-Hermitian quantum walk. Phys Rev Lett 102:065703
44. Hatano N, Nelson DR (1996) Localization transitions in non-Hermitian quantum mechanics. Phys Rev Lett 77:570–573
45. Fukui T, Kawakami N (1998) Breakdown of the Mott insulator: Exact solution of an asymmetric Hubbard model. Phys Rev B 58:16051–16056
46. Oka T, Aoki H (2010) Dielectric breakdown in a Mott insulator: many-body Schwinger-Landau-Zener mechanism studied with a generalized Bethe ansatz. Phys Rev B 81:033103
47. Shen H, Zhen B, Fu L (2018) Topological band theory for non-Hermitian hamiltonians. Phys Rev Lett 120:146402
48. Yoshida T, Peters R, Kawakami N (2018) Non-Hermitian perspective of the band structure in heavy-fermion systems. Phys Rev B 98:035141
49. Esposito M, Harbola U, Mukamel S (2009) Nonequilibrium fluctuations, fluctuation theorems, and counting statistics in quantum systems. Rev Mod Phys 81:1665–1702
50. Ren J, Hänggi P, Li B (2010) Berry-phase-induced heat pumping and its impact on the fluctuation theorem. Phys Rev Lett 104:170601
51. Sagawa T, Hayakawa H (2011) Geometrical expression of excess entropy production. Phys Rev E 84:051110

52. Nelson DR, Shnerb NM (1998) Non-Hermitian localization and population biology. Phys Rev E 58:1383–1403
53. Amir A, Hatano N, Nelson DR (2016) Non-Hermitian localization in biological networks. Phys Rev E 93:042310
54. Murugan A, Vaikuntanathan S (2017) Topologically protected modes in non-equilibrium stochastic systems. Nat Commun 8:13881
55. McGrath T, Jones NS, ten Wolde PR, Ouldridge TE (2017) Biochemical machines for the interconversion of mutual information and work. Phys Rev Lett 118:028101
56. Fisher ME (1978) Yang-Lee edge singularity and ϕ^3 field theory. Phys Rev Lett 40:1610–1613
57. Bender CM, Boettcher S (1998) Real spectra in non-Hermitian hamiltonians having PT symmetry. Phys Rev Lett 80:5243–5246
58. El-Ganainy R, Makris KG, Khajavikhan M, Musslimani ZH, Rotter S, Christodoulides DN (2018) Non-Hermitian physics and PT symmetry. Nat Phys 14:11
59. Zhao H, Feng L (2018) Parity-time symmetric photonics. Natl Sci Rev 5:183–199
60. Kato T (1966) Perturbation theory for linear operators. Springer, New York
61. Mostafazadeh A (2009) Spectral singularities of complex scattering potentials and infinite reflection and transmission coefficients at real energies. Phys Rev Lett 102:220402
62. Bender CM, Berry MV, Mandilara A (2002) Generalized PT symmetry and real spectra. J Phys A 35:L467
63. Makris KG, El-Ganainy R, Christodoulides DN, Musslimani ZH (2008) Beam dynamics in $\mathcal{PT}$ symmetric optical lattices. Phys Rev Lett 100:103904
64. Regensburger A, Bersch C, Miri M-A, Onishchukov G, Christodoulides DN, Peschel U (2012) Parity-time synthetic photonic lattices. Nature 488:167–171
65. Feng L, Xu Y-L, Fegadolli WS, Lu M-H, Oliveira JEB, Almeida VR, Chen Y-F, Scherer A (2013) Experimental demonstration of a unidirectional reflectionless parity-time metamaterial at optical frequencies. Nat Mater 12:108–113
66. Peng B, Özdemir SK, Lei F, Monifi F, Gianfreda M, Long GL, Fan S, Nori F, Bender CM, Yang L (2014) Parity-time-symmetric whispering-gallery microcavities. Nat Phys 10:394–398
67. Chtchelkatchev NM, Golubov AA, Baturina TI, Vinokur VM (2012) Stimulation of the fluctuation superconductivity by PT symmetry. Phys Rev Lett 109:150405
68. Peng P, Cao W, Shen C, Qu W, Jianming W, Jiang L, Xiao Y (2016) Anti-parity-time symmetry with flying atoms. Nat Phys 12:1139–1145
69. Jing H, Özdemir SK, Lü X-Y, Zhang J, Yang L, Nori F (2014) $\mathcal{PT}$-symmetric phonon laser. Phys Rev Lett 113:053604
70. Landau LD (1959) On the theory of the fermi liquid. Sov Phys JETP 35:70–74
71. Haldane FDM (1981) Effective harmonic-fluid approach to low-energy properties of one-dimensional quantum fluids. Phys Rev Lett 47:1840–1843
72. Bour E (1862) Théorie de la déformation des surfaces. J. École Impériale Polytech 19:1
73. Ablowitz MJ, Clarkson PA (1991) Solitons, nonlinear evolution equations and inverse scattering. Cambridge University Press, Cambridge, England
74. Berezinskii VL (1971) Destruction of long-range order in one-dimensional and two-dimensional systems having a continuous symmetry group I. Classical systems. Sov Phys JETP 32:493–500
75. Berezinskii VL (1972) Destruction of long-range order in one-dimensional and two-dimensional systems possessing a continuous symmetry group II. Quantum systems. Sov Phys JETP 34:610–616
76. Kosterlitz JM, Thouless DJ (1973) Ordering, metastability and phase transitions in two-dimensional systems. J Phys C 6:1181–1203
77. Giamarchi T (2004) Quantum physics in one dimension. Oxford University Press, Oxford
78. Amit DJ, Goldschmidt YY, Grinstein S (1980) Renormalisation group analysis of the phase transition in the 2D Coulomb gas, Sine-Gordon theory and XY-model. J Phys A 13:585
79. Kaushal RS (2002) Quantum mechanics of complex hamiltonian systems in one dimension. J Phys A 35:8743

References 83

80. Jannussis A, Brodimas G, Baskoutas S, Leodaris A (2003) Non-Hermitian harmonic oscillator with discrete complex or real spectrum for non-unitary squeeze operators. J Phys A 36:2507
81. Furukawa S, Kim YB (2011) Entanglement entropy between two coupled Tomonaga-Luttinger liquids. Phys Rev B 83:085112
82. Polkovnikov A, Altman E, Demler E (2006) Interference between independent fluctuating condensates. Proc Natl Acad Sci USA 103:6125–6129
83. Hofferberth S, Lesanovsky I, Schumm T, Gritsev V, Demler E, Schmiedmayer J (2008) Probing quantum and thermal noise in an interacting many-body system. Nat Phys 4:489–495
84. Song HF, Rachel S, Le Hur K (2010) General relation between entanglement and fluctuations in one dimension. Phys Rev B 82:012405
85. Syassen N, Bauer DM, Lettner M, Volz T, Dietze D, García-Ripoll JJ, Cirac JI, Rempe G, Dürr S (2008) Strong dissipation inhibits losses and induces correlations in cold molecular gases. Science 320:1329–1331
86. Garcia-Ripoll JJ, Dürr S, Syassen N, Bauer DM, Lettner M, Rempe G, Cirac JI (2009) Dissipation-induced hard-core boson gas in an optical lattice. New J Phys 11:013053
87. Cazalilla MA, Citro R, Giamarchi T, Orignac E, Rigol M (2011) One dimensional bosons: from condensed matter systems to ultracold gases. Rev Mod Phys 83:1405–1466
88. Büchler HP, Blatter G, Zwerger W (2003) Commensurate-incommensurate transition of cold atoms in an optical lattice. Phys Rev Lett 90:130401
89. Cazalilla MA (2004) Bosonizing one-dimensional cold atomic gases. J Phys B 37:S1
90. Beige A, Braun D, Tregenna B, Knight PL (2000) Quantum computing using dissipation to remain in a decoherence-free subspace. Phys Rev Lett 85:1762–1765
91. Facchi P, Pascazio S (2002) Quantum Zeno subspaces. Phys Rev Lett 89:080401
92. Fisher MPA, Weichman PB, Grinstein G, Fisher DS (1989) Boson localization and the superfluid-insulator transition. Phys Rev B 40:546–570
93. Mostafazadeh A (2002) Pseudo-Hermiticity versus PT symmetry: the necessary condition for the reality of the spectrum of a non-Hermitian hamiltonian. J Math Phys 43:205–214
94. Bender CM, Jones HF, Rivers RJ (2005) Dual PT-symmetric quantum field theories. Phys Lett B 625:333–340
95. Zamolodchikov AB (1986) Irreversibility of the flux of the renormalization group in a 2-d field theory. JETP Lett 43:730–732
96. Seiberg N (1990) Notes on quantum Liouville theory and quantum gravity. Prog Theor Phys Supp 102:319–349
97. Ikhlef Y, Jacobsen JL, Saleur H (2016) Three-point functions in $c \leq 1$ Liouville theory and conformal loop ensembles. Phys Rev Lett 116:130601
98. Castro-Alvaredo OA, Doyon B, Ravanini F (2017) Irreversibility of the renormalization group flow in non-unitary quantum field theory. J Phys A 50:424002
99. Lourenco JAS, Eneias RL, Pereira RG (2018) Kondo effect in a $\mathcal{PT}$-symmetric non-Hermitian hamiltonian. Phys Rev B 98:085126
100. Nakagawa M, Kawakami N, Ueda M (2018) Non-Hermitian kondo effect in ultracold alkaline-earth atoms. Phys Rev Lett 121:203001
101. Affleck I, Ludwig AWW (1991) Universal noninteger "ground-state degeneracy" in critical quantum systems. Phys Rev Lett 67:161–164
102. Nomura K (1995) Correlation functions of the 2D sine-Gordon model. J Phys A 28:5451
103. Kitazawa A, Nomura K, Okamoto K (1996) Phase diagram of $s = 1$ bond-alternating XXZ chains. Phys Rev Lett 76:4038–4041
104. Heiss WD, Steeb W-H (1991) Avoided level crossings and riemann sheet structure. J Math Phys 32:3003–3007
105. Vidal G (2007) Classical simulation of infinite-size quantum lattice systems in one spatial dimension. Phys Rev Lett 98:070201
106. Bender CM, Berntson BK, Parker D, Samuel E (2013) Observation of PT phase transition in a simple mechanical system. Am J Phys 81:173–179
107. Wetterich C (1991) Average action and the renormalization group equations. Nucl Phys B 352:529

108. Wetterich C (1993) Exact evolution equation for the effective potential. Phys Lett B 301:90
109. Nándori I (2013) Functional renormalization group with a compactly supported smooth regulator function. J High Energy Phys 2013:150
110. Defenu N, Bacsó V, Márián IG, Nándori I, Trombettoni A (2017) Criticality of models interpolating between the sine- and the sinh-Gordon Lagrangians. arXiv:1706.01444
111. Lieb EH, Liniger W (1963) Exact analysis of an interacting Bose gas. I. The general solution and the ground state. Phys Rev 130:1605–1616
112. Oberthaler MK, Abfalterer R, Bernet S, Keller C, Schmiedmayer J, Zeilinger A (1999) Dynamical diffraction of atomic matter waves by crystals of light. Phys Rev A 60:456–472
113. Turlapov A, Tonyushkin A, Sleator T (2003) Optical mask for laser-cooled atoms. Phys Rev A 68:023408
114. Johnson KS, Thywissen JH, Dekker NH, Berggren KK, Chu AP, Younkin R, Prentiss M (1998) Localization of metastable atom beams with optical standing waves: nanolithography at the Heisenberg limit. Science 280:1583–1586
115. Sarchi D, Carusotto I, Wouters M, Savona V (2008) Coherent dynamics and parametric instabilities of microcavity polaritons in double-well systems. Phys Rev B 77:125324
116. Carusotto I, Ciuti C (2005) Spontaneous microcavity-polariton coherence across the parametric threshold: quantum Monte Carlo studies. Phys Rev B 72:125335
117. Wouters M, Carusotto I (2006) Absence of long-range coherence in the parametric emission of photonic wires. Phys Rev B 74:245316
118. Gladilin VN, Ji K, Wouters M (2014) Spatial coherence of weakly interacting one-dimensional nonequilibrium bosonic quantum fluids. Phys Rev A 90:023615
119. Trotzky S, Chen Y-A, Flesch A, McCulloch IP, Schollwöck U, Eisert J, Bloch I (2012) Probing the relaxation towards equilibrium in an isolated strongly correlated one-dimensional Bose gas. Nat Phys 8:325–330
120. Jaksch D, Bruder C, Cirac JI, Gardiner CW, Zoller P (1998) Cold bosonic atoms in optical lattices. Phys Rev Lett 81:3108–3111
121. Greiner M, Mandel O, Esslinger T, Hänsch TW, Bloch I (2002) Quantum phase transition from a superfluid to a Mott insulator in a gas of ultracold atoms. Nature 415:39–44
122. Sachdev S (2001) Quantum phase transitions. Cambridge University Press, Cambridge, England
123. Daley AJ, Taylor JM, Diehl S, Baranov M, Zoller P (2009) Atomic three-body loss as a dynamical three-body interaction. Phys Rev Lett 102:040402
124. Zhu B, Gadway B, Foss-Feig M, Schachenmayer J, Wall ML, Hazzard KRA, Yan B, Moses SA, Covey JP, Jin DS, Ye J, Holland M, Rey AM (2014) Suppressing the loss of ultracold molecules via the continuous quantum Zeno effect. Phys Rev Lett 112:070404
125. Yan B, Moses SA, Gadway B, Covey JP, Hazzard KR, Rey AM, Jin DS, Ye J (2013) Observation of dipolar spin-exchange interactions with lattice-confined polar molecules. Nature 501:521–525
126. Freericks JK, Monien H (1996) Strong-coupling expansions for the pure and disordered Bose-Hubbard model. Phys Rev B 53:2691–2700
127. Lacroix C, Mendels P, Mila F (eds) (2011) Introduction to frustrated magnetism: materials, experiments, theory. Springer, Berlin
128. Preiss PM, Ma R, Tai ME, Simon J, Greiner M (2015) Quantum gas microscopy with spin, atom-number, and multilayer readout. Phys Rev A 91:041602
129. Tomita T, Nakajima S, Danshita I, Takasu Y, Takahashi Y (2017) Observation of the Mott insulator to superfluid crossover of a driven-dissipative Bose-Hubbard system. Sci Adv 3:e1701513
130. Ashida Y, Ueda M (2015) Diffraction-unlimited position measurement of ultracold atoms in an optical lattice. Phys Rev Lett 115:095301
131. Pichler H, Daley AJ, Zoller P (2010) Nonequilibrium dynamics of bosonic atoms in optical lattices: decoherence of many-body states due to spontaneous emission. Phys Rev A 82:063605

132. Fukuhara T, Kantian A, Endres M, Cheneau M, Schauß P, Hild S, Bellem D, Schollwöck U, Giamarchi T, Gross C, Bloch I, Kuhr S (2013) Quantum dynamics of a mobile spin impurity. Nat Phys 9:235–241
133. Bloch I, Dalibard J, Nascimbène S (2012) Quantum simulations with ultracold quantum gases. Nat Phys 8:267–276
134. Lee TE, Häffner H, Cross MC (2012) Collective quantum jumps of Rydberg atoms. Phys Rev Lett 108:023602
135. Brennecke F, Mottl R, Baumann K, Landig R, Donner T, Esslinger T (2013) Real-time observation of fluctuations at the driven-dissipative dicke phase transition. Proc Natl Acad Sci USA 110:11763–11767
136. Endres M, Fukuhara T, Pekker D, Cheneau M, Schauß P, Gross C, Demler E, Kuhr S, Bloch I (2012) The Higgs amplitude mode at the two-dimensional superfluid/Mott insulator transition. Nature 487:454–458
137. Uetake S, Murakami R, Doyle JM, Takahashi Y (2012) Spin-dependent collision of ultracold metastable atoms. Phys Rev A 86:032712
138. Trotzky S, Pollet L, Gerbier F, Schnorrberger I, Bloch U, Prokofev NV, Svistunov B, Troyer M (2010) Suppression of the critical temperature for superfluidity near the mott transition. Nat Phys 6:998–1004
139. Kheruntsyan KV, Gangardt DM, Drummond PD, Shlyapnikov GV (2005) Finite-temperature correlations and density profiles of an inhomogeneous interacting one-dimensional Bose gas. Phys Rev A 71:053615
140. Cardy JL (1985) Conformal invariance and the Yang-Lee edge singularity in two dimensions. Phys Rev Lett 54:1354–1356
141. Pasquier V, Saleur H (1990) Common structures between finite systems and conformal field theories through quantum groups. Nucl Phys B 330:521–556
142. Gong Z, Ashida Y, Kawabata K, Takasan K, Higashikawa S, Ueda M (2018) Topological phases of non-Hermitian systems. Phys Rev X 8:031079
143. Ashida Y, Ueda M (2016) Precise multi-emitter localization method for fast super-resolution imaging. Opt Lett 41:72–75
144. Facchi P, Pascazio S (2008) Quantum Zeno dynamics: mathematical and physical aspects. J Phys A 41:493001
145. Reiter F, Sørensen AS (2012) Effective operator formalism for open quantum systems. Phys Rev A 85:032111

Chapter 4
Out-of-Equilibrium Quantum Dynamics

Abstract We present three general frameworks to describe the dynamics of quantum many-particle systems under continuous observation. In Sect. 4.1, we formulate the dynamics conditioned on the number of quantum jump events, which we term as the full-counting dynamics. We apply it to an exactly solvable model of noninteracting fermions and analyze its out-of-equilibrium dynamics after the quench. We find nonlocal and chiral propagation of correlations beyond the Lieb-Robinson bound. The unique features originate from the non-Hermiticity of the continuously monitored dynamics and do not appear in the corresponding closed systems or the ensemble-averaged dissipative dynamics. In Sect. 4.2, we formulate the thermalization and heating dynamics in generic many-body systems under measurements. Employing the eigenstate thermalization hypothesis, we show that a generic (nonintegrable) many-body system will thermalize at a single-trajectory level under continuous observation. We provide numerical evidence of our findings by studying specific nonintegrable models that are relevant to state-of-the-art experimental setups in ultracold gases. In Sect. 4.3, we formulate the diffusive dynamics under a minimally destructive spatial observation. We derive the many-body stochastic Schrödinger equation for indistinguishable particles under continuous position measurement. We show that the measurement indistinguishability of particles results in complete suppression of relative positional decoherence, leading to persistent correlations in transport dynamics under measurement. We apply the theory of minimally destructive spatial observation to a setup of ultracold atoms in an optical lattice. In Sect. 4.4, we discuss possible experimental realizations of our theoretical studies presented in this chapter. Finally, we conclude this chapter with an outlook in Sect. 4.5.

Keywords Nonequilibrium quantum dynamics · Quench dynamics · Thermalization · The Lieb-Robinson bound · Diffusive dynamics

4.1 Propagation of Correlations and Entanglement Under Measurement

4.1.1 Introduction

In the previous chapter, we have studied how the equilibrium many-body properties, in particular, quantum critical phenomena are modified under the influences of measurement backaction from continuous observation. We have focused on the simplest continuously monitored evolution without quantum jumps, which is governed by the effective non-Hermitian Hamiltonian. This consideration can be justified in a short-time regime during which occurrences of quantum jumps are not very significant or can be eliminated by postselections. Yet, when we move onto the question on the full out-of-equilibrium dynamics of quantum systems under continuous observation, the backaction from quantum jump events will be increasingly important in a longer time regime. This motivates us to develop a theoretical framework to take into account all the possible quantum trajectories.

The aim of this chapter is to develop common frameworks for studying out-of-equilibrium dynamics of quantum many-particle systems under continuous observation. We formulate the frameworks in general ways and apply them to particular models to elucidate the underlying physics unique to many-particle dynamics under measurements.

In this section, by generalizing the notion of full-counting statistics to open many-particle systems, we develop a theoretical formalism to analyze out-of-equilibrium dynamics under continuous observation with an arbitrary number of quantum jumps. The full-counting statistics [1–4] (i.e., the probability distribution of the number of detected signals) has been originally introduced to characterize the underlying nonequilibrium dynamics of mesoscopic devices. Examples include quantum dots, where the electron exchanged with the external environment can be detected at the single-electron resolution [5–8], and photoemissions from atoms, which can be measured individually over a wide range of frequencies [9]. Related ideas have been also applied to one-dimensional Bose gases [10–12] and measurements of entanglement between electron leads [13]. However, most of the previous developments in this direction have been achieved for quantum systems having small degrees of freedom. Here, we consider an open many-particle system and develop a notion of the full-counting dynamics that gives the time evolution of the density matrix conditioned on the number of quantum jumps being observed. We apply our formalism to an exactly solvable model and study its nonequilibrium dynamics under continuous observation. We reveal the emergent unique features such as nonlocal and chiral propagation of correlations, which accompany an entanglement growth with oscillations. In particular, we show that correlations can propagate beyond the Lieb-Robinson bound [14], which is the conventional maximal speed limit on propagation of correlations, at the cost of probabilistic nature of quantum measurement. We identify the origin of these phenomena as the non-Hermiticity of the continuously monitored dynamics, which becomes most prominent at an exceptional point of an effective Hamiltonian.

In our solvable model, this singular point coincides with a spectrum transition point in the parity-time (PT) symmetric non-Hermitian Hamiltonian [15].

4.1.2 General Idea: The Full-Counting Many-Particle Dynamics

We first formulate our idea in a general manner. Consider an open quantum many-particle system whose dissipative dynamics is described by the following master equation

$$\frac{d\hat{\rho}(t)}{dt} = -i\left(\hat{H}_{\text{eff}}\hat{\rho} - \hat{\rho}\hat{H}_{\text{eff}}^\dagger\right) + \mathcal{J}[\hat{\rho}], \quad (4.1)$$

where $\hat{\rho}(t)$ is the density matrix, $\hat{H}_{\text{eff}} = \hat{H} - (i/2)\sum_a \hat{L}_a^\dagger \hat{L}_a$ is an effective non-Hermitian Hamiltonian, $\hat{L}_a$ is a jump operator, and $\mathcal{J}[\hat{\rho}] = \sum_a \hat{L}_a \hat{\rho} \hat{L}_a^\dagger$ describes quantum jump processes [16–19]. We set $\hbar = 1$. We consider continuously monitored dynamics during the time interval $[0, t]$ and count the number of quantum jumps that have occurred during that interval. Let n be the number n of observed quantum jumps. We then introduce the full-counting many-particle dynamics as follows:

$$\hat{\rho}^{(n)}(t) = \frac{\hat{\mathcal{P}}_n \hat{\rho}(t) \hat{\mathcal{P}}_n}{P_n(t)}, \quad (4.2)$$

where $\hat{\mathcal{P}}_n$ is a projection operator onto the subspace with n jumps and $P_n(t) = \text{Tr}[\hat{\mathcal{P}}_n \hat{\rho}(t) \hat{\mathcal{P}}_n]$ provides the probability of finding n jumps during the interval $[0, t]$. To be concrete, we assume that the jump process $\hat{L}_a$ causes a one-body loss. Then, one can in practice access the information about $\hat{\rho}^{(n)}(t)$ by preparing N particles at the initial time, letting the system evolve until time t, and measuring the total number of particles after the evolution by, e.g., quantum gas microscopy. If the measurement outcome is $N - n$ particles, one can know that n jumps have occurred during the time interval. Collecting data for realizations with n jumps, one can reconstruct the full-counting density matrix $\hat{\rho}^{(n)}(t)$. In experiments, we remark that similar postselective operations have been already realized in ultracold atoms [20–23].

We can obtain a formal expression of the full-counting dynamics by taking the average over all possible occurrences of n jump events. Classifying the trajectory dynamics according to the number of quantum jumps $\hat{\rho} = \sum_{n=0}^{N} \hat{\varrho}^{(n)}$, an unnormalized conditional density matrix $\hat{\varrho}^{(n)} = \hat{\mathcal{P}}_n \hat{\rho} \hat{\mathcal{P}}_n$ can be solved as

$$\hat{\varrho}^{(n)}(t) = \sum_{\{a_k\}_{k=1}^n} \int_0^t dt_n \cdots \int_0^{t_2} dt_1 \prod_{k=1}^n \left[\hat{\mathcal{U}}_{\text{eff}}(\Delta t_k)\hat{L}_{a_k}\right] \hat{\mathcal{U}}_{\text{eff}}(t_1)\hat{\rho}(0)\hat{\mathcal{U}}_{\text{eff}}^\dagger(t_1) \prod_{k=1}^n \left[\hat{L}_{a_k}^\dagger \hat{\mathcal{U}}_{\text{eff}}^\dagger(\Delta t_k)\right], \quad (4.3)$$

where $\Delta t_k = t_{k+1} - t_k$ with $t_{n+1} \equiv t$, and $\hat{\mathcal{U}}_{\text{eff}}(t) = e^{-i\hat{H}_{\text{eff}}t}$. For the unconditional dissipative dynamics $\hat{\rho}(t) = \sum_n \hat{\varrho}^{(n)}(t)$, the speed at which correlations build up between distant particles is known to be bounded by the generalized Lieb-Robinson (LR) velocity [24, 25] if the Liouvillian of Eq. (4.1) is the sum of local operators (see Appendix A for a brief review of the LR bound). This is similar to closed systems [26–34], where the (original) LR bound [14] sets the maximal speed limit on propagation of correlations, leading to an effective light-cone causal structure in locally interacting many-body systems. For the full-counting dynamics $\hat{\rho}^{(n)}(t)$ that is conditioned on the measurement outcome, one can no longer expect that the propagation speed obeys the LR velocity due to the nonlocal nature of the measurement acting on an entire many-particle system. The aim of this section is to elucidate the nature of such previously unexplored nonequilibrium dynamics. To analyze the emergent nonequilibrium phenomena in the full-counting dynamics, we hereafter focus on a simple exactly solvable model.

4.1.3 System: Atoms Subject to a Spatially Modulated Loss

As a concrete example, we consider a noninteracting version of the lattice model introduced in the previous chapter analyzing the unconventional critical behavior. We here study its out-of-equilibrium dynamics after the quench by taking into account all the possible quantum jumps.

For the sake of completeness, we briefly review the derivation of the time-evolution equation of the model. We start from the continuum model of one-dimensional spin-polarized noninteracting fermions subject to two weak standing waves with wavelength λ. One standing wave is far detuned from the atomic resonance and thus creates a shallow real potential $h_0 \cos(2\pi x/d)$, where h_0 is the potential depth and $d = \lambda/2$ is the lattice spacing. The other is near resonant to the atomic resonance and creates a weak dissipative potential that leads to a one-body loss.[1] These beams are superimposed and displaced from each other by $d/4$ (see Fig. 4.1). Then, after adiabatically eliminating the dynamics of excited states, the time evolution of ground-state atoms can be described by the following Lindblad master equation:

$$\frac{d\hat{\rho}}{dt} = -i\left(\hat{\mathcal{H}}_{\text{eff}}\hat{\rho} - \hat{\rho}\hat{\mathcal{H}}_{\text{eff}}^\dagger\right) + 2\gamma_0 \int dx \left[1 + \sin\left(\frac{2\pi x}{d}\right)\right] \hat{\Psi}(x)\hat{\rho}\,\hat{\Psi}^\dagger(x), \quad (4.4)$$

where

$$\hat{\mathcal{H}}_{\text{eff}} \equiv \int dx\, \hat{\Psi}^\dagger(x)\left(-\frac{\nabla^2}{2m} + V_{\text{eff}}(x) - i\gamma_0\right)\hat{\Psi}(x) \quad (4.5)$$

---

[1] While the wavelengths for the two potentials are slightly different due to the detuning, the difference can be estimated to be negligibly small ($\sim 0.1\%$).

4.1 Propagation of Correlations and Entanglement Under Measurement 91

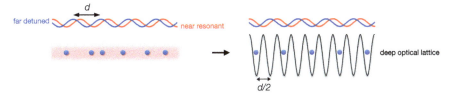

Fig. 4.1 Schematic figures illustrating the model considered. (Left panel) The continuum model of one-dimensional ultracold atoms subject to two shallow optical lattices, one being far detuned (blue) and the other being near-resonant (red) to an atomic resonance. (Right panel) Superimposing a deep optical potential (black) having the half periodicity $d/2$ and employing the tight-binding approximation, we obtain the lattice model analyzed in this section. Reproduced from Fig. S2 of Ref. [35]. Copyright © 2018 by the American Physical Society

is an effective non-Hermitian Hamiltonian, $\hat{\Psi}(x)$ denotes the fermonic field operator, γ_0 characterizes the strength of the dissipation that is determined by the intensity of the near-resonant light. We also introduce

$$V_{\text{eff}}(x) = h_0 \cos\left(\frac{2\pi x}{d}\right) - i\gamma_0 \sin\left(\frac{2\pi x}{d}\right) \quad (4.6)$$

as a complex effective potential [36]. We then superimpose a deep lattice potential with half periodicity $d/2$ (see Fig. 4.1). Employing the standard procedure of the tight-binding approximation for the atomic field [22], we obtain the following master equation:

$$\frac{d\hat{\rho}(t)}{dt} = -i\left(\hat{H}_{\text{eff}}\hat{\rho} - \hat{\rho}\hat{H}_{\text{eff}}^\dagger\right) + \mathcal{J}[\hat{\rho}], \quad (4.7)$$

where

$$\hat{H}_{\text{eff}} = -\sum_{l=0}^{L-1}\left[\left(J + (-1)^l i\gamma\right)\left(\hat{c}_{l+1}^\dagger \hat{c}_l + \hat{c}_l^\dagger \hat{c}_{l+1}\right) + (-1)^l h\hat{c}_l^\dagger \hat{c}_l\right] - 2i\gamma w\hat{N} \equiv \hat{H}_{\text{PT}} - 2i\gamma w\hat{N} \quad (4.8)$$

is a tight-binding version of the effective Hamiltonian (which is equivalent to the noninteracting limit $\Delta \to 0$ of the Hamiltonian in the previous chapter), and

$$\mathcal{J}[\hat{\rho}] = 2\gamma \sum_{l=0}^{L-1}[2w\hat{c}_l\hat{\rho}\hat{c}_l^\dagger + (-1)^l(\hat{c}_l\hat{\rho}\hat{c}_{l+1}^\dagger + \hat{c}_{l+1}\hat{\rho}\hat{c}_l^\dagger)] \quad (4.9)$$

is a quantum jump process. Here, $\hat{c}_l$ ($\hat{c}_l^\dagger$) is the annihilation (creation) operator of the atom at site l, J (γ) is the real (imaginary) hopping parameter, h is the staggered on-site potential (see Fig. 4.2a), w is a factor determined by the depth of the deep lattice, and $\hat{N} = \sum_l \hat{c}_l^\dagger \hat{c}_l$ is the total atom-number operator. To ensure that the dynamical map (4.7) is completely positive trace-preserving (CPTP) (i.e., the dynamics is Markovian), we must impose the condition $w \geq 1$. For the sake of concreteness, we assume $w = 1$ below though the specific choice of its value is irrelevant to our main results. We remark that both of the effective Hamiltonian $\hat{H}_{\text{eff}}$ and the jump process $\mathcal{J}[\hat{\rho}]$ consist of local operators and thus the generalized Lieb-Robinson bound [24, 25] should hold in the unconditional dynamics.

We assume that the initial state is half-filled $N = L/2$, where the length L of the lattice is assumed to be even, impose the periodic boundary conditions, and denote the first part of $\hat{H}_{\text{eff}}$ in Eq. (4.8) as $\hat{H}_{\text{PT}}$ since it satisfies the parity-time (PT) symmetry [15]. Since the effective Hamiltonian $\hat{H}_{\text{PT}}$ is quadratic, it can be diagonalized as

$$\hat{H}_{\text{PT}} = -\sum_{l=0}^{L-1}\left[\left(J+(-1)^l i\gamma\right)\left(\hat{c}_{l+1}^\dagger \hat{c}_l + \hat{c}_l^\dagger \hat{c}_{l+1}\right) + (-1)^l h \hat{c}_l^\dagger \hat{c}_l\right]$$
$$= \sum_{0 \leq k < 2\pi} \sum_{\lambda = \pm} \epsilon_\lambda(k) \hat{g}_{\lambda k}^\dagger \hat{f}_{\lambda k}, \qquad (4.10)$$

where $\epsilon_\pm(k) = \pm\sqrt{h^2 - 4\gamma^2 + 2J'^2(1 + \cos(k))}$ are eigenvalues with $J' = \sqrt{J^2 + \gamma^2}$ and $k = 2\pi n/(L/2)$ ($n = 0, 1, \ldots, L/2 - 1$). The operators $\hat{g}^\dagger$ and $\hat{f}$ create the right and left eigenvectors, i.e., $\hat{H}_{\text{PT}} \hat{g}_{\lambda k}^\dagger |0\rangle = \epsilon_\lambda(k) \hat{g}_{\lambda k}^\dagger |0\rangle$ and $\langle 0| \hat{f}_{\lambda k} \hat{H}_{\text{PT}} = \langle 0| \hat{f}_{\lambda k} \epsilon_\lambda(k)$, and they obey an anticommutation relation

$$\{\hat{f}_{\lambda k}, \hat{g}_{\lambda' k'}^\dagger\} = \delta_{k,k'} \delta_{\lambda, \lambda'}. \qquad (4.11)$$

An important consequence of the non-Hermiticity in the effective Hamiltonian is the nonorthogonality of the eigenstates. It manifests itself as an unusual commutation relation satisfied by $\hat{g}$ and $\hat{g}^\dagger$:

$$\{\hat{g}_{\lambda k}, \hat{g}_{\lambda' k'}^\dagger\} = \delta_{k,k'} \Delta_{\lambda \lambda'}(k), \qquad (4.12)$$

where $\Delta_{\lambda \lambda'}(k)$ are the 2×2 matrices. Their nonzero off-diagonal elements represent the nonorthogonality between the right eigenvectors of different bands at mode k (see Appendix B for details). In the PT-unbroken regime $\gamma < h/2$, the band spectrum is real and gapped. The gap closes at $k = \pi$ for $\gamma = h/2$ (see Fig. 4.2b). This point corresponds to an exceptional point [37], i.e., two $k = \pi$ eigenstates of the upper and lower bands coalesce into a single one. In the PT-broken regime $\gamma > h/2$, eigenstates around $k = \pi$ turn out to have complex pairs of pure imaginary eigenvalues.

4.1 Propagation of Correlations and Entanglement Under Measurement 93

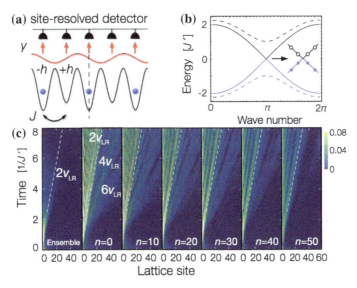

Fig. 4.2 **a** Schematic illustration of an exactly solvable model for the full-counting dynamics. **b** Energy bands $\epsilon_\pm(k)$ for Eq. (4.10) with $\gamma/h = 1/2$ (solid curves) and $\gamma = 0, h = J$ (dashed curves). **c** Spatiotemporal dynamics of equal-time correlations for the unconditional evolution (leftmost panel) and the full-counting evolutions (other panels) conditioned on the number n of quantum jumps. We use $\gamma = h/2 = 0.5J$ and $N = L/2 = 61$. Reproduced from Fig. 1 of Ref. [35]. Copyright © 2018 by the American Physical Society

4.1.4 Nonequilibrium Dynamics of Correlations and Entanglement

Nonlocal propagation of correlations

We next discuss emergent nonequilibrium phenomena unique to the full-counting dynamics. Substituting the diagonalized effective Hamiltonian (4.10) into a general solution (4.3), we can obtain an exact solution of the full-counting dynamics $\hat{\rho}^{(n)}(t)$ (see Appendix B for details). To be concrete, we consider the following quench operation. Initially, we set h and γ to be zero and prepare the ground state of $\hat{H}$. At time $t = 0$, h and γ are suddenly switched on, and the system evolves in time after which the site-resolved measurement is performed to determine the number n of quantum jumps, i.e., the number of atoms lost. Classifying the realizations according to n, one can study the full-counting dynamics $\hat{\rho}^{(n)}(t)$. We set $\gamma = h/2$ such that the postquench Hamiltonian $\hat{H}_{\text{PT}}$ is at the verge of the spectrum transition point, leading to the linear dispersion around $k = \pi$ (see Fig. 4.2b).

Let us first consider the unconditional case $\hat{\rho}(t) = \sum_{n=0}^{N} P_n(t)\hat{\rho}^{(n)}(t)$. The leftmost panel in Fig. 4.2c shows its spatiotemporal dynamics of an equal-time correlation

$$C(l, t) = \text{Tr}[\hat{\rho}(t)\hat{c}_l^\dagger \hat{c}_0]. \tag{4.13}$$

Since the Liouvillian of the master equation consists of local operators, it is expected that correlations can propagate no faster than twice the LR velocity $2v_{\text{LR}}$ [24, 25, 38], where v_{LR} is determined from the maximum group velocity[2] $|\partial \epsilon_\pm(k)/\partial(k/2)|_{k=\pi} = 2J'$. We remark that a similar blurred light cone has also been recently found in numerical simulations of the dissipative Bose-Hubbard model [39].

A qualitatively different propagation can be found in the full-counting dynamics $\hat{\rho}^{(n)}(t)$. Figure 4.2c plots an equal-time correlation

$$C^{(n)}(l, t) = \text{Tr}[\hat{\rho}^{(n)}(t)\hat{c}_l^\dagger \hat{c}_0] \tag{4.14}$$

for the full-counting dynamics conditioned on different values of n. There, nonlocal propagations faster than the LR velocity of the corresponding unconditional dynamics can be found. We also find that such supersonic modes propagate at the velocities that are integer multiples of $2v_{\text{LR}}$. We identify the origin of these nonlocal propagations as the non-Hermiticity in the underlying continuously monitored dynamics. To elucidate this, we focus on the simplest trajectory dynamics with no jumps being observed [40, 41]

$$\hat{\rho}^{(0)}(t) = \frac{e^{-i\hat{H}_{\text{PT}}t} \hat{\rho}(0) e^{i\hat{H}_{\text{PT}}^\dagger t}}{\text{Tr}[e^{-i\hat{H}_{\text{PT}}t} \hat{\rho}(0) e^{i\hat{H}_{\text{PT}}^\dagger t}]}. \tag{4.15}$$

Because an initially pure state remains pure [41] in Eq. (4.15), an unnormalized time-dependent state can be written as $|\Psi_t\rangle = e^{-i\hat{H}_{\text{PT}}t}|\Psi_0\rangle$, where we denote the initial state as $\hat{\rho}(0) = |\Psi_0\rangle\langle\Psi_0|$. Let $\psi_{\lambda k}$'s be expansion coefficients of the initial state with right eigenstates: $|\Psi_0\rangle = \prod_k [\sum_\lambda \psi_{\lambda k} \hat{g}_{\lambda k}^\dagger]|0\rangle$. Introducing the unequal-time correlation $\tilde{C}^{(0)}(l, t) = \langle \Psi_0 | \hat{c}_l^\dagger(t) \hat{c}_0(0) | \Psi_0 \rangle / \langle \Psi_t | \Psi_t \rangle$ with $\hat{c}_l^\dagger(t) = e^{i\hat{H}_{\text{PT}}t} \hat{c}_l^\dagger e^{-i\hat{H}_{\text{PT}}t}$ and substituting the diagonalized $\hat{H}_{\text{PT}}$ into it, we obtain

$$\tilde{C}^{(0)}(l, t) = \frac{2}{L} \sum_k \sum_{\lambda=\pm} \begin{Bmatrix} \alpha_{\lambda k} \\ \beta_{\lambda k} \end{Bmatrix} \frac{\psi_{\lambda k}^* e^{i\epsilon_\lambda(k)t - ik\lceil l/2 \rceil}}{\mathcal{N}_k(t)}. \tag{4.16}$$

Here $\alpha_{\lambda k}$ and $\beta_{\lambda k}$ are coefficients chosen according to the parity of l, $\lceil \cdot \rceil$ is the ceiling function, and we introduce the norm factor by

---

[2] A factor of two in the group velocity results from our choice of wavevectors in the Fourier transforms of the sublattices (see e.g., the phase factor in Eq. 4.16).

4.1 Propagation of Correlations and Entanglement Under Measurement

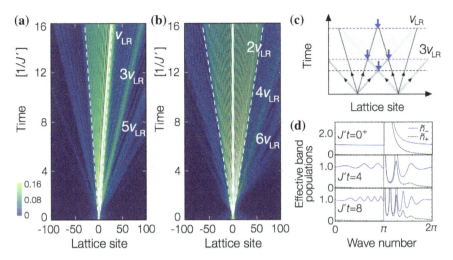

Fig. 4.3 The panels **a** and **b** show the unequal- and equal-time correlations in the non-Hermitian quench dynamics, respectively. We use $\gamma = h/2 = 0.5J$ and $N = L/2 = 61$. The Lieb-Robinson bound is indicated by the white dashed lines. **c** The time needed to establish the correlations is determined by pairs of quasiparticles propagating at velocities v_{LR} and $3v_{\text{LR}}$. **d** Time evolutions of effective band populations at $J't = 0^+$ (postquench state), 4, and 8. Reproduced from Fig. 2 of Ref. [35]. Copyright © 2018 by the American Physical Society

$$\mathcal{N}_k(t) = \sum_{\lambda\lambda'=\pm} \psi^*_{\lambda k}(t)\Delta_{\lambda\lambda'}(k)\psi_{\lambda' k}(t), \tag{4.17}$$

where $\psi_{\lambda k}(t) = \psi_{\lambda k}e^{-i\epsilon_\lambda(k)t}$ (see Eq. (B.24) in Appendix B). The total norm of an unnormalized quantum state $|\Psi_t\rangle$ is given by the product of these factors $\langle \Psi_t | \Psi_t \rangle = \prod_k \mathcal{N}_k(t)$.

The crucial point here is that due to the nonorthogonality of eigenvectors ($\Delta_{+-} = \Delta^*_{-+} \neq 0$) the norm $\mathcal{N}_k(t)$ oscillates at the frequency $2\epsilon_+(k)$. Thus, $\tilde{C}^{(0)}(l,t)$ in Eq. (4.16) involves terms that oscillate at frequencies $\epsilon_\lambda(k), 3\epsilon_\lambda(k), 5\epsilon_\lambda(k), \ldots$, leading to the propagations of quasiparticles at velocities $v_{\text{LR}}, 3v_{\text{LR}}, 5v_{\text{LR}}, \ldots$ (see Fig. 4.3a). As a result, the equal-time correlation $C^{(0)}(l,t)$ involves the propagations at velocities $2v_{\text{LR}}, 4v_{\text{LR}}, 6v_{\text{LR}}, \ldots$ (see Fig. 4.3b) since it originates from quasiparticle pairs propagating in opposite directions [26] (see Fig. 4.3c). We stress that the emergence of these supersonic modes is a consequence of the unique interplay between the many-particle nature of the system and the non-Hermiticity in the continuously monitored dynamics. Supersonic modes result from the nonlinearity due to the oscillating norm factors $\mathcal{N}_k(t)$ in the denominator in Eq. (4.16). This is a direct consequence of the total norm of a many-particle state, which is given by the product of $\mathcal{N}_k(t)$ (cf. the derivation given in Appendix B). For the (one-body) classical non-Hermitian systems, such oscillating factors do not appear inside the summation over k in Eq. (4.16) since the total norm is given by the sum of $\mathcal{N}_k(t)$ rather than their product [42]. Thus, our finding is a genuine quantum effect in the sense that the supersonic modes do not appear in the single-particle sector or the mean-field non-Hermitian dynamics observed in optics [43, 44] or dissipative matter waves [45–48].

In parallel with discussions in closed systems [26–34], the quantities $\tilde{n}_{\lambda k}(t) = |\psi_{\lambda k}|^2 / \mathcal{N}_k(t)$ can be understood as an effective band population in the full-counting dynamics. The conventional band population does not change during the time evolution after the quench in noninteracting closed systems [26, 32–34]. Meanwhile, the effective population $\tilde{n}_{\lambda k}(t)$ can oscillate in time (see Fig. 4.3d). This is a nontrivial consequence of the nonorthogonality between two eigenstates at mode k.

Finally, we make a remark on the manifestation of the Lieb-Robinson bound in the unconditional dynamics. In the full-counting dynamics $\hat{\rho}^{(n)}(t)$ (see e.g., Fig. 4.3a and b), there exist the robust supersonic modes propagating with velocities $3v_{\text{LR}}, 5v_{\text{LR}}, \ldots$ for unequal-time correlations and with velocities $4v_{\text{LR}}, 6v_{\text{LR}}, \ldots$ for equal-time correlations. These modes clearly violate the LR bound since the supersonic modes will eventually protrude into the tail beyond the light cone allowed by the bounds (cf. Eqs. (A.1) and (A.2) in Appendix A). Yet, the propagation of correlations in the unconditional dynamics $\hat{\rho}(t) = \sum_n P_n(t) \hat{\rho}^{(n)}(t)$ is still limited by the LR velocity (see the left-most panel in Fig. 4.2c). Here, the LR bound manifests itself as an exponential suppression of the supersonic modes due to the exponentially decaying probability factor $P_n(t)$ which multiplies the full-counting dynamics $\hat{\rho}^{(n)}(t)$. To see this, in Fig. 4.4 we plot (a) a typical profile of the correlation function in the full-counting dynamics and (b) the values of the supersonic contribution for different numbers of quantum jumps n. The latter shows that the supersonic contribution dwindles very rapidly (faster than exponential decrease) as n increases. Thus, the major contributions of the supersonic modes come from the trajectories with a relatively

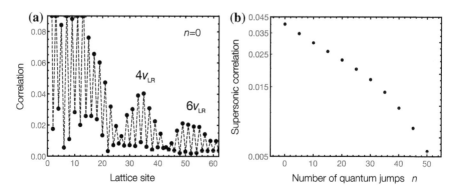

Fig. 4.4 **a** Absolute value of the equal-time correlation $|C^{(n)}(l, t_0)| = |\text{Tr}[\hat{\rho}^{(n)}(t)\hat{c}_l^\dagger \hat{c}_0]|$ for the null quantum jump $n = 0$ at time $t_0 = 5/J'$ plotted against lattice site l. All the parameters and the quench protocol are the same as those in Fig. 4.2c. **b** Equal-time correlation $C^{(n)}(l_0, t_0)$ associated with the supersonic modes propagating with the velocity $4v_{\text{LR}}$ plotted against the number of quantum jumps. The lattice site is chosen to be $l_0 = 35$ at which the supersonic contribution is maximal (see also the panel **a**). Reproduced from Fig. S1 of Ref. [35]. Copyright © 2018 by the American Physical Society

small number n of jumps (i.e., atomic loss). Meanwhile, the occurrence probability of such trajectories will eventually be suppressed exponentially as a function of time t. It is this exponential suppression that recovers the LR bound in the overall unconditional density matrix $\hat{\rho}(t) = \sum_n P_n(t) \hat{\rho}^{(n)}(t)$.

Chirality in propagation of correlations

The observed propagation also exhibits the chirality, i.e., it violates the left-right symmetry. The gain-loss structure of $\hat{H}_{\text{PT}}$ allows one to intuitively understand the observed enhancement of propagation of correlations in the right direction (see Fig. 4.5). The bond with positive (negative) imaginary hopping corresponds to the "gain" ("loss") bond at which particles are injected (removed). Due to the staggered potential, a majority of the injected particles flow into the deeper potential in the right direction and they are removed at the loss bond. As a result, the flow in the right direction overweighs the reverse one.

This chirality becomes most prominent at the transition point $\gamma = h/2$. Figure 4.5a plots time-averaged values of the current $iJ \sum_{l=0}^{L-1} (\hat{c}_l^\dagger \hat{c}_{l+1} - \hat{c}_{l+1}^\dagger \hat{c}_l)$ for various h and γ in a steady-state regime. At the threshold $\gamma = h/2$, the current takes the maximum value, which originates from the strong nonorthogonality due to the exceptional point at $k = \pi$ (see Fig. 4.2b) [49]. Around this point, two eigenvectors of different bands coalescence into the one having positive group velocities $\partial \epsilon / \partial (k/2) > 0$ (see Fig. 4.6). This confluent band structure causes the imbalanced band populations $\tilde{n}_{\lambda k}(t)$ in Fig. 4.3d (especially in a regime $k < \pi$), leading to a pronounced propagation in the right direction. We stress that the chirality here is particularly prominent

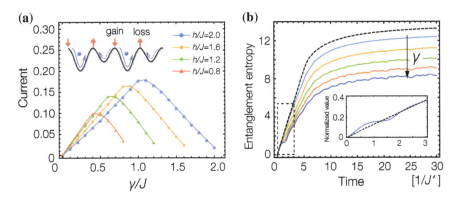

Fig. 4.5 **a** The particle current is plotted against dissipation strength γ with different staggered potentials h. We calculate the current by taking the average over a nearly steady-state regime (from time $t = 15/J'$ to $20/J'$). The inset schematic figure indicates a gain-loss structure leading to a net particle current in the right direction. **b** Dynamics of entanglement entropy. We choose the initial state to be a product state that is the ground state of $\hat{H}$ with $h = \infty$ and consider a subregion of a length of 20 sites. We plot the results for different postquench parameters γ from 0.0 to 0.5 (top to bottom) with step 0.1 and $\gamma = h/2$ held fixed. The oscillatory behavior shown in the inset originates from the time-dependence of $\tilde{n}_{\lambda k}$. Reproduced from Fig. 3 of Ref. [35]. Copyright © 2018 by the American Physical Society

due to the formation of the Fermi sea, making contrast to the single-particle sector [50–55] that reduces to classical dynamics.

To further elucidate the origin of the chiral structure, we can make simple analytical arguments based on the eigenvectors in the vicinity of the exceptional point. Let us start from the initial (Hermitian) Hamiltonian, i.e., the one with $\gamma = h = 0$. Without loss of generality, we set $J = 1$ throughout this subsection. There are two band dispersions: one has a positive group velocity ($\epsilon_>(k) = 2\sin(k/2)$) and the other has a negative group velocity ($\epsilon_<(k) = -2\sin(k/2)$). The corresponding eigenvectors are given by diagonalizing the 2×2 Hermitian matrices with $\gamma = h = 0$ and $J = 1$ (cf. Eq. (B.3) in Appendix B). In the vicinity of the gapless point at $k = \pi$, we obtain the results

$$c_>(\delta k) = \frac{1}{\sqrt{2}} \begin{pmatrix} -i - \frac{\delta k}{2} \\ 1 \end{pmatrix} + O\left((\delta k)^2\right), \tag{4.18}$$

$$c_<(\delta k) = \frac{1}{\sqrt{2}} \begin{pmatrix} i + \frac{\delta k}{2} \\ 1 \end{pmatrix} + O\left((\delta k)^2\right), \tag{4.19}$$

where $\delta k = k - \pi$ is the displacement satisfying $|\delta k| \ll 1$ and $c_{>(<)}$ is the eigenvector of the band dispersion having a positive (negative) group velocity in the basis of the two sublattices. Since the lower band ($\epsilon(k) < 0$) is filled in the initial ground state, the eigenvector $c_>(\delta k)$ ($c_<(\delta k)$) is populated for $\delta k < 0$ ($\delta k > 0$) at the initial time (see the shaded region in Fig. 4.6).

We next consider the postquench Hamiltonian, i.e., the non-Hermitian Hamiltonian with nonzero γ satisfying $h = 2\gamma$. In this case, two eigenvectors of different bands coalesce at the exceptional point at $k = \pi$. In the vicinity of the exceptional point, we obtain simple analytical expressions of right eigenvectors as follows:

$$c_>^R(\delta k) = \frac{1}{\sqrt{2}} \begin{pmatrix} -i \\ 1 \end{pmatrix} + \frac{1}{4\sqrt{2}\gamma} \begin{pmatrix} -i(1 - 2i\gamma - \sqrt{1+\gamma^2}) \\ \sqrt{1+\gamma^2} - 1 \end{pmatrix} \delta k + O\left((\delta k)^2\right), \tag{4.20}$$

$$c_<^R(\delta k) = \frac{1}{\sqrt{2}} \begin{pmatrix} -i \\ 1 \end{pmatrix} + \frac{1}{4\sqrt{2}\gamma} \begin{pmatrix} -i(1 - 2i\gamma + \sqrt{1+\gamma^2}) \\ -\sqrt{1+\gamma^2} - 1 \end{pmatrix} \delta k + O\left((\delta k)^2\right), \tag{4.21}$$

which are valid for $|\delta k| \ll \min(\gamma, 1)$. Here $c_{>(<)}^R$ is the right eigenvector of the band dispersion having the positive (negative) group velocity. Note that in the limit of $\delta k \to 0$ these two eigenvectors coalesce into the one having a positive group velocity given in Eq. (4.18) (see the left panel in Fig. 4.6). This coalescence of eigenvectors near the exceptional point leads to the imbalanced effective populations of quasiparticles having positive group velocities in $k < \pi$ (see Fig. 4.3d), resulting in the pronounced propagation of correlations in the positive direction as we demonstrated above. We remark that if the gain-loss structure is reversed, i.e., if we set $h = -2\gamma$, the two right eigenvectors can be shown to coalesce into the one having a negative group velocity given in Eq. (4.19) (see the right panel in Fig. 4.6), leading to the pronounced propagation in the negative direction.

4.1 Propagation of Correlations and Entanglement Under Measurement   99

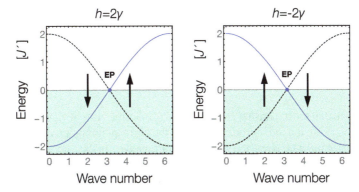

Fig. 4.6 Coalescence of eigenvectors near the exceptional point. (Left panel) When the postquench Hamiltonian is near the spectral transition point, the gapless point at $k = \pi$ forms an exceptional point (EP). In the vicinity of EP, the two eigenvectors in different bands coalesce into the one having positive group velocities (blue solid curve). Since the lower band is initially populated (shaded region), this coalescence leads to an imbalanced effective population in the band having positive group velocities in $k < \pi$, resulting in the pronounced propagation of correlations in the positive direction as demonstrated above. (Right panel) In contrast, if the gain-loss structure is reversed (if we set $h = -2\gamma$), the band having negative group velocities (blue solid curve) is dominantly populated in $k > \pi$. This results in the pronounced propagation in the negative direction. Reproduced from Fig. S3 of Ref. [35]. Copyright © 2018 by the American Physical Society

Finally, we mention the origin of the discontinuity and the divergence of the effective band populations appearing at $k = \pi$ (see Fig. 4.3d). They are caused by singularities in left eigenvectors at the exceptional point:

$$c_>^L(\delta k) = \frac{2\gamma}{\sqrt{2(1+\gamma^2)}} \begin{pmatrix} i \\ 1 \end{pmatrix} \frac{1}{\delta k} + \frac{1}{2\sqrt{2(1+\gamma^2)}} \begin{pmatrix} -i(1+2i\gamma+\sqrt{1+\gamma^2}) \\ \sqrt{1+\gamma^2}+1 \end{pmatrix} + O(\delta k), \quad (4.22)$$

$$c_<^L(\delta k) = -\frac{2\gamma}{\sqrt{2(1+\gamma^2)}} \begin{pmatrix} i \\ 1 \end{pmatrix} \frac{1}{\delta k} + \frac{1}{2\sqrt{2(1+\gamma^2)}} \begin{pmatrix} -i(-1-2i\gamma+\sqrt{1+\gamma^2}) \\ \sqrt{1+\gamma^2}-1 \end{pmatrix} + O(\delta k), \quad (4.23)$$

where $c_{>(<)}^L$ are left eigenvectors of the band dispersions having the positive (negative) group velocity. The divergence of the left eigenvectors in the limit $\delta k \to 0$ originates from the vanishing inner product between the right and left eigenvectors at the exceptional point [44, 49]. To understand how this divergence leads to the discontinuity of the effective band populations $\tilde{n}_{\lambda k}$, we recall that $\tilde{n}_{\lambda k}$ is proportional to the square of the expansion coefficient $\psi_{\lambda k}$ of the initial ground state in terms of the right eigenvectors for the postquench non-Hermitian matrices, i.e., $n_{\lambda k} \propto |\psi_{\lambda k}|^2$ with $|\Psi_0\rangle = \prod_k [\sum_\lambda \psi_{\lambda k} \hat{g}_{\lambda k}^\dagger]|0\rangle$. We then obtain the following relations:

For $\delta k < 0$:
$$\begin{cases} \tilde{n}_{+,k} \propto \left|c_<^{\dagger L}(\delta k) \cdot c_>(\delta k)\right|^2 \sim 0; \\ \tilde{n}_{-,k} \propto \left|c_>^{\dagger L}(\delta k) \cdot c_>(\delta k)\right|^2 \sim 1. \end{cases} \quad (4.24)$$

For $\delta k > 0$:
$$\begin{cases} \tilde{n}_{+,k} \propto \left|c_>^{\dagger L}(\delta k) \cdot c_<(\delta k)\right|^2 \simeq \frac{\gamma^2}{1+\gamma^2}\frac{1}{\delta k^2}; \\ \tilde{n}_{-,k} \propto \left|c_<^{\dagger L}(\delta k) \cdot c_<(\delta k)\right|^2 \simeq \frac{\gamma^2}{1+\gamma^2}\frac{1}{\delta k^2}. \end{cases} \quad (4.25)$$

Here we note that the expansion coefficients $\psi_{\lambda k}$ in terms of right eigenvectors can be obtained by taking the inner product between the corresponding left eigenvectors and the initial ground state since the left and right eigenvectors satisfy the orthonormal condition $\{\hat{f}_{\lambda k}, \hat{g}^{\dagger}_{\lambda',k'}\} = \delta_{k,k'}\delta_{\lambda,\lambda'}$. Equation (4.24) shows that the populations $\tilde{n}_{\pm,k}$ remain finite if we approach the exceptional point $k = \pi$ from below. This is because the diverging contribution $(i, 1)^T$ in the left eigenvectors in Eqs. (4.22) and (4.23) is orthogonal to the leading contribution $(-i, 1)^T$ of $c_>(\delta k)$ in Eq. (4.18). In contrast, if we approach the exceptional point from above, the populations $\tilde{n}_{\pm k}$ diverge in $\delta k \to 0$ as shown in Eq. (4.25) since the diverging contribution $(i, 1)^T$ in Eqs. (4.22) and (4.23) is parallel to the leading contribution of $c_<(\delta k)$ in Eq. (4.19).

Entanglement dynamics

The chirality found in the continuously monitored dynamics also alters the evolution of the entanglement growth. Figure 4.5b shows the entanglement entropy $S_A[\hat{\rho}^{(0)}(t)]$ [56] after the quench with varying measurement strength γ. We choose a subregion A to be an interval of 20 sites. Increasing γ, the entanglement entropy decreases due to the enhanced chirality. We can understand this as follows. After the quench, pairs of entangled quasiparticles are generated and they propagate in opposite directions [26]. The number of correlated pairs with one quasiparticle being inside and the other being outside of A determines the entanglement entropy. The chirality thus leads to a decrease of the entanglement entropy because such correlated pairs propagating in opposite directions cannot be generated via the chiral (i.e., unidirectional) modes. The time-dependent band populations $\tilde{n}_{\lambda k}(t)$ cause an oscillation on the linear increase of the entanglement entropy (inset in Fig. 4.5b). In light of the simple dispersion in the Hamiltonian, this oscillatory behavior also gives an intriguing feature in open systems; such a oscillation can appear only if the band spectrum possess multiple local maxima in closed integrable systems [32].

4.2 Thermalization and Heating Dynamics Under Measurement

4.2.1 Introduction

In the previous section, we have studied the peculiar out-of-equilibrium dynamics in many-particle systems under measurement. Meanwhile, after such transient dynamics, it is natural to expect that an interacting many-body system will ultimately

4.2 Thermalization and Heating Dynamics Under Measurement

equilibrate, especially in light of recent developments in our understanding of equilibration and thermalization in quantum many-body systems [28, 57–78]. In this section, we ask the question of whether or not statistical mechanics [79–86] remains valid as a description of such open many-body systems under quantum measurement.

In general, statistical mechanics provides a universal description to characterize thermodynamic properties of an equilibrated physical system. To address the problem of thermalization in quantum systems, we can classify them as follows: (i) systems coupled to thermal environments, (ii) isolated systems, and (iii) systems in contact with nonthermal environments. In class (i), it has been known that thermalization to the Gibbs ensemble can be formulated by the phenomenological Lindblad master equation [87–92]. There, the detailed balance condition for the Lindblad operators ensures that the master equation always possesses the Gibbs ensemble at the temperature of the thermal bath as a solution of the master equation. The last decade has been devoted to advancing our understanding of thermalization in class (ii) [28, 57–78]. This is mainly motivated by recent experimental developments in ultracold atoms [93–97] that allow one to prepare quantum many-body systems in nearly isolated situations. As a generic and possible mechanism of thermalization under unitary dynamics, the eigenstate thermalization hypothesis (ETH) [68–78, 84, 85] has been proposed. Numerical calculations for several nonintegrable many-body Hamiltonians [68–78] have verified the ETH, while it does not hold for integrable [98–104] or many-body localized systems [105, 106].

An important remaining issue is to address a possible thermalization mechanism in systems falling into class (iii). In this class, the dynamics is intrinsically nonunitary while the detailed balanced condition is violated since a coupling to nonthermal environment permits general nonunitary processes such as continuous measurements [107–124] and engineered dissipation [35, 45, 47, 125–137]. In such a nonthermal environment, the bath temperature does not exist a priori and it is not obvious whether or not the system still thermalizes and how a (possibly) thermalized state can be related to the Gibbs ensemble. While related numerical studies have been done for specific examples [19, 107, 118, 138, 139], a unified (i.e., model-independent) understanding of thermalization in open generic many-body systems is still lacking. In this section, we aim to extend the framework of thermalization to quantum many-body systems coupled to generic environments permitted by controlled dissipations and quantum measurements [140]. In experiments, our study is related to recent realizations of a variety of open many-body systems [45, 47, 134–137]. The results presented in this section are applicable not only to many-body systems under continuous observation, but also to dissipative Lindblad dynamics of many-body systems coupled to environments (that are not necessarily thermal) [35, 45, 47, 126–138] or under noisy unitary operations [141–150]. The results presented in this section thus provide yet another insight into why thermodynamics emerges so universally.

4.2.2 Statistical Ensemble Under Minimally Destructive Observation

Suppose that a generic (typically, non integrable) quantum many-body system is subject to continuous observation. The initial state is assumed to be a thermal equilibrium state $\hat{\rho}_{eq}$ whose mean energy is E_0 and the corresponding temperature is $T_0 = 1/\beta_0$. We set $k_B = 1$ and $\hbar = 1$ in this section. We consider the quantum trajectory dynamics under continuous observation as formulated in Chap. 2. The dynamics is characterized by a sequence of measurement outcomes, i.e., types $\mathcal{M} = (m_1, \ldots m_n)$ and occurrence times $\mathcal{T} = (t_1, \ldots t_n)$ of quantum jumps, and is represented as

$$\hat{\varrho}_\mathcal{M}(t; \mathcal{T}) = \hat{\Pi}^\mathcal{M}_{t;\mathcal{T}} \hat{\rho}_{eq} \hat{\Pi}^{\dagger\mathcal{M}}_{t;\mathcal{T}}, \tag{4.26}$$

where we introduce the nonunitary operator $\hat{\Pi}^\mathcal{M}_{t;\mathcal{T}}$ as

$$\hat{\Pi}^\mathcal{M}_{t;\mathcal{T}} = \prod_{i=1}^{n} [\hat{\mathcal{U}}(\Delta t_i) \sqrt{\gamma} \hat{L}_{m_i}] \hat{\mathcal{U}}(t_1). \tag{4.27}$$

We recall that $\hat{\mathcal{U}}(\tau) = e^{-i\hat{H}_{eff}\tau}$ with $\hat{H}_{eff} = \hat{H} - i\hat{\Gamma}/2$ and $\hat{\Gamma} = \gamma \sum_m \hat{L}_m^\dagger \hat{L}_m$ describes the non-Hermitian evolution during no-jump events, γ characterizes a measurement strength, $\hat{L}_m$ is a jump operator associated with a measurement outcome m, and we denote $\Delta t_i = t_{i+1} - t_i$ and $t_{n+1} = t$. A jump operator $\hat{L}_m$ is assumed to be either a local operator or the sum of local operators and conserves the total number of particles.

To extract generic features of the many-body trajectory dynamics, we assume the ETH and the limit of minimally destructive observation. The former states that the expectation values of arbitrary few-body observables with respect to a high-energy eigenstate $|E_a\rangle$ of $\hat{H}$ coincide with those of the corresponding Gibbs ensemble [76], as numerically supported in a number of nonintegrable many-body Hamiltonians [68–78]. The latter indicates a situation in which a waiting time of quantum jumps is much longer than the equilibration time of the many-body dynamics during a no-jump process. More specifically, we consider the limit $\gamma \to 0$ with keeping $\gamma t = \mu$ finite. Physically, this limit means that one typically observes a finite number of quantum jumps during $[0, t]$ while the system still does not reach an ultimate steady state, which can typically be a trivial state such as an infinite-temperature state. When a waiting time exceeds the equilibration time in the many-body dynamics, it is expected that the memory of the occurrence times $\mathcal{T}$ will eventually be lost due to the fast decay of the rapidly oscillating, time-dependent terms in the density matrix [60]. Thus, long-time values of expectation values can be studied in terms of the time-averaged density matrix [60]:

$$\hat{\varrho}_\mathcal{M}(t) = \int_0^t dt_n \cdots \int_0^{t_2} dt_1 \hat{\varrho}_\mathcal{M}(t; \mathcal{T}). \tag{4.28}$$

4.2 Thermalization and Heating Dynamics Under Measurement 103

Assuming the minimally destructive limit and the ETH, we can achieve several simplifications in Eq. (4.28). To begin with, we rewrite it by expanding a non-Hermitian effective Hamiltonian $\hat{H}_{\text{eff}}$ as $\hat{H}_{\text{eff}} = \sum_a \Lambda_a |\Lambda_a^R\rangle\langle\Lambda_a^L|$, where Λ_a is a complex eigenvalue, and the right (left) eigenstates $|\Lambda_a^R\rangle$ ($|\Lambda_a^L\rangle$) satisfy the orthonormal condition $\langle\Lambda_a^R|\Lambda_b^L\rangle = \delta_{ab}$. Inserting the relation $\hat{\mathcal{U}}(\tau) = e^{-i\hat{H}_{\text{eff}}\tau} = \sum_a e^{-i\Lambda_a\tau}|\Lambda_a^R\rangle\langle\Lambda_a^L|$ into Eq. (4.28), we obtain

$$\hat{\varrho}_M(t) = \sum_{\{a_i\}\{b_i\}} \mathcal{F}(t; \{a_i\}, \{b_i\}) \prod_{i=1}^{n}[\sqrt{\gamma}(V_{m_i})_{a_{i+1}a_i}](P_{\text{eq}})_{a_1 b_1} \prod_{i=1}^{n}[\sqrt{\gamma}(V_{m_i}^\dagger)_{b_i b_{i+1}}]|\Lambda_{a_{n+1}}^R\rangle\langle\Lambda_{b_{n+1}}^L|, \tag{4.29}$$

where we introduce the matrices V_m and P_{eq} as

$$(V_m)_{ab} = \langle\Lambda_a^L|\hat{L}_m|\Lambda_b^R\rangle, \quad (P_{\text{eq}})_{ab} = \langle\Lambda_a^L|\hat{\rho}_{\text{eq}}|\Lambda_b^L\rangle, \tag{4.30}$$

and $\mathcal{F}$ involves the time integration of the exponential factor

$$\mathcal{F}(t; \{a_i\}, \{b_i\}) = e^{-i(\Lambda_{a_{n+1}} - \Lambda_{b_{n+1}}^*)t} \int_0^t dt_n \cdots \int_0^{t_2} dt_1 e^{-i\sum_{i=1}^n \Delta_i t_i} \tag{4.31}$$

with

$$\Delta_i = \Lambda_{a_i} - \Lambda_{b_i}^* - (\Lambda_{a_{i+1}} - \Lambda_{b_{i+1}}^*). \tag{4.32}$$

In the minimally destructive limit (leading to $t \to \infty$), we have only to take into account the leading (nonoscillating) contributions in the integral $\mathcal{F}$, i.e., the terms with $a_i = b_i$ with $i = 1, 2, \ldots, n+1$. Also, the eigenstates $|\Lambda_a^{R,L}\rangle$ can be replaced by those $|E_a\rangle$ of the system Hamiltonian $\hat{H}$, as the minimally destructive limit ensures the vanishingly small non-Hermiticity $\gamma \to 0$ in the effective Hamiltonian. Accordingly, to express an imaginary part Γ_a of an eigenvalue Λ_a, we can use the perturbative result $\Gamma_a = \gamma\langle E_a|\sum_m \hat{L}_m^\dagger\hat{L}_m|E_a\rangle$. These simplifications lead to

$$\hat{\varrho}_M(t) \simeq \sum_{\{a_i\}} e^{-\mu\tilde{\Gamma}_{a_{n+1}}} \int_0^\mu d\mu_n \cdots \int_0^{\mu_2} d\mu_1 e^{-\sum_{i=1}^n \delta\tilde{\Gamma}_{a_i}\mu_i} (\mathcal{V}_{m_n})_{a_{n+1}a_n} \cdots (\mathcal{V}_{m_1})_{a_2 a_1} (p_{\text{eq}})_{a_1} |E_{a_1}\rangle\langle E_{a_1}|, \tag{4.33}$$

where $\mathcal{V}_m$ and p_{eq} are matrices whose elements are defined by

$$(\mathcal{V}_m)_{ab} = |\langle E_a|\hat{L}_m|E_b\rangle|^2, \quad (p_{\text{eq}})_a = \langle E_a|\hat{\rho}_{\text{eq}}|E_a\rangle, \tag{4.34}$$

and we introduce variables

$$\mu_i \equiv \gamma t_i, \tag{4.35}$$

$$\tilde{\Gamma}_a \equiv \Gamma_a/\gamma = \langle E_a|\sum_m \hat{L}_m^\dagger\hat{L}_m|E_a\rangle, \tag{4.36}$$

$$\delta\tilde{\Gamma}_{a_i} \equiv \tilde{\Gamma}_{a_i} - \tilde{\Gamma}_{a_{i+1}}. \tag{4.37}$$

To attain a further simplification, we note the fact that the off-diagonal elements $(\mathcal{V}_m)_{ab}$ vanish exponentially fast with the energy difference $\omega = |E_a - E_b|$ [66, 67]. Thus, the dominant contributions to Eq. (4.33) are made from matrix-vector products for the elements a_i and a_{i+1} that are close in energy, i.e., the elements satisfying $|E_{a_i} - E_{a_{i+1}}|/|E_{a_i} + E_{a_{i+1}}| \ll 1$. For such elements, the ETH guarantees that the fluctuation of the decay rate is strongly suppressed $|\Gamma_{a_i} - \Gamma_{a_{i+1}}|/|\Gamma_{a_i} + \Gamma_{a_{i+1}}| \ll 1$, as we consider physical jump operators $\hat{L}_m$ consisting of few-body operators. We thus neglect $\delta\tilde{\Gamma}_a$'s in Eq. (4.33), leading to

$$\hat{\varrho}_\mathcal{M}(t) \simeq \frac{\mu^n}{n!} \sum_a e^{-\mu\tilde{\Gamma}_a} [\mathcal{V}_{m_n} \cdots \mathcal{V}_{m_1} p_{\text{eq}}]_a |E_a\rangle\langle E_a|. \qquad (4.38)$$

Finally, while successive multiplications of matrices $\mathcal{V}_m$ on the initial distribution p_{eq} can eventually change the mean energy $\overline{E}$ by an extensive amount, they still keep the energy fluctuation subextensive. This follows from the cluster decomposition property [101, 151, 152] of thermal eigenstates for local operators $\hat{O}_{x,y}$ (see Appendix C for details):

$$\lim_{|x-y|\to\infty} \text{Tr}[\hat{O}_x \hat{O}_y \hat{P}_a] - \text{Tr}[\hat{O}_x \hat{P}_a]\text{Tr}[\hat{O}_y \hat{P}_a] = 0, \qquad (4.39)$$

where $\hat{P}_a = |E_a\rangle\langle E_a|$. In other words, the energy distribution is strongly peaked around the mean value during each time interval between jump events. The ETH then guarantees that the distribution of the detection rate $\hat{\Gamma}$ is also strongly peaked and its fluctuation around the mean value is vanishingly small in the thermodynamic limit (see e.g., the top panel in Fig. 4.8c below). We thus replace $\tilde{\Gamma}_a$ in Eq. (4.38) by its mean value $\overline{\tilde{\Gamma}}$ in the final distribution $(\prod_{i=1}^n \mathcal{V}_{m_i}) p_{\text{eq}}$, and arrive at the following simple expression of the density matrix:

$$\hat{\varrho}_\mathcal{M}(t) \simeq \frac{\mu^n}{n!} e^{-\mu\overline{\tilde{\Gamma}}} \sum_a [\mathcal{V}_{m_n} \cdots \mathcal{V}_{m_1} p_{\text{eq}}]_a |E_a\rangle\langle E_a|. \qquad (4.40)$$

After taking the normalization, we can rewrite $\hat{\varrho}_\mathcal{M}$ as

$$\hat{\rho}_\mathcal{M} = \frac{\Lambda_\mathcal{M}[\hat{\rho}_{\text{eq}}]}{Z(\mathcal{M})} = \frac{1}{Z(\mathcal{M})} \sum_a [\mathcal{V}_{m_n} \cdots \mathcal{V}_{m_1} p_{\text{eq}}]_a \hat{P}_a, \qquad (4.41)$$

where we define $\Lambda_\mathcal{M} = \prod_{i=1}^n (\Lambda \circ \mathcal{L}_{m_i} \circ \Lambda)$ with $\mathcal{L}_m[\hat{O}] = \hat{L}_m \hat{O} \hat{L}_m^\dagger$ and $\Lambda[\hat{O}] = \sum_a \hat{P}_a \hat{O} \hat{P}_a$, and $Z(\mathcal{M})$ is a normalization constant. Physically, the dephasing channel Λ originates from the ergodic, relaxation dynamics during the no-count process. We remark that, in finite-size systems, the distribution in the energy basis is often rather broad. Hence, in practice it can also be useful to use the expression (4.38) especially when the diagonal elements of the detection rate Γ_a vary significantly as a function of energy.

4.2 Thermalization and Heating Dynamics Under Measurement

Because of the subextensiveness of the standard deviation of energy in $\hat{\rho}_M$ (see Appendix C), the energy distribution of $\hat{\rho}_M$ has a sharp peak around the mean $\overline{E}_M$ as mentioned above. An effective temperature β_{eff}^M of the system can thus be introduced from the relation $\overline{E}_M = \text{Tr}[\hat{H}\hat{\rho}_{\beta_{\text{eff}}^M}]$, where $\hat{\rho}_\beta = e^{-\beta\hat{H}}/Z_\beta$ is the Gibbs ensemble. It is then guaranteed by the ETH that $\hat{\rho}_M$ is indistinguishable from the Gibbs ensemble if we focus on an expectation value of a few-body observable $\hat{O}$:

$$\text{Tr}[\hat{O}\hat{\rho}_M] \simeq \text{Tr}[\hat{O}\hat{\rho}_{\beta_{\text{eff}}^M}]. \tag{4.42}$$

From now on, $\simeq$ is understood to be the equality in the thermodynamic limit. In this sense, a generic quantum many-body system under continuous observation thermalizes at the single-trajectory level.

Several remarks are in order. Firstly, we emphasize that it is highly nontrivial to precisely determine the temperature of an open many-body system. We usually need to rely on ad hoc techniques for each problem. Our approach based on the matrix-vector product ensemble (MVPE) (see Eq. (4.40)) offers a general approach to extract its effective temperature. In principle, any physical quantities can then be obtained from the Gibbs ensemble at an extracted temperature provided that the ETH holds true. Secondly, as the system Hamiltonian $\hat{H}$ is assumed to be nonintegrable, it should obey $[\hat{H}, \hat{L}_m] \neq 0$. This noncommutativity indicates that the multiplication of the matrix $\mathcal{V}_m$ alters the temperature of the system. Finally, we remark on similarities and differences between Eq. (4.40) and the density matrix of isolated systems reached after a sudden quench [68, 153–155] or slow unitary operations [76]. Both density matrices are diagonal in the energy basis and coefficients are expressed by the product of matrices and the vector. In the slow unitary operations, the matrix can be interpreted as the transition matrix and it obeys the doubly stochastic condition $\sum_a (\mathcal{V})_{ab} = \sum_b (\mathcal{V})_{ab} = 1$ that inevitably leads to heating of the system [66, 67, 156]. Yet, in the MVPE obtained for the open-system dynamics, one cannot interpret the matrix $\mathcal{V}$ as the transition matrix and, in particular, it violates the doubly stochastic condition. One can thus perform the cooling of the system if artificial (typically non-Hermitian) measurement operators $\hat{L}_m$ are implemented [126, 127].

4.2.3 Numerical Simulations in Nonintegrable Open Many-Body Systems

Nonintegrable systems under local and global measurements

We demonstrate our general approach formulated above by studying a Hamiltonian $\hat{H} = \hat{K} + \hat{U}$ with nearest- and next-nearest-neighbor hopping and on-site interaction:

$$\hat{K} = -\sum_{l}(t_{\rm h}\hat{b}_{l}^{\dagger}\hat{b}_{l+1}+t_{\rm h}'\hat{b}_{l}^{\dagger}\hat{b}_{l+2}+{\rm H.c.}), \qquad (4.43)$$

$$\hat{U} = \sum_{l}(U\hat{n}_{l}\hat{n}_{l+1}+U'\hat{n}_{l}\hat{n}_{l+2}), \qquad (4.44)$$

where $\hat{b}_l$ ($\hat{b}_l^{\dagger}$) annihilates (creates) a hard-core boson on site l and $\hat{n}_l = \hat{b}_l^{\dagger}\hat{b}_l$. We assume that hard-core bosons are trapped in an open chain. It has been well established that this Hamiltonian obeys the ETH [69, 73, 76, 78]. The system size and the total number of bosons are set to be $L_s = 18$ and $N = 6$. We study two types of measurements: (i) the site-resolved density measurement $\hat{L}_l = \hat{n}_l$, where a jump operator acts on a local region and a jump is labeled by site l, and (ii) a global measurement $\hat{L} = \sum_l (-1)^l \hat{n}_l$, in which a type of jump is unique. To be concrete, we choose the initial state to be an eigenstate[3] $|E_0\rangle$ with temperature $T_0 = 3t_{\rm h}$. We set the initial time $t = 0$ to be the first detection time of a quantum jump.

Figure 4.7 shows a typical trajectory dynamics under the local measurement with $\hat{L}_l = \hat{n}_l$. After each detection of a quantum jump, measurement backaction localizes an atom on the detected lattice site, which subsequently spreads over and quickly relaxes toward an equilibrium density (see Fig. 4.7a). Figure 4.7b shows the corresponding dynamics of the kinetic energy $\langle \hat{K} \rangle$ (top) and the occupation $\langle \hat{n}_{k=0} \rangle$ at zero momentum (bottom), and compares them with the predictions from the MVPE $\hat{\rho}_M$ (red chain) and the Gibbs ensemble $\hat{\rho}_{\beta_{\rm eff}^M}$ (green dashed). For each time interval, the dynamical values agree with the MVPE predictions within time-dependent fluctuations. To gain further insights, Fig. 4.7c plots the diagonal matrix elements of each observable in the energy basis (top two panels) and energy distributions after every quantum jump (the other panels). Small eigenstate-to-eigenstate fluctuations in the observables and the remarkable agreements in the energy distributions explain the success of the MVPE description in Fig. 4.7b. It is notable that only a few jumps are sufficient to smear out the initial memory of a single eigenstate after which the distribution is almost indistinguishable from that of the corresponding Gibbs ensemble, thus validating the relation (4.42). Small fluctuations after the first jump (see Fig. 4.7b) indicate that even a single quantum jump generates a sufficiently large effective dimension to make the system equilibrate, which can be understood from the substantial delocalization of an energy eigenstate in the Fock basis [157]. A discrepancy of the Gibbs ensemble from $\langle \hat{n}_{k=0} \rangle (t)$ in Fig. 4.7b can be attributed to the small system size as detailed below. In this respect, it may be advantageous to use $\hat{\rho}_M$ rather than $\hat{\rho}_{\beta_{\rm eff}^M}$ for small systems that can be prepared in experiments.

Figure 4.8 shows trajectory dynamics under the global measurement. Figure 4.8a shows dynamics of the distribution of the jump operator $\hat{L} = \sum_l (-1)^l \hat{n}_l$. After each jump, the measurement backaction localizes the distribution, leading to a cat-like post-measurement state with relatively large weights on $\hat{L} = \pm 4$. The peaks rapidly collapse and the distribution relaxes to a thermal one due to the noncommutativity between $\hat{H}$ and $\hat{L}$. Figure 4.8b plots the corresponding time evolution of $\langle \hat{n}_{k=0} \rangle$ in

---

[3] We remark that results for a general initial equilibrium distribution $p_{\rm eq}$ can be merely given as a linear sum of the results for single eigenstates.

4.2 Thermalization and Heating Dynamics Under Measurement

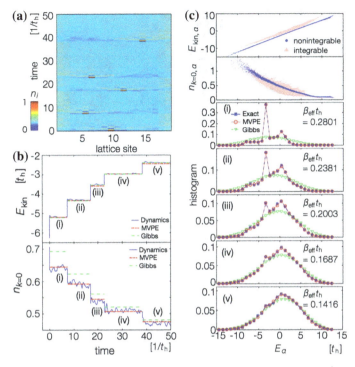

Fig. 4.7 Typical quantum trajectory dynamics under the local measurement with $\hat{L}_l = \hat{n}_l$. **a** Spatiotemporal dynamics of the particle-number density. Every time an atom is detected, a high-density region appears which subsequently diffuses away toward an equilibrium value. **b** Time evolution of the kinetic energy $\hat{K}$ and that of the occupation $\hat{n}_{k=0}$ at zero momentum. Superimposed are the values corresponding to the Gibbs ensemble $\hat{\rho}_{\beta_{\text{eff}}^M}$ (green dashed) and the MVPE $\hat{\rho}_M$ (red dashed). **c** Top two panels plot diagonal elements of the observables in the energy basis. The other panels plot the energy distributions after each jump. We use $t_h = U = t'_h = U' = 1$ and $\gamma = 0.02$ except for the integrable case in (**c**) for which $t_h = U = 1$ and $t'_h = U' = 0$. Reproduced from Fig. S1 of Ref. [140]. Copyright © 2018 by the American Physical Society

comparison with the values obtained from the Gibbs ensemble $\hat{\rho}_{\beta_{\text{eff}}^M}$ and the MVPE $\hat{\rho}_M$. The top panel in Fig. 4.8c plots the diagonal values Γ_a of $\hat{\Gamma}$ while the other panels show energy distributions after each jump. We again observe an excellent agreement between the MVPE and the exact dynamical values.

Finite-size scaling analyses

Figure 4.9 shows the results of the finite-size scaling analysis to test the precision of the predictions from the MVPE $\hat{\rho}_M$ and the corresponding Gibbs ensemble $\hat{\rho}_{\beta_{\text{eff}}^M}$ with an effective temperature. We calculate the relative deviations from the time-averaged value:

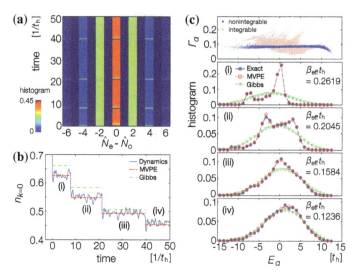

Fig. 4.8 Typical quantum trajectory dynamics under the global measurement with $\hat{L}=\sum_l(-1)^l\hat{n}_l$. Dynamics of **a** the distribution of $\hat{L}$ and **b** $\hat{n}_{k=0}$. **c** Top panel shows diagonal elements of the detection rate $\hat{\Gamma}$ in the energy basis. The other panels show the energy distributions after each jump. We use the parameters as in Fig. 4.7. Reproduced from Fig. 1 of Ref. [140]. Copyright © 2018 by the American Physical Society

$$r_{\hat{\rho}} = \frac{\left|\overline{\hat{O}(t)} - \langle\hat{O}\rangle_{\hat{\rho}}\right|}{\left|\overline{\hat{O}(t)} + \langle\hat{O}\rangle_{\hat{\rho}}\right|}, \quad (4.45)$$

where $\overline{\hat{O}(t)}$ denotes the time-averaged value of an observable $\hat{O}$ over the trajectory dynamics during the time interval involving t between quantum jumps, $\langle\cdot\rangle_{\hat{\rho}} = \mathrm{Tr}[\cdot\hat{\rho}]$ with $\hat{\rho}$ being chosen to be either the MVPE or the Gibbs ensemble. As an observable $\hat{O}$, we use the kinetic energy $\hat{K}$ or the occupation number $\hat{n}_{k=0}$ at zero momentum. We set the filling $N/L = 1/3$ with N and L being the total number of atoms and the system size, and vary N from 3 to 6. The results are presented in Fig. 4.9. The top (bottom) panels show the relative deviations $r_{\hat{\rho}}$ for each time interval after the n-th jump event in the trajectory dynamics with the local (global) measurement processes. These finite-size scaling analyses indicate that the relative errors of the MVPE predictions (filled circles) converge almost exponentially to zero in the thermodynamic limit. We find that the convergence of the corresponding Gibbs ensemble predictions (open circles) is slower than that of the MVPE. This fact can be attributed to a combination of broad energy distributions of finite-size systems and large fluctuations in diagonal elements of observables (see e.g., Fig. 4.7c). It is worthwhile to mention that a similar slow convergence of the observable $\hat{n}_{k=0}$ to the equilibrium value due to finite-size effects has also been found in the time-dependent density-matrix

4.2 Thermalization and Heating Dynamics Under Measurement

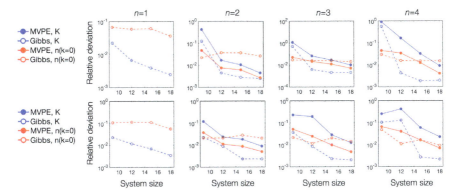

Fig. 4.9 Finite-size scaling analyses of the relative deviations of the predictions of the MVPE and the corresponding Gibbs ensemble from the time-averaged value. The top (bottom) panels show the relative deviations of the observables for each ensemble from their time-averaged values in the trajectory dynamics after the n-th quantum jump with the local (global) measurement. The relative deviations of the MVPE values from the time-averaged values are plotted as the blue (red) filled circles for the kinetic energy $\hat{K}$ (the occupation number at zero momentum $\hat{n}_{k=0}$). In the same way, the deviations of the corresponding Gibbs ensemble are plotted as open circles. We remark that the deviations of the MVPE predictions after the first jump $n = 1$ are negligibly small and not shown in the plots. We set the parameters as in Fig. 4.7. Reproduced from Fig. S3 of Ref. [140]. Copyright © 2018 by the American Physical Society

renormalization-group calculations of the Bose-Hubbard model with spontaneous emissions [139].

Numerical results on integrable systems

An isolated integrable many-body system often fails to thermalize because the Gibbs ensemble is not sufficient to fix distributions of an extensive number of local conserved quantities. Here we present numerical results of trajectory dynamics with an integrable system Hamiltonian. To be specific, we consider a local measurement process $\hat{L}_l = \hat{n}_l$ and choose the parameters as $t_h = U = 1$ and $t'_h = U' = 0$. For the sake of comparison, in Fig. 4.10 we present the results for the trajectory dynamics with the same occurrence times and types of quantum jumps as realized in the nonintegrable results presented in Fig. 4.7. The initial state is again chosen to be the energy eigenstate $|E_0\rangle$ of the integrable many-body Hamiltonian having an energy corresponding to the temperature $T_0 = 3t_h$.

Figure 4.10a shows the spatiotemporal dynamics of the atom number at each lattice site. Measurement backaction localizes an atom at the site of detection and the density waves propagate ballistically through the system. The induced density fluctuations are significantly larger than those found in the corresponding nonintegrable results, and the relaxation to the equilibrium profile seems to be not reached during each time interval between quantum jumps in the integrable case. Also, the ballistically propagating density waves are reflected back at the boundaries and can disturb the density; the finite-size effects can be more significant in the integrable

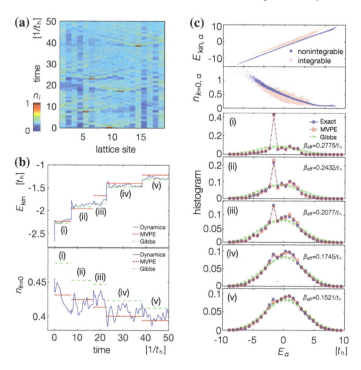

Fig. 4.10 Numerical results on the trajectory dynamics for which the system Hamiltonian is integrable and the measurement is local. **a** Spatiotemporal dynamics of the atom-number distribution at each lattice site and **b** the corresponding dynamics of the kinetic energy $\hat{K}$ (top panel) and the occupation number at zero momentum $\hat{n}_{k=0}$ (bottom panel). **c** Top two panels show the diagonal values of each observable in the energy basis for the integrable case (red triangle) and the nonintegrable case (blue circle). Other panels show the changes of energy distributions after each jump. We set the parameters as in Fig. 4.7. Reproduced from Fig. S4 of Ref. [140]. Copyright © 2018 by the American Physical Society

case than the corresponding nonintegrable one. It merits further study to identify an equilibration time scale in an integrable many-body system under continuous measurement. Figure 4.10b shows the corresponding dynamics of the kinetic energy $\hat{K}$ and the occupation number $\hat{n}_{k=0}$ at zero momentum. Relatively small (large) time-dependent fluctuations in the kinetic energy (the zero-momentum occupation) can be attributed to the small (large) fluctuations of its diagonal values in the energy basis (see the top two panels in Fig. 4.10c). Other panels in Fig. 4.10c show the corresponding changes of energy distributions after each quantum jump. In a similar manner as in the nonintegrable results presented in Fig. 4.7, a few jumps are enough to smear out the initial memory as a single energy eigenstate. This fact implies that the biased weight on a possible nonthermal eigenstate admitted by the weak variant of the ETH [70, 158] will disappear after observing a few number of quantum jumps. The jump operator $\hat{L}_l$ acts as a weak-integrability breaking (nonunitary) perturbation and should eventually make the system thermalize.

4.2 Thermalization and Heating Dynamics Under Measurement

The thermalization behavior can be also inferred from the eventual agreement between the time-dependent values of the observables and the predictions from the Gibbs ensemble after several jumps (see Fig. 4.10b). Nevertheless, it is still evident that a largely biased weight on an initial (possibly nonthermal) state can survive when the number of jumps is small (see e.g., the panels (i) and (ii) in Fig. 4.10c), and thus the generalized Gibbs ensemble can be a suitable description in such a case. To make concrete statements, we need a larger system size and more detailed analyses with physically plausible initial conditions. We leave it as an interesting open question to examine to what extent the initial memory of a possible nonthermal state can be kept under the integrability-breaking continuous measurement process.

4.2.4 Application to Many-Body Lindblad Dynamics

Finally, we discuss an application of the MVPE to the Lindblad dynamics. Aside from the description of continuously monitored systems, the quantum trajectory dynamics also provides a method to numerically solve the Lindblad master equation [19, 92]:

$$\frac{d\hat{\rho}}{dt} = \mathcal{L}[\hat{\rho}] = -i(\hat{H}_{\text{eff}}\hat{\rho} - \hat{\rho}\hat{H}_{\text{eff}}^{\dagger}) + \gamma \sum_m \hat{L}_m \hat{\rho} \hat{L}_m^{\dagger}, \qquad (4.46)$$

where $\hat{\rho}(t) = \sum_\mathcal{M} \hat{\varrho}_\mathcal{M}(t)$ is the ensemble-averaged density matrix. The Lindblad equation (4.46) can be used to study the dynamics of an open system weakly coupled to an environment [92] or a system subject to noisy unitary operations [141–144, 150]. However, taking the ensemble average is often very demanding especially for a many-body system due to a large number of possible trajectories.

When we focus on a (physically plausible) case of a translationally invariant $\hat{H}$ and $\hat{L}_m$, the MVPE offers a simple approach to overcome this difficulty. The translational invariance and the locality of $\hat{L}_m$ lead to the matrix $\mathcal{V}_m$ that is independent of a spatial label m. Therefore, the MVPE is characterized only by the number n of jumps rather than their sequence[4]: $\hat{\rho}_n \propto \sum_a [\mathcal{V}^n p_{\text{eq}}]_a \hat{P}_a$. Since the detection rate $\hat{\Gamma}$ is assumed to consist of few-body observables, the ETH guarantees that the distribution of the number n of quantum jumps has a sharp peak around the mean $\bar{n}$. We thus obtain

$$\text{Tr}[\hat{O}e^{\mathcal{L}t}\hat{\rho}_{\text{eq}}] \simeq \text{Tr}[\hat{O}\hat{\rho}_{\bar{n}_t}] \simeq \text{Tr}[\hat{O}\hat{\rho}_{\beta_{\text{eff}}^{\bar{n}_t}}], \qquad (4.47)$$

where $\beta_{\text{eff}}^{\bar{n}_t}$ is the corresponding effective temperature and $\bar{n}_t$ is the mean number of n during $[0, t]$, which can be calculated from the implicit relation $t \simeq \sum_{n=0}^{\bar{n}_t} 1/\Gamma_n$ with $\Gamma_n = \text{Tr}[\hat{\Gamma}\hat{\rho}_n]$. Equation (4.47) indicates that expectation values of few-body observables in the Lindblad many-body dynamics agree with those obtained from

---

[4] In fact, $\hat{\rho}_n$ is nothing but the MVPE approximation of the full-counting dynamics introduced in the previous section.

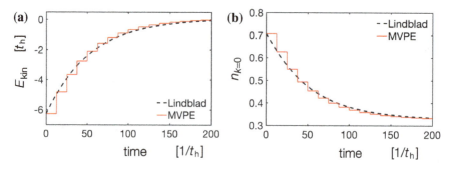

Fig. 4.11 Comparisons between the matrix-vector product ensemble (MVPE) predictions (red solid line) and the Lindblad dynamics (black dashed curve) for **a** the kinetic energy $\hat{K}$ and **b** the occupation number at zero momentum $\hat{n}_{k=0}$. The global continuous measurement is performed with a jump operator $\hat{L} = \sum_l (-1)^l \hat{n}_l$. We set the parameters as in Fig. 4.7. The panel (**b**) is reproduced from Fig. 2c of Ref. [140]. Copyright © 2018 by the American Physical Society

the typical MVPE or the Gibbs ensemble at an appropriate effective temperature. Given the fact that the analysis of Eq. (4.46) requires the diagonalization of a $D^2 \times D^2$ Liouvillean (where D is the dimension of the Hilbert space), our approach (4.47) can significantly simplify the analysis.

As an example, we have applied our approach to the nonintegrable lattice model discussed in the previous subsections. To be specific, we consider the global measurement $\hat{L} = \sum_l (-1)^l \hat{n}_l$. Since the MVPE $\hat{\rho}_n$ is characterized solely by the number n of quantum jumps, it suffices to determine its typical number $\bar{n}_t$ as a function of time t to apply the MVPE to the Lindblad dynamics. This can be achieved by using the relation $\bar{n}_t = \sum_{m=0}^{\infty} \Theta(t - \sum_{n=0}^{m} 1/\Gamma_n)$, where $\Theta(x)$ is the Heaviside step function. Figure 4.11 compares the MVPE predictions (red solid line) with the Lindblad dynamics (black dashed curve) corresponding to the heating dynamics of (a) the kinetic energy and (b) the occupation number at zero momentum under the global measurement. The stepwise behavior in the MVPE prediction originates from the discreteness of $\bar{n}_t$, which will diminish in the thermodynamic limit. The agreement found in Fig. 4.11 thus demonstrates the relation (4.47) aside from the stepwise finite-size contributions.

4.3 Diffusive Quantum Dynamics Under Measurement

4.3.1 Introduction

As we have reviewed in Chap. 2, there are two types of continuous measurements depending on different underlying stochastic processes. The first one is a quantum jump process that associates with a discontinuous change of a quantum state described by the Poisson-like stochastic process. This type of measurement is what

4.3 Diffusive Quantum Dynamics Under Measurement

we have so far considered as continuous observation. Here, we study the other type of measurement associated with the diffusive stochastic process known as the Wiener process. To study diffusive dynamics of many-particle systems under continuous position measurement, we consider the limit of a weak spatial resolution of *in-site* measurements of quantum gases [116, 118, 134, 159–164]. This limit is of practical importance as it will offer a nondestructive way of real-time observation of many-particle systems, which may allow us to apply measurement-based feedback control [165, 166] to quantum many-body systems. In contrast, the conventional techniques of quantum gas microscopy [167] have been so far limited to a destructive single-shot imaging [168].

In this section, we develop a theoretical framework to describe quantum dynamics of multiple particles under a minimally destructive spatial observation. While the stochastic Schrödinger equation of distinguishable particles under the spatial measurement has been obtained as a straightforward generalization of the result for the single-particle case [169–173], to our knowledge the derivation for indistinguishable particles has long been unknown [174]. We here achieve the latter by taking the appropriate limit of strong atom-light coupling and weak spatial resolution. The resulting equation indicates that measurement indistinguishability of particles completely suppresses the decoherence in the relative positions, resulting in the unique quantum transport dynamics. In previous works on the site-resolved position measurement [107, 130, 131, 175, 176] and continuous position measurement of a single quantum particle [169–173], we remark that the indistinguishability does not play such a nontrivial role. We numerically demonstrate our findings for the minimal model consisting of noninteracting two particles.

4.3.2 System: Atoms Under Spatial Observation

We consider the setup of our previous work [162] that discusses the collapse of many-particle wavefunction of atoms in an optical lattice due to the spatial measurement. While we have neglected the internal quantum dynamics (i.e., hopping of atoms) in Ref. [162] (which can be justified for a deep optical lattice), we are here interested in a situation in which both the internal dynamics and the measurement backaction play nontrivial roles.

We consider N atoms trapped in an optical lattice, which are described by the Bose-Hubbard Hamiltonian $\hat{H} = -J\sum_m (\hat{b}_m^\dagger \hat{b}_{m+1} + \text{H.c.}) + (U/2)\sum_m \hat{n}_m(\hat{n}_m - 1)$. Here, J is the hopping rate and U denotes the on-site interaction, $\hat{b}_m^\dagger$ ($\hat{b}_m$) creates (annihilates) an atom at site m, and $\hat{n}_m = \hat{b}_m^\dagger \hat{b}_m$. Here and henceforth, we set $U = 0$ for the sake of simplicity. We discuss the spatial measurement of atoms by an off-resonant probe light (see Fig. 4.12). The detection of scattered photons causes the collapse of the many-body wavefunction. The jump operator corresponding to this

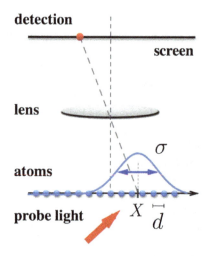

Fig. 4.12 Schematic figure of spatially resolved measurement of atoms on a lattice whose spacing is d. Trapped atoms off-resonantly scatter probe light. A lens aperture diffracts the scattered photons. The detection of photons on the screen induces the reduction of the wavefunction according to the outcome X and a spatial resolution σ. Reproduced from Fig. 2 of Ref. [119]. Copyright © 2017 by the American Physical Society

continuous position measurement can be obtained as follows (see the Supplementary Material of Ref. [162] for microscopic derivations):

$$\hat{M}(X) = \sqrt{\gamma} \sum_m f(X - md) \hat{b}_m^\dagger \hat{b}_m, \tag{4.48}$$

where X is a measurement outcome, γ denotes the measurement strength. We represent an amplitude of a general point spread function by f, which satisfies the normalization condition $\int f^2(X)dX = 1$ and the parity symmetry $f(X) = f(-X)$. As we show later, the resulting time-evolution equation derived in the weak-resolution limit depends on f only through its effective spatial resolution σ defined as

$$\int dX f(X - md) f(X - ld) \simeq 1 - \frac{(m-l)^2 d^2}{4\sigma^2}. \tag{4.49}$$

For instance, it is just the standard deviation if f is chosen to be a Gaussian point spread function $f(X) = e^{-X^2/(2\sigma^2)}/(\sigma^2 \pi)^{1/4}$.

We start from formulating the dynamics under the above spatial measurement following a general theory reviewed in Chap. 2.[5] The measurement process corresponds to a photodetection on the screen and the quantum jump process is described by the discrete stochastic process $dN(X; t)$ that depends on X. The time evolution is described by the following stochastic many-body Schrödinger equation:

---

[5] It is straightforward to generalize our discussion on the lattice system presented here to the continuum space.

4.3 Diffusive Quantum Dynamics Under Measurement

$$d|\psi\rangle = -\frac{i}{\hbar}\hat{H}|\psi\rangle dt - \frac{1}{2}\int dX\left(\hat{M}^\dagger(X)\hat{M}(X) - \langle\hat{M}^\dagger(X)\hat{M}(X)\rangle\right)|\psi\rangle dt$$
$$+ \int dX\left(\frac{\hat{M}(X)|\psi\rangle}{\sqrt{\langle\hat{M}^\dagger(X)\hat{M}(X)\rangle}} - |\psi\rangle\right)dN(X;t), \qquad (4.50)$$

where $\langle\cdots\rangle$ is an expectation value with respect to $|\psi\rangle$. In general, the time evolution is described by the non-Hermitian Hamiltonian in the no-jump process. In fact, for a single-particle case, this dynamics reduces to the unitary evolution since the term $-1/2\int dX\hat{M}^\dagger(X)\hat{M}(X)$ is simply proportional to the identity operator. In the case of multiple particles, the no-count evolution is in general intrinsically distinct from the unitary evolution as the interference among particles leads to a change in the detection rate. The second line in Eq. (4.50) corresponds to a jump process caused by a detection of scattered photons on the screen. Mathematically, the stochastic process $dN(X;t)$ is known as a marked point process [177]

$$dN(X;t)dN(Y;t) = \delta(X-Y)dN(X;t), \quad E[dN(X;t)] = \langle\hat{M}^\dagger(X)\hat{M}(X)\rangle dt, \qquad (4.51)$$

where $E[\cdots]$ denotes taking an ensemble average over measurement outcomes. Physically, $E[dN(X;t)]/dt$ provides the intensity of photons detected on position X at time t.

4.3.3 Minimally Destructive Spatial Observation

To perform continuous spatial observation of a quantum gas in a nondestructive manner, one has to implement a weak and frequent measurement. The reason is that substantial heating and loss of atoms are caused by the conventional single-shot observations [168, 178–181]. From this perspective, we consider the limit of weak spatial resolution ($\sigma \gg d$) and strong atom-light coupling ($\gamma \gg J/\hbar$) while keeping γ/σ^2 finite. Taking such a limit is essential to describe nontrivial dynamics under measurement which is continuous in time [169–173, 182]; otherwise, the quantum Zeno effect would completely suppress the quantum dynamics [134].

To achieve the limiting procedure, we first consider the scaling of the quantity $\tilde{N}(t) = \int N(t;X)dX$, which is the number of photons detected during a time interval $[0, t]$. The mean of its rate of change is (at the leading order)

$$\left\langle\frac{d\tilde{N}(t)}{dt}\right\rangle = \int dX\langle\hat{M}^\dagger(X)\hat{M}(X)\rangle \simeq \gamma\sum_{m,l}\left(1 - \frac{(m-l)^2 d^2}{4\sigma^2}\right)\langle\hat{n}_m\hat{n}_l\rangle \simeq \gamma N^2, \qquad (4.52)$$

where the first approximate equality follows from Eq. (4.49). Since the number of detections goes to infinity in the frequent measurement limit $\gamma \to \infty$, one can approximate the fluctuation of stochastic intensity as

$$\frac{d\tilde{N}(t) - \int dX \langle \hat{M}^\dagger(X)\hat{M}(X)\rangle dt}{N\sqrt{\gamma}} \simeq dW(t), \qquad (4.53)$$

where we use the central limit theorem and employ the fact that the standard deviation of $\langle d\tilde{N}(t)/dt\rangle$ can be approximated by the square root of Eq. (4.52). We introduce $dW(t)$ as the Wiener stochastic process, which is characterized by

$$E[dW(t)] = 0, \quad (dW(t))^2 = dt. \qquad (4.54)$$

In a similar way, we can approximate the term including $dN(X;t)$ as

$$\frac{dN(X;t) - \langle \hat{M}^\dagger(X)\hat{M}(X)\rangle dt}{\sqrt{\langle \hat{M}^\dagger(X)\hat{M}(X)\rangle}} \simeq dW(X;t), \qquad (4.55)$$

where $dW(X;t)$ is the spatially dependent Wiener process that is characterized by

$$E[dW(X;t)] = 0, \quad dW(X;t)dW(Y;t) = \delta(X-Y)dt. \qquad (4.56)$$

4.3.4 Many-Body Stochastic Schrödinger Equations

In this subsection, we discuss the fundamental time-evolution equations of quantum many-particle systems under the minimally destructive spatial observation. While the equation for distinguishable particles can be straightforwardly obtained from that for a single-particle case, to our knowledge the equation for indistinguishable particles had not been known before our derivation.

Indistinguishable particles

We first study the case for indistinguishable particles. We rewrite Eq. (4.50) in terms of the density matrix of a pure state $\hat{\rho} = |\psi\rangle\langle\psi|$. Then, we take the limit of weak spatial resolution and strong atom-light coupling by using Eqs. (4.53) and (4.55) (see Appendix D for technical details). The resulting equation is the following diffusive stochastic many-body Schrödinger equation:

$$d\hat{\rho} = -\frac{i}{\hbar}\left[\hat{H},\hat{\rho}\right]dt - \frac{N^2\gamma d^2}{4\sigma^2}\left[\hat{X}_{\text{c.m.}},\left[\hat{X}_{\text{c.m.}},\hat{\rho}\right]\right]dt + \sqrt{\frac{N^2\gamma d^2}{2\sigma^2}}\left\{\hat{X}_{\text{c.m.}} - \langle\hat{X}_{\text{c.m.}}\rangle,\hat{\rho}\right\}dW(t), \qquad (4.57)$$

where we denote the center-of-mass operator as

$$\hat{X}_{\text{c.m.}} = \frac{\sum_m m\hat{n}_m}{N}. \qquad (4.58)$$

A salient feature of this evolution equation is the absence of the relative positional decoherence term, which does exist for distinguishable particles [170, 183] (see the

4.3 Diffusive Quantum Dynamics Under Measurement

second line in Eq. (4.64) below). This absence qualitatively alters quantum transport dynamics as demonstrated in the next subsection. We remark that, for practically indistinguishable particles in measurement (but fundamentally distinguishable), the same suppression should occur.[6] In experiments, such measurement distinguishability would be relevant when one performs the state-selective imaging [21, 161, 184–187] or the polarization measurement. We may interpret the suppression of the relative positional decoherence as an emergent decoherence-free subspace (DFS) resulting from the symmetry in the measurement operator under the exchange of indistinguishable particles. To our best knowledge, Eq. (4.57) presents the first derivation of a model of continuous position measurement for quantum many-body systems. We remark that the degrees of freedom of relative positions are *not* frozen in our consideration unlike in a rigid system [183, 188, 189].

Distinguishable particles

We next mention the case for distinguishable particles. As for N noninteracting distinguishable particles, we introduce the associated measurement operators $\hat{M}_i(X)$ as

$$\hat{M}_i(X) = \sqrt{\gamma_i} \sum_x f(X - xd) |x\rangle_{ii}\langle x|, \tag{4.59}$$

where i labels a particle, γ_i is a detection rate associated with particle i, f represents a general point spread function, and x denotes a lattice site. Then, the time-evolution equation for distinguishable particles is written as

$$d|\psi\rangle = -\frac{i}{\hbar}\hat{H}|\psi\rangle dt - \frac{1}{2}\sum_{i=1}^{N}\int dX_i \left(\hat{M}_i^\dagger(X_i)\hat{M}_i(X_i) - \langle\hat{M}_i^\dagger(X_i)\hat{M}_i(X_i)\rangle\right)|\psi\rangle dt$$

$$+ \sum_{i=1}^{N}\int dX_i \left(\frac{\hat{M}_i(X_i)|\psi\rangle}{\sqrt{\langle\hat{M}_i^\dagger(X_i)\hat{M}_i(X_i)\rangle}} - |\psi\rangle\right) dN_i(X_i;t), \tag{4.60}$$

where $dN_i(X;t)$ are the marked point processes corresponding to particle i that satisfy

$$dN_i(X;t)dN_j(X;t) = \delta_{ij}\delta(X-Y)dN_i(X;t). \tag{4.61}$$

Taking the limit of weak spatial resolution in Eq. (4.60), the following diffusive stochastic time-evolution equation of distinguishable particles can be obtained (see Appendix E for details)

---

[6] A coherent light scattering from isotopes is a typical example.

$$d\hat{\rho} = -\frac{i}{\hbar}\left[\hat{H}, \hat{\rho}\right]dt - \sum_{i=1}^{N} \frac{\gamma_i d^2}{4\sigma^2}\left[\hat{x}_i, [\hat{x}_i, \hat{\rho}]\right]dt + \sum_{i=1}^{N} \sqrt{\frac{\gamma_i d^2}{2\sigma^2}}\left\{\hat{x}_i - \langle\hat{x}_i\rangle, \hat{\rho}\right\}dW_i(t),$$
(4.62)

where $\hat{x}_i = \sum_m m|m\rangle_{ii}\langle m|$ is the position operator of particle i, and $dW_i(t)$'s are independent Wiener processes that obey

$$E[dW_i(t)] = 0, \quad dW_i(t)dW_j(t) = \delta_{i,j}dt.$$
(4.63)

For the sake of simplicity, let the scattering rate $\gamma_i = \gamma$ be independent of i. We can then simplify Eq. (4.62) as

$$d\hat{\rho} = -\frac{i}{\hbar}\left[\hat{H}, \hat{\rho}\right]dt - \frac{N\gamma d^2}{4\sigma^2}\left[\hat{X}_{\text{c.m.}}, [\hat{X}_{\text{c.m.}}, \hat{\rho}]\right]dt + \sqrt{\frac{N\gamma d^2}{2\sigma^2}}\left\{\hat{X}_{\text{c.m.}} - \langle\hat{X}_{\text{c.m.}}\rangle, \hat{\rho}\right\}dW(t)$$
$$- \frac{\gamma d^2}{4\sigma^2}\sum_{i=1}^{N}\left[\hat{r}_i, [\hat{r}_i, \hat{\rho}]\right]dt + \sqrt{\frac{\gamma d^2}{2\sigma^2}}\sum_{i=1}^{N}\left\{\hat{r}_i - \langle\hat{r}_i\rangle, \hat{\rho}\right\}dW_i(t),$$
(4.64)

where we introduce the center-of-mass operator $\hat{X}_{\text{c.m.}} = \sum_{i=1}^{N}\hat{x}_i/N$, the relative coordinate $\hat{r}_i = \hat{x}_i - \hat{X}_{\text{c.m.}}$, and the linear superposition of Wiener processes $dW(t) \equiv \sqrt{1/N}\sum_i dW_i(t)$ (which is also a Wiener process). We remark that Eqs. (4.62) and (4.64) are consistent with the known result [170, 183] that can be obtained by applying a single-particle model [169–173].

Implications to quantum transport dynamics

The distinct role of the measurement distinguishability discussed above can lead to a unique quantum transport dynamics. In particular, this can be understood from the peculiar feature of the decoherence rate of the off-diagonal term $\langle\{n_m\}|E[\hat{\rho}]|\{n'_m\}\rangle$ of the density matrix for indistinguishable particles:

$$\Gamma_{\{n_m\},\{n'_m\}} = \frac{N^2\gamma d^2}{4\sigma^2}\left(X_{\text{c.m.}} - X'_{\text{c.m.}}\right)^2.$$
(4.65)

This expression results from the second term of the right-hand side of Eq. (4.57). Here, we represent the center-of-mass coordinate (CMC) of the Fock state $\{n_m\}$ ($\{n'_m\}$) as $X_{\text{c.m.}}$ ($X'_{\text{c.m.}}$). From the expression (4.65), we can infer the three distinct time regimes. In the first regime, the coherence between states with different CMCs is rapidly lost. As a result, the many-particle wavefunction collapses into a state with a well-localized CMC. The time required for the collapse can be estimated as $4\sigma^2/(N^2\gamma d^2 L^2)$ with L being a typical distance between the CMCs of the superposed Fock states. In the second regime, although the CMC is localized, the coherence within the subspace of the Fock states taking similar CMCs can be preserved. This results from the suppression of the relative positional decoherence in Eq. (4.57). We term this regime as the inertial regime, as indistinguishable particles exhibit ballistic transports. Finally, in the long-time regime $t \gg 4\sigma^2/(\gamma d^2)$, the coherence between the nearest Fock states with the minimal CMC difference $\delta X_{\text{c.m.}} = 1/N$ is eventually

4.3 Diffusive Quantum Dynamics Under Measurement

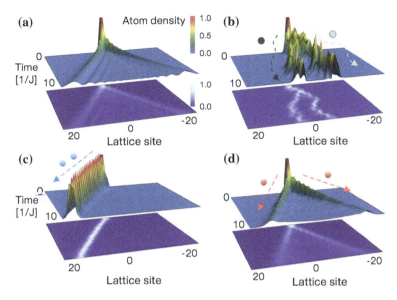

Fig. 4.13 Numerical simulations of dynamics of two noninteracting particles under minimally destructive spatial observation. **a** Unitary dynamics ($\Gamma = 0$). The evolution of the density distribution is almost not affected by quantum statistics. **b–d** Quantum dynamics under the spatial observation for **b** distinguishable particles, **c** bosons, and **d** fermions, calculated for a relatively weak measurement strength with $\Gamma = 2.0$. The absence of the relative positional decoherence results in the **c** bunched and **d** anti-correlated ballistic motions, while distinguishable particles show (**b**) uncorrelated diffusive behavior. Reproduced from Fig. 3 of Ref. [119]. Copyright © 2017 by the American Physical Society

lost, resulting in a diffusive behavior. The diffusion constant depends on quantum statistics (see Fig. 4.14a below).

4.3.5 Numerical Demonstrations

We numerically demonstrate the general properties discussed above by focusing on the minimal case of two noninteracting particles. We calculate the single-trajectory dynamics by Eq. (4.57) for indistinguishable particles and by Eq. (4.64) for distinguishable particles.

Inertial regime

A particularly interesting time regime is the second regime mentioned above, in which the coherence between relative positions can be preserved, resulting in ballistic quantum walks of two particles. To be specific, we choose the initial state to be two particles that are localized at adjacent sites. For the unitary dynamics, the evolution of the density distribution is almost not affected by quantum statistics (see Fig. 4.13a), where the peaks of the density ballistically propagate with $\langle x^2 \rangle = 2J^2t^2/\hbar^2$. In contrast, under the spatial observation, the quantum dynamics strongly depends on quantum statistics. Although distinguishable particles show uncorrelated diffusive transport (see Fig. 4.13b), indistinguishable particles show ballistic and correlated transport (see Fig. 4.13c, d). The correlation can be understood as the standard multi-particle interference between identical particles, which are also known as bunching (antibunching) of bosons (fermions).

In Fig. 4.13c, two bosons transport in the same direction due to constructive interference. Because of rapid collapse of the CMC, two bosons form a localized wave packet. The rapid collapse is characterized by a time scale $t_{\text{col}} \simeq (3\sqrt{2}\hbar^2/(\Gamma J^2))^{1/3}$. In contrast, in Fig. 4.13d, two fermions move in the opposite directions and the CMC is localized around zero. In this sense, the weakly resolved spatial observation does not appreciably alter the quantum dynamics of fermions in comparison with the unitary dynamics (see Fig. 4.13a). This sharply contrasts with the site-resolved measurement [175], in which atoms inevitably transport diffusively due to the strong measurement backaction.

Diffusive regime

Figure 4.14 shows our numerical results for the diffusive regime of the quantum dynamics under the spatial measurement, where we set the measurement strength Γ to be a larger value than the one used above. Figure 4.14a plots the square σ_r^2 of relative distance between two particles. For distinguishable particles, it is characterized by the diffusion constant (see Appendix F for its derivation)

$$D_c = \frac{16J^2\sigma^2}{\gamma\hbar^2 d^2}. \tag{4.66}$$

Quantum statistics acts as an effective attractive (repulsive) interaction for bosons (fermions), leading to a diffusion constant smaller (larger) than D_c. Typical diffusive trajectory dynamics for bosons and fermions are shown in Fig. 4.14b and c.

While we have so far focused on the simplest two-particle case, our general arguments below Eq. (4.65) hold true for many-particle cases as long as atom interactions are negligibly weak. We leave it as an interesting open question to study the dynamics of interacting many-body systems under the weak spatial observation formulated here. At this point, it is worthwhile to remark that, taking the high-resolution limit $\sigma \ll d$ and the ensemble average of Eq. (4.50), we can reproduce the following Lindblad equation:

$$\frac{d E[\hat{\rho}]}{dt} = -\frac{i}{\hbar}[\hat{H}, \hat{\rho}] - \frac{\gamma}{2} \sum_m [\hat{n}_m, [\hat{n}_m, E[\hat{\rho}]]], \tag{4.67}$$

4.3 Diffusive Quantum Dynamics Under Measurement 121

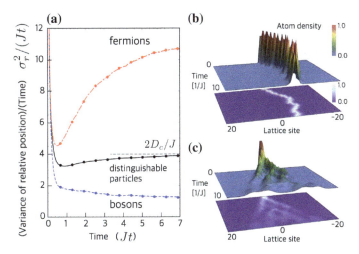

Fig. 4.14 Diffusive dynamics under the spatial observation with a relatively strong measurement strength. **a** The diffusion constant for distinguishable particles (solid black curve), bosons (blue dashed curve) and fermions (red dash-dotted curve). Typical diffusive trajectory dynamics is plotted for **b** bosons and **c** fermions. We use $\Gamma = 16.0$ and the results in (**a**) are obtained by taking the average over 10^3 realizations. Reproduced from Fig. 4 of Ref. [119]. Copyright © 2017 by the American Physical Society

which has been used to study the dissipative dynamics under the site-resolved measurement [107, 130, 131, 176].[7] In this dissipative dynamics, if the interaction is absent, it has been known that the rapid decoherence occurs with the time $\sim 1/\gamma$, resulting in diffusive behavior characteristic of classical random walk [190]. Yet, an exotic transport such as anomalous diffusion can still appear in the presence of a nonzero interaction [130, 131]. It would be intriguing to explore such anomalous behavior in a many-body system under the continuous spatial observation discussed in this section.

4.4 Experimental Situations in Ultracold Gases

We have developed three general formalisms to discuss the emergent out-of-equilibrium phenomena in quantum many-body systems under continuous observation. We have applied them to several specific examples that can be realized in ultracold atoms. We here briefly discuss experimental situations to implement those models. Firstly, as a possible experimental implementation of the full-counting dynamics studied in Sect. 4.1, we propose to realizing our exactly solvable model in quantum gas microscopy [167] to count the number of quantum jumps. We remark

---

[7]Note that the jump operator $\hat{L}_m = \hat{n}_m$ coincides with the one used for the local measurement discussed in our study on the thermalization in Sect. 4.2.

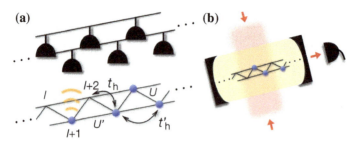

Fig. 4.15 Quantum many-body systems under **a** local and **b** global continuous measurements. **a** Hard-core bosons on a lattice are subject to site-resolved position measurement via light scattering. **b** The cavity photon field is coupled to a collective mode of atoms and photons emanating from the cavity are continuously monitored

that one can simultaneously measure the total number of particles and the site-resolved density-density correlation. In noninteracting models, the density-density correlation can be expressed as the product of the equal-time correlations and thus both correlations share the same information. One can realize the one-body loss via a weak resonant beam. The parameters γ, J, h in the solvable model can be experimentally controlled by tuning the intensities of external beams. In practice, we propose to use ^{6}Li atoms with a beam resonant to the $^2S_{1/2} \to {}^2P_{3/2}$ transition [191].

Secondly, as for our study of thermalization presented in Sect. 4.2, we remark that both local and global measurement processes assumed for numerical simulations are experimentally realizable with state-of-the-art techniques of ultracold atoms. The former associates with the jump operator $\hat{L}_l = \hat{n}_l$ and corresponds to a site-resolved position measurement of ultracold atoms via light scattering, as realized in quantum gas microscopy (see Fig. 4.15a) [167]. The latter associates with the jump operator $\hat{L} = \sum_l (-1)^l \hat{n}_l$ acting on an entire region of the system, which can be realized by continuously monitoring photons leaking out of a cavity coupled to a certain collective atomic mode (see Fig. 4.15b) [118].

Finally, we make several remarks on experimental situations in our study on the minimally destructive spatial observation in Sect. 4.3. In the discussions above, while a unit collection efficiency of photons has been implicitly assumed, we note that it is straightforward to generalize our theory to take into account contributions from uncollected signals. Heating effects can often be suppressed in the deep Lamb-Dicke regime or by Raman sideband cooling. When heating becomes substantial, one has to take into account higher bands of an optical lattice. The resolution σ and the collection efficiency can be controlled via a numerical aperture of lens.

4.5 Conclusions and Outlook

In this chapter, we have developed three general formalisms to study the out-of-equilibrium dynamics of many-body systems under continuous observation and applied them to analyze the dynamics of specific models that can be realized with ultracold atoms.

In Sect. 4.1, we have introduced the class of open many-body dynamics conditioned on measurement outcomes, which we termed as the full-counting dynamics. We have shown that the ability to measure individual quanta can reveal previously unexplored types of many-particle dynamics. In particular, we have demonstrated that, at the cost of the probabilistic nature of quantum measurement, correlations can propagate faster than the LR bound. We have identified the origin of the nonlocal propagation as the nonorthogonality of eigenstates due to the non-Hermiticity of the continuously monitored dynamics. Given that the nonorthogonality is a ubiquitous feature in non-Hermitian systems, we expect that the findings presented here can also emerge in other types of open many-particle systems. The observed features become most significant when the nonorthogonality is maximally strong due to an exceptional point (e.g., in PT-symmetric systems as we demonstrated in the solvable model). It remains an interesting open question to elucidate roles of interactions or nonintegrability in the full-counting dynamics. It would be interesting to develop a possible field-theoretic argument on the quench dynamics of quasiparticles [26, 27]. In fact, the low-energy field theory of the effective Hamiltonian $\hat{H}_{\rm PT}$ in Eq. (4.10) is a special case of the generalized sine-Gordon model introduced in the previous Chapter. It is also worthwhile to mention that this class of field theory includes quantum Liouville theory [192], which has been studied especially in high-energy physics.

In Sect. 4.2, combining the eigenstate thermalization hypothesis and quantum measurement theory, we have shown that a generic quantum many-body system under continuous observation thermalizes at a single trajectory level. The effective temperature and the dynamics of open many-body systems can be quantitatively described by the matrix-vector product ensemble (4.40). Also, it can be widely applied to systems described by a many-body Lindblad master equation such as dissipative or noisy systems. Our general results are supported by numerical analyses of nonintegrable systems subject to local and global observations, which can be experimentally realized by quantum gas microscopy and in atom-cavity systems. There exist several interesting future directions. Firstly, when the Hamiltonian is integrable, quantum jumps act as weak integrability-breaking perturbations. It would be interesting to study thermalization at the trajectory level in such a situation, where one can expect the appearance of prethermalization if the effects from quantum jumps are insignificant. When quantum jumps sufficiently mix the distribution, (possibly) biased probability weights on nonthermal rare states admitted in the weak variant of ETH will vanish and ultimate thermalization should happen. While our first attempt to elucidate this picture has been provided in Sect. 4.2.3, we leave detailed analyses as interesting future problems. It also merits further study to analyze effects of measurement on a many-body localized system, where even the weak ETH can be violated. Secondly, there may exist connections between the random unitary circuit dynamics [147] and the nonintegrable open trajectory dynamics studied here. Both dynamics share important common features; they have no energy conservation, obey the Lindblad master equation upon the ensemble average and satisfy the locality. In particular, it merits further study to explore the Kardar-Parisi-Zhang universal behavior [146] or the predicted scrambling dynamics [147] in the present setup.

In Sect. 4.3, we derived the diffusive type of the stochastic Schrödinger equation (4.57) that governs quantum many-particle dynamics under the weak spatial observation. Taking the weak and frequent measurement limit, an overall fluctuation in the measurement outcomes has been taken into account as a simple Wiener stochastic process in the resulting equation. As a consequence, the indistinguishability of particles leads to a complete suppression of relative positional decoherence. Analyzing the minimal example with two particles, we have demonstrated that this suppression results in unique transport dynamics. In particular, there emerges the time regime in which the coherence between two particles persists even under measurement. In the long-time regime, our results indicate that particles finally exhibit diffusive behavior, in which particle species can alter the diffusion constants. Our study suggests several interesting future directions. Since relevant decoherence operators in Eq. (4.57) now reduce to the center-of-mass coordinate alone, it would be possible to apply theory of quantum feedback control to quantum many-body systems under the spatial observation. In this sense, a nondestructive observation discussed here can be a starting point for realizing real-time feedback control of quantum many-body systems.

It remains an interesting open question to elucidate a role of interactions in many-particle systems under the weak spatial observation. For example, we can use Eq. (4.57) to investigate effects of interactions by switching on the on-site interaction term U in the system Hamiltonian $\hat{H}$. Also, our results may be applicable to preserving coherence in quantum control. Besides such a practical importance, a minimally destructive observation of many-body systems will also offer interesting possibilities from a fundamental point of views. For instance, it may provide a tool to study effects of measurements on strongly correlated phenomena such as quantum critical phenomena, as we have discussed in the previous chapter.

Appendix A: The Lieb-Robinson Bound

Here we briefly summarize the statement of the Lieb-Robinson bound and describe its relation to the full-counting dynamics discussed in Sect. 4.1. Lieb and Robinson (LR) have shown that, for a unitary time evolution in nonrelativistic quantum spins (or fermonic particles for spin-1/2 case) on a lattice, there exists a finite group velocity v_{LR} which bounds the velocity of propagation of information in the system. Specifically, they have shown the bound

4.5 Conclusions and Outlook

$$\left\| [\hat{O}_A(t), \hat{O}_B(0)] \right\| \leq c \min(|A|, |B|) \left\| \hat{O}_A \right\| \left\| \hat{O}_B \right\| \exp\left(-\frac{L - v_{\text{LR}} t}{\xi}\right), \quad \text{(A.1)}$$

where $\hat{O}_A$ and $\hat{O}_B$ are local operators acing on two subsystems A and B that are separated by the distance L, $\|\cdot\|$ is the operator norm, $|A|$ ($|B|$) denotes the volume of A (B), v_{LR} is the LR velocity, ξ characterizes the size of the tail in the effective light cone. The operator $\hat{O}(t)$ denotes the Heisenberg representation. We note that values of constants c, v_{LR} and ξ cannot be given by the bound in general. Physically, the bound (A.1) shows that a signal given in region B at $t = 0$ cannot be transferred to region A faster than the velocity v_{LR}. Bravyi, Hastings and Verstraete have used the inequality (A.1) to obtain the bound on the connected equal-time correlation functions after the quench:

$$\langle \Psi_t | \hat{O}_A \hat{O}_B | \Psi_t \rangle < c'(|A| + |B|) \exp\left(-\frac{L - 2v_{\text{LR}} t}{\chi}\right), \quad \text{(A.2)}$$

where c' and χ are constants. These relations play crucial roles especially in quantum information science and have laid the cornerstone in studies of gapped many-body ground states. While the relations (A.1) and (A.2) have been originally derived for closed quantum systems, later they have been generalized to the open quantum dynamics described by the Lindblad master equation, where the Liouvillian is assumed be the sum of local operators acting on the density matrix. In particular, this locality condition is satisfied in the exactly solvable model discussed in Sect. 4.1 as inferred from the non-Hermitian term $-\sum_l [(-1)^l i\gamma(\hat{c}_{l+1}^\dagger \hat{c}_l + \hat{c}_l^\dagger \hat{c}_{l+1}) + 2i\gamma \hat{c}_l^\dagger \hat{c}_l]$ in the effective Hamiltonian and the jump term $\mathcal{J}[\hat{\rho}] = 2\gamma \sum_l [2\hat{c}_l \hat{\rho} \hat{c}_l^\dagger + (-1)^l (\hat{c}_l \hat{\rho} \hat{c}_{l+1}^\dagger + \hat{c}_{l+1} \hat{\rho} \hat{c}_l^\dagger)]$; both of them consist of only local operators.

Appendix B: The Exact Solution of the Lattice Model

We provide technical details on the exact solution of the full-counting dynamics for the lattice model discussed in Sect. 4.1. To begin with, we diagonalize the effective Hamiltonian $\hat{H}_{\text{PT}}$ in Eq. (4.10). We divide the periodic chain of even length L into two sublattices ($\hat{a}_l = \hat{c}_{2l}$ and $\hat{b}_l = \hat{c}_{2l+1}$ with $l = 0, 1, \ldots, L/2 - 1$) and introduce their Fourier transforms by

$$\hat{a}_l = \sqrt{\frac{2}{L}} \sum_{0 \leq k < 2\pi} \hat{a}_k e^{ikl}, \quad \hat{b}_l = \sqrt{\frac{2}{L}} \sum_{0 \leq k < 2\pi} \hat{b}_k e^{ikl}, \quad k = \frac{2\pi n}{(L/2)} \left(n = 0, 1, \ldots, \frac{L}{2} - 1 \right). \quad \text{(B.1)}$$

Using these operators, we can rewrite $\hat{H}_{\text{PT}}$ as follows:

$$\hat{H}_{\text{PT}} = -\sum_{l=0}^{L-1}\left[\left(J+(-1)^l i\gamma\right)\left(\hat{c}_{l+1}^\dagger \hat{c}_l + \hat{c}_l^\dagger \hat{c}_{l+1}\right) + (-1)^l h \hat{c}_l^\dagger \hat{c}_l\right] \tag{B.2}$$

$$= \sum_{0\le k<2\pi}\left(\hat{a}_k^\dagger \ \hat{b}_k^\dagger\right)\begin{pmatrix} -h & -J-i\gamma+(-J+i\gamma)e^{-ik} \\ -J-i\gamma+(-J+i\gamma)e^{ik} & h \end{pmatrix}\begin{pmatrix}\hat{a}_k \\ \hat{b}_k\end{pmatrix}. \tag{B.3}$$

Diagonalizing the 2×2 matrix for each mode k, we obtain

$$\hat{H}_{\text{PT}} = \sum_k \sum_{\lambda=\pm} \epsilon_\lambda(k)\hat{g}_{\lambda k}^\dagger \hat{f}_{\lambda k}, \quad \epsilon_\pm(k) = \pm\sqrt{h^2 - 4\gamma^2 + 2J'^2(1+\cos k)}, \quad \hat{g}_{\lambda k}^\dagger = \alpha_\lambda^R(k)\hat{a}_k^\dagger + \beta_\lambda^R(k)\hat{b}_k^\dagger, \tag{B.4}$$

where $\epsilon_\lambda(k)$ ($\lambda = \pm$) are two eigenvalues for each mode k with $J' = \sqrt{J^2+\gamma^2}$, $\hat{g}_{\lambda k}^\dagger$ creates a right eigenvector of $\hat{H}_{\text{PT}}$, i.e., $\hat{H}_{\text{PT}}\hat{g}_{\lambda k}^\dagger|0\rangle = \epsilon_\lambda(k)\hat{g}_{\lambda k}^\dagger|0\rangle$, and $(\alpha_\lambda^R(k), \beta_\lambda^R(k))^T$ are components of the corresponding right eigenvector of the 2×2 non-Hermitian matrix in Eq. (B.3). The operator $\hat{f}_{\lambda k}$ creates a left eigenvector of $\hat{H}_{\text{PT}}$, i.e., $\langle 0|\hat{f}_{\lambda k}\hat{H}_{\text{PT}} = \langle 0|\hat{f}_{\lambda k}\epsilon_\lambda(k)$, and its form is uniquely determined when we impose a generalized anticommutation relation $\{\hat{f}_{\lambda k}, \hat{g}_{\lambda' k'}^\dagger\} = \delta_{k,k'}\delta_{\lambda,\lambda'}$.

We can similarly diagonalize the jump term $\mathcal{J}[\hat{\rho}]$ in Eq. (4.9), obtaining

$$\mathcal{J}[\hat{\rho}] = 4\sum_k \sum_{\lambda=\pm} \gamma_\lambda(k)\hat{d}_{\lambda k}\hat{\rho}\,\hat{d}_{\lambda k}^\dagger, \quad \gamma_\pm(k) = \gamma\left(1\pm\left|\sin\left(\frac{k}{2}\right)\right|\right), \tag{B.5}$$

where we introduce the operators $\hat{d}_{+,k} = (-ie^{\frac{ik}{2}}\hat{a}_k + \hat{b}_k)/\sqrt{2}$ and $\hat{d}_{-,k} = (ie^{\frac{ik}{2}}\hat{a}_k + \hat{b}_k)/\sqrt{2}$.

We next derive the solution of the full-counting dynamics. To do so, we introduce the interaction picture by

$$\tilde{\hat{\rho}}(t) = e^{i\hat{H}_{\text{eff}}t}\hat{\rho}(t)e^{-i\hat{H}_{\text{eff}}^\dagger t}, \quad \tilde{\hat{d}}_{\lambda k}(t) = e^{i\hat{H}_{\text{eff}}t}\hat{d}_{\lambda k}e^{-i\hat{H}_{\text{eff}}t}. \tag{B.6}$$

Then, the time-evolution equation becomes

$$\frac{d\tilde{\hat{\rho}}(t)}{dt} = 4\sum_k\sum_{\lambda=\pm}\gamma_\lambda(k)\tilde{\hat{d}}_{\lambda k}(t)\tilde{\hat{\rho}}(t)\tilde{\hat{d}}_{\lambda k}^\dagger(t). \tag{B.7}$$

For the sake of concreteness, we assume that $N = L/2$ particles are present at time $t=0$. Introducing a projector $\hat{\mathcal{P}}_n$ onto the subspace containing $N-n$ particles, we denote an unnormalized density matrix accompanying n quantum jumps by $\hat{\varrho}^{(n)}(t) = \hat{\mathcal{P}}_n\hat{\rho}(t)\hat{\mathcal{P}}_n$. Integrating equation (B.7) with the initial condition $\hat{\varrho}^{(n)}(0) = 0$ for $n < N$ and noting the relation $[\hat{\mathcal{P}}_n, \hat{H}_{\text{eff}}] = 0$, we obtain the following recursion relation:

$$\tilde{\hat{\varrho}}^{(n)}(t) = 4\int_0^t d\tau \sum_{k,\lambda}\gamma_\lambda(k)\tilde{\hat{d}}_{\lambda k}(\tau)\tilde{\hat{\varrho}}^{(n-1)}(\tau)\tilde{\hat{d}}_{\lambda k}^\dagger(\tau). \tag{B.8}$$

4.5 Conclusions and Outlook

Solving the recursion relation (B.8) iteratively, we obtain the formal solution of $\hat{\varrho}^{(n)}(t)$ as

$$\hat{\varrho}^{(n)}(t) = 4^n \int_0^t dt_n \cdots \int_0^{t_2} dt_1$$
$$\sum_{k_1\lambda_1,\ldots,k_n\lambda_n} \gamma_{\lambda_1}(k_1)\cdots\gamma_{\lambda_n}(k_n)\hat{\tilde{d}}_{\lambda_n k_n}(t_n)\cdots\hat{\tilde{d}}_{\lambda_1 k_1}(t_1)\hat{\varrho}^{(0)}(t_1)\hat{\tilde{d}}^{\dagger}_{\lambda_1 k_1}(t_1)\cdots\hat{\tilde{d}}^{\dagger}_{\lambda_n k_n}(t_n)$$

$$= \frac{4^n}{n!} \int_0^t dt_n \cdots \int_0^t dt_1$$
$$\sum_{k_1\lambda_1,\ldots,k_n\lambda_n} \gamma_{\lambda_1}(k_1)\cdots\gamma_{\lambda_n}(k_n)\overrightarrow{T}\left[\hat{\tilde{d}}_{\lambda_n k_n}(t_n)\cdots\hat{\tilde{d}}_{\lambda_1 k_1}(t_1)\right]\hat{\rho}(0)\overleftarrow{T}\left[\hat{\tilde{d}}^{\dagger}_{\lambda_1 k_1}(t_1)\cdots\hat{\tilde{d}}^{\dagger}_{\lambda_n k_n}(t_n)\right], \quad \text{(B.9)}$$

where we use $\hat{\varrho}^{(0)}(t_1) = \hat{\rho}(0)$ in obtaining the second equality, and $\overrightarrow{T}$ ($\overleftarrow{T}$) denotes the time-ordering (anti-time-ordering) operator. To perform the time integration, let us simplify the following time-dependent operator $\hat{\tilde{d}}_{\lambda k}(t)$:

$$\hat{\tilde{d}}_{\lambda k}(t) = e^{i\hat{H}_{\text{eff}}t}\hat{d}_{\lambda k}e^{-i\hat{H}_{\text{eff}}t} = e^{-2\gamma t}e^{i\hat{H}_{\text{PT}}t}\hat{d}_{\lambda k}e^{-i\hat{H}_{\text{PT}}t}. \quad \text{(B.10)}$$

Because $\hat{H}_{\text{PT}}$ is quadratic in fermionic operators (see Eq. (B.3)), we can solve Eq. (B.10) by introducing an eigenoperator $\hat{\Lambda}_{\eta k}$ satisfying the relation $[\hat{\Lambda}_{\eta k}, \hat{H}_{\text{PT}}] = \epsilon_{\eta}(k)\hat{\Lambda}_{\eta k}$ ($\eta = \pm$). We thus obtain

$$\hat{\tilde{d}}_{\lambda k}(t) = e^{-2\gamma t}\sum_{\eta=\pm} c_{\eta\lambda}\hat{\Lambda}_{\eta k}e^{-i\epsilon_{\eta}(k)t}, \quad \text{(B.11)}$$

where $c_{\eta\lambda}$ are the expansion coefficients of $\hat{d}_{\lambda k}$ with respect to $\hat{\Lambda}_{\eta k}$'s. Using a right eigenvector $(\alpha_{\lambda}^{R}(k), \beta_{\lambda}^{R}(k))^T$ of $\hat{H}_{\text{PT}}$ (see Eq. (B.4)), an explicit expression of the eigenoperators can be given as $\hat{\Lambda}_{\eta k} = \alpha_{\eta}^{R}(-k)\hat{a}_k + \beta_{\eta}^{R}(-k)\hat{b}_k$. We then consider the following time integration:

$$\int_0^t d\tau \cdots \hat{\tilde{d}}_{\lambda k}(\tau)\cdots\hat{\tilde{d}}^{\dagger}_{\lambda k}(\tau)\cdots = \int_0^t d\tau \sum_{\eta\eta'} c_{\eta\lambda}c^*_{\eta'\lambda}e^{-4\gamma\tau - i\epsilon_{\eta}(k)\tau + i\epsilon_{\eta'}(k)\tau}\cdots\hat{\Lambda}_{\eta k}\cdots\hat{\Lambda}^{\dagger}_{\eta' k}\cdots$$

$$= \sum_{\eta\eta'} c_{\eta\lambda}c^*_{\eta'\lambda}\frac{1 - e^{-4\gamma t - i\epsilon_{\eta}(k)t + i\epsilon_{\eta'}(k)t}}{4\gamma + i(\epsilon_{\eta}(k) - \epsilon_{\eta'}(k))}\cdots\hat{\Lambda}_{\eta k}\cdots\hat{\Lambda}^{\dagger}_{\eta' k}\cdots.$$

(B.12)

Using Eq. (B.9) and introducing the 2×2 matrix $\gamma^c_{\eta\eta'}(k) = \sum_{\lambda}\gamma_{\lambda}(k)c_{\eta\lambda}c^*_{\eta'\lambda}$, we obtain

$$\hat{\varrho}^{(n)}(t) = \sum_{\eta_1\eta'_1 k_1\cdots\eta_n\eta'_n k_n} \frac{1}{n!}\left[\prod_{i=1}^n \gamma^c_{\eta_i\eta'_i}(k_i)\frac{1 - e^{-4\gamma t - i\epsilon_{\eta_i}(k_i)t + i\epsilon_{\eta'_i}(k_i)t}}{\gamma + i(\epsilon_{\eta_i}(k_i) - \epsilon_{\eta'_i}(k_i))/4}\right]\hat{\Lambda}_{\eta_n k_n}\cdots\hat{\Lambda}_{\eta_1 k_1}\hat{\rho}(0)\hat{\Lambda}^{\dagger}_{\eta'_1 k_1}\cdots\hat{\Lambda}^{\dagger}_{\eta'_n k_n}.$$

(B.13)

Transforming back to the Schrödinger picture by using

$$\hat{\varrho}^{(n)}(t) = e^{-i\hat{H}_{\text{eff}}t}\hat{\tilde{\varrho}}^{(n)}(t)e^{i\hat{H}_{\text{eff}}^\dagger t} = e^{-4\gamma(N-n)t}e^{-i\hat{H}_{\text{PT}}t}\hat{\varrho}^{(n)}(t)e^{i\hat{H}_{\text{PT}}^\dagger t} \quad (B.14)$$

and the relation $e^{-i\hat{H}_{\text{PT}}t}\hat{\Lambda}_{\eta k}e^{i\hat{H}_{\text{PT}}t} = \hat{\Lambda}_{\eta k}e^{i\epsilon_\eta(k)t}$, we obtain the solution of the full-counting dynamics:

$$\hat{\varrho}^{(n)}(t) = \sum_{\eta_1\eta'_1 k_1 \cdots \eta_n \eta'_n k_n} \frac{e^{-4\gamma(N-n)t}}{n!}\left[\prod_{i=1}^n \mathcal{D}_{\eta_i \eta'_i}(k_i;t)\right]\hat{\Lambda}_{\eta_n k_n}\cdots\hat{\Lambda}_{\eta_1 k_1}e^{-i\hat{H}_{\text{PT}}t}\hat{\rho}(0)e^{i\hat{H}_{\text{PT}}^\dagger t}\hat{\Lambda}^\dagger_{\eta'_1 k_1}\cdots\hat{\Lambda}^\dagger_{\eta'_n k_n}, \quad (B.15)$$

where we introduce the 2 × 2 Hermitian matrix $\mathcal{D}_{\eta\eta'}$ by

$$\mathcal{D}_{\eta\eta'}(k;t) = \gamma^c_{\eta\eta'}(k)\frac{e^{i\epsilon_\eta(k)t - i\epsilon_{\eta'}(k)t} - e^{-4\gamma t}}{\gamma + i(\epsilon_\eta(k) - \epsilon_{\eta'}(k))/4}. \quad (B.16)$$

In practice, to calculate the nonequilibrium properties of the system such as correlation functions, we proceed as follows. First, we diagonalize the operators $\hat{\Lambda}_{\eta k}$ and $\hat{\Lambda}^\dagger_{\eta' k}$ in Eq. (B.15) with respect to the indices η and η'. To this end, for each time t and wavevector k, we numerically diagonalize the following 2 × 2 Hermitian matrix:

$$\begin{pmatrix} \sum_{\eta\eta'}\mathcal{D}_{\eta\eta'}(k;t)\alpha^R_\eta(-k)\alpha^{*R}_{\eta'}(-k) & \sum_{\eta\eta'}\mathcal{D}_{\eta\eta'}(k;t)\beta^R_\eta(-k)\alpha^{*R}_{\eta'}(-k) \\ \sum_{\eta\eta'}\mathcal{D}_{\eta\eta'}(k;t)\alpha^R_\eta(-k)\beta^{*R}_{\eta'}(-k) & \sum_{\eta\eta'}\mathcal{D}_{\eta\eta'}(k;t)\beta^R_\eta(-k)\beta^{*R}_{\eta'}(-k) \end{pmatrix}. \quad (B.17)$$

Using its two real eigenvalues $\lambda_{\pm,k}(t)$ and the corresponding orthonormal eigenvectors $\hat{v}_{\pm,k}(t)$, we can simplify Eq. (B.15) as follows:

$$\hat{\varrho}^{(n)}(t) = \frac{e^{-4\gamma(N-n)t}}{n!}\sum_{\eta_1 k_1 \cdots \eta_n k_n}\left[\prod_{i=1}^n \lambda_{\eta_i k_i}(t)\right]\hat{v}_{\eta_n k_n}(t)\cdots\hat{v}_{\eta_1 k_1}(t)e^{-i\hat{H}_{\text{PT}}t}\hat{\rho}(0)e^{i\hat{H}_{\text{PT}}^\dagger t}\hat{v}^\dagger_{\eta_1 k_1}(t)\cdots\hat{v}^\dagger_{\eta_n k_n}(t). \quad (B.18)$$

The time evolution $e^{-i\hat{H}_{\text{PT}}t}\hat{\rho}(0)e^{i\hat{H}_{\text{PT}}^\dagger t}$ can be calculated by using Eq. (B.4). Denoting the initial state as $\hat{\rho}(0) = |\Psi_0\rangle\langle\Psi_0|$ and expanding it in terms of the right eigenvectors $|\Psi_0\rangle = \prod_k[\sum_\lambda \psi_{\lambda k}\hat{g}^\dagger_{\lambda k}]|0\rangle$, the time evolution is given by

$$|\Psi_t\rangle = e^{-i\hat{H}_{\text{PT}}t}|\Psi_0\rangle = \prod_k\left[\sum_\lambda \psi_{\lambda k}e^{-i\epsilon_\lambda(k)t}\hat{g}^\dagger_{\lambda k}\right]|0\rangle = \prod_k\left[\sum_\eta \psi^v_\eta(k;t)\hat{v}^\dagger_{\eta k}(t)\right]|0\rangle. \quad (B.19)$$

In obtaining the last equality, we have expanded the time-dependent state in the basis of $\hat{v}_{\pm,k}(t)$ and introduced the corresponding expansion coefficients $\psi^v_\eta(k;t)$. Combining Eqs. (B.18) and (B.19), we can obtain the following expression for the trace of an unnormalized density matrix:

$$\text{Tr}\left[\hat{\varrho}^{(n)}(t)\right] = e^{-4\gamma(N-n)t}\left[\prod_k \mathcal{N}_k(t)\right]\sigma_n\left(\{f^v_k(t)\}\right), \quad (B.20)$$

4.5 Conclusions and Outlook

where we introduce the time-dependent norm factor $\mathcal{N}_k(t) = \sum_\eta |\psi^\nu_{\eta k}(t)|^2$ of each mode k, and σ_n denotes the n-th symmetric polynomial of $f^\nu_k(t) \equiv \sum_\eta \lambda_{\eta k}(t)|\psi^\nu_{\eta k}(t)|^2/\mathcal{N}_k(t)$:

$$\sigma_n\left(\{f^\nu_k(t)\}\right) = \frac{(-1)^n}{(N-n)!}\frac{d^{N-n}}{dx^{N-n}}\bigg|_{x=0} \prod_k \left(x - f^\nu_k(t)\right). \tag{B.21}$$

We note that $\{f^\nu_k(t)\}$ forms a set of N variables with $k = 0, 2\pi/N, \ldots, 2\pi(N-1)/N$.

The full-counting equal-time correlation is now given as

$$\begin{aligned}C^{(n)}(l,t) &= \frac{\text{Tr}\left[\hat{c}^\dagger_l \hat{c}_0 \hat{\varrho}^{(N-n)}(t)\right]}{\text{Tr}\left[\hat{\varrho}^{(N-n)}(t)\right]} \\ &= \frac{2}{L}\sum_k e^{-ik\lceil l/2 \rceil}\left[\frac{\sum_\lambda \psi^*_{\lambda k}(t) O^{*R}_\lambda(k) \times \sum_\lambda \psi_{\lambda k}(t) \alpha^R_\lambda(k)}{\mathcal{N}_k(t)}\right] \frac{\sigma_n\left(\{f^\nu_{k'}(t)\}_{k' \neq k}\right)}{\sigma_n\left(\{f^\nu_{k'}(t)\}\right)},\end{aligned} \tag{B.22}$$

where we choose $O = \alpha$ (β) when l is even (odd) and introduce $\psi_{\lambda k}(t) = \psi_{\lambda k} e^{-i\epsilon_\lambda(k)t}$ and

$$\sigma_n\left(\{f^\nu_{k'}(t)\}_{k' \neq k}\right) = \frac{(-1)^n}{(N-1-n)!}\frac{d^{N-1-n}}{dx^{N-1-n}}\bigg|_{x=0} \prod_{k' \neq k} \left(x - f^\nu_{k'}(t)\right). \tag{B.23}$$

Finally, for the null-jump case $n = 0$, we can further simplify the expressions of correlation functions. For example, the unequal-time correlation defined by $\tilde{C}^{(0)}(l,t) = \langle \Psi_0|\hat{c}^\dagger_l(t)\hat{c}_0(0)|\Psi_0\rangle/\langle \Psi_t|\Psi_t\rangle$ with $\hat{c}^\dagger_l(t) \equiv e^{i\hat{H}_{\text{PT}}t}\hat{c}^\dagger_l e^{-i\hat{H}_{\text{PT}}t}$ can be expressed as

$$\tilde{C}^{(0)}(l,t) = \frac{2}{L}\sum_k \sum_{\lambda=\pm} O_{\lambda k} \frac{\psi^*_{\lambda k} e^{i\epsilon_\lambda(k)t - ik\lceil l/2 \rceil}}{\mathcal{N}_k(t)} \quad \text{with} \quad O_{\lambda k} = O^{*R}_\lambda(k)\sum_\eta \psi_{\eta k}\alpha^R_\eta(k), \tag{B.24}$$

which gives Eq. (4.16) in Sect. 4.1.

Appendix C: Subextensive Energy Fluctuation in the Matrix-Vector Product Ensemble

We here show the subextensiveness of the energy fluctuation in the matrix-vector product ensemble introduced in Sect. 4.2:

$$\hat{\rho}_M = \frac{\Lambda_M[\hat{\rho}_{\text{eq}}]}{Z(M)} = \frac{1}{Z(M)} \sum_a [\mathcal{V}_{m_n} \cdots \mathcal{V}_{m_1} p_{\text{eq}}]_a |E_a\rangle\langle E_a|. \quad (\text{C.1})$$

We recall that $\Lambda_M = \prod_{i=1}^n (\Lambda \circ \mathcal{L}_{m_i} \circ \Lambda)$ with $\mathcal{L}_m[\hat{O}] = \hat{L}_m \hat{O} \hat{L}_m^\dagger$, $\Lambda[\hat{O}] = \sum_a \hat{P}_a \hat{O} \hat{P}_a$ and $\hat{P}_a = |E_a\rangle\langle E_a|$, and $Z(M) = \sum_a [\mathcal{V}_{m_n} \cdots \mathcal{V}_{m_1} p_{\text{eq}}]_a$ is a normalization constant. We assume that thermal eigenstates satisfy the cluster decomposition property (CDP) for local operators $\hat{O}_{x,y}$:

$$\lim_{|x-y|\to\infty} \text{Tr}[\hat{O}_x \hat{O}_y \hat{P}_a] - \text{Tr}[\hat{O}_x \hat{P}_a]\text{Tr}[\hat{O}_y \hat{P}_a] = 0, \quad (\text{C.2})$$

which is a fundamental property that lies at the heart of quantum many-body theory and should hold true for any physical states (with only a few exceptions such as (long-lived) macroscopic superposition states). From this assumption together with the ETH, it follows that any diagonal ensemble with a strongly peaked energy distribution satisfies the CDP in the thermodynamic limit. In particular, the initial thermal equilibrium state $\hat{\rho}_{\text{eq}}$ also satisfies the CDP since its energy fluctuation (i.e., the standard deviation) is subextensive by definition. Below we show that if an energy fluctuation in an ensemble $\hat{\rho}$ diagonal in the energy basis is subextensive and thus $\hat{\rho}$ satisfies the CDP, then a post-measurement ensemble $\hat{\rho}_m \propto \Lambda_m[\hat{\rho}]$ after a single quantum jump with $\Lambda_m = \Lambda \circ \mathcal{L}_m \circ \Lambda$ also satisfies these conditions. By induction it then follows that an energy fluctuation of the density matrix (C.1) is also subextensive.

Local measurement

We first show the subextensiveness of the energy fluctuation in the post-measurement ensemble $\hat{\rho}_m$ for a local measurement, in which a measurement operator $\hat{L}_m$ acts on a local spatial region. The variance of energy is given as

$$(\Delta E)^2 = \text{Tr}[\hat{H}^2 \hat{\rho}_m] - (\text{Tr}[\hat{H}\hat{\rho}_m])^2 = \frac{1}{(Z(m))^2}\left[Z(m)\text{Tr}[\hat{H}^2 \mathcal{L}_m[\hat{\rho}]] - \left(\text{Tr}[\hat{H}\mathcal{L}_m[\hat{\rho}]]\right)^2\right]. \quad (\text{C.3})$$

where $Z(m) = \text{Tr}[\Lambda_m[\hat{\rho}]]$ is a normalization constant. We express the Hamiltonian and measurement operators as sums of local operators:

$$\hat{H} = \sum_x \hat{h}_x, \quad \hat{L}_m = \sum_{x\in\mathcal{D}_m} \hat{l}_x, \quad (\text{C.4})$$

where $\mathcal{D}_m$ denotes a local spatial region on which $\hat{L}_m$ acts. To rewrite Eq. (C.3), we calculate the quantity

$$\Delta_{xy} \equiv \frac{1}{(Z(m))^2}\left[Z(m)\text{Tr}[\hat{h}_x \hat{h}_y \mathcal{L}_m[\hat{\rho}]] - \text{Tr}[\hat{h}_x \mathcal{L}_m[\hat{\rho}]]\text{Tr}[\hat{h}_y \mathcal{L}_m[\hat{\rho}]]\right] \quad (\text{C.5})$$

$$= \frac{1}{(\langle \hat{L}_m^\dagger \hat{L}_m\rangle)^2}\left[\langle \hat{L}_m^\dagger \hat{L}_m\rangle\langle \hat{L}_m^\dagger \hat{h}_x \hat{h}_y \hat{L}_m\rangle - \langle \hat{L}_m^\dagger \hat{h}_x \hat{L}_m\rangle\langle \hat{L}_m^\dagger \hat{h}_y \hat{L}_m\rangle\right], \quad (\text{C.6})$$

4.5 Conclusions and Outlook

where we denote $\text{Tr}[\cdot \hat{\rho}] = \langle \cdot \rangle$. Using the condition $[\hat{l}_x, \hat{h}_y] = 0$ for $x \neq y$, we obtain in the limit $|x - y| \to \infty$

$$\langle \hat{L}_m^\dagger \hat{h}_x \hat{h}_y \hat{L}_m \rangle = \langle \hat{L}_m^\dagger \hat{L}_m \hat{h}_x \hat{h}_y \rangle \simeq \langle \hat{L}_m^\dagger \hat{L}_m \rangle \langle \hat{h}_x \rangle \langle \hat{h}_y \rangle, \tag{C.7}$$

$$\langle \hat{L}_m^\dagger \hat{h}_x \hat{L}_m \rangle \langle \hat{L}_m^\dagger \hat{h}_y \hat{L}_m \rangle \simeq \langle \hat{L}_m^\dagger \hat{L}_m \rangle^2 \langle \hat{h}_x \rangle \langle \hat{h}_y \rangle. \tag{C.8}$$

Here, we use the CDP of $\hat{\rho}$ in deriving the last expressions. We thus obtain $\lim_{|x-y|\to\infty} \Delta_{xy} = 0$. It follows that the standard deviation of energy is subextensive:

$$\lim_{V \to \infty} \frac{\Delta E}{V} = \lim_{V \to \infty} \frac{\sqrt{\sum_{xy} \Delta_{xy}}}{\sum_{xy} 1} = 0. \tag{C.9}$$

In particular, it is physically plausible to assume that Δ_{xy} decays exponentially fast or at least faster than V^{-1} in the thermodynamic limit. Under this condition, we obtain the square-root scaling:

$$\Delta E \simeq \sqrt{\sum_x \Delta_{xx}} \propto O(\sqrt{V}). \tag{C.10}$$

Global measurement

We next consider a global measurement, in which a measurement operator acts on an entire region of the system. As $\mathcal{D}_m$ is independent of a label m, we abbreviate a label and denote a measurement operator as $\hat{L} = \sum_z \hat{l}_z$ for the sake of simplicity. It turns out that we need to discuss the two different cases separately depending on whether or not an expectation value $\lim_{V \to \infty} \langle \hat{L} \rangle / V$ vanishes in the thermodynamic limit.

We first consider the case in which $\langle \hat{L} \rangle$ scales as

$$\langle \hat{L} \rangle = \langle \sum_z \hat{l}_z \rangle \propto O(V), \tag{C.11}$$

so that the expectation value $\lim_{V \to \infty} \langle \hat{L} \rangle / V$ does not vanish. From the CDP of $\hat{\rho}$, the leading term in Δ_{xy} defined in Eq. (C.5) can be estimated in the limit $|x - y| \to \infty$ as

$$\Delta_{xy} = \frac{1}{\langle \hat{L}^\dagger \hat{L} \rangle^2} \left(\langle \sum_{z,w} \hat{l}_z^\dagger \hat{l}_w \rangle \langle \sum_{z,w} \hat{l}_z^\dagger \hat{h}_x \hat{h}_y \hat{l}_w \rangle - \langle \sum_{z,w} \hat{l}_z^\dagger \hat{h}_x \hat{l}_w \rangle \langle \sum_{z,w} \hat{l}_z^\dagger \hat{h}_y \hat{l}_w \rangle \right) \tag{C.12}$$

$$\simeq \frac{1}{|\langle \hat{L} \rangle|^4} \left[(\langle \hat{l}_x^\dagger \rangle \langle \hat{h}_x \rangle - \langle \hat{l}_x^\dagger \hat{h}_x \rangle)(\langle \hat{l}_y^\dagger \rangle \langle \hat{h}_y \rangle - \langle \hat{l}_y^\dagger \hat{h}_y \rangle) \langle \hat{L} \rangle^2 + \text{c.c.} \right] \propto O\left(\frac{1}{V^2}\right). \tag{C.13}$$

We thus conclude that the standard deviation of energy is subextensive:

$$\Delta E = \sqrt{\sum_x \Delta_{xx} + \sum_{x\neq y} \Delta_{xy}} \simeq \sqrt{\sum_x \Delta_{xx}} \propto O(\sqrt{V}). \quad (C.14)$$

We next consider the other case in which an expectation value of $\hat{L}/V$ vanishes in the thermodynamic limit. To be specific, we impose the following condition

$$\langle \hat{L} \rangle = \langle \sum_z \hat{l}_z \rangle \propto o(\sqrt{V}). \quad (C.15)$$

For instance, in the numerical example presented in Sect. 4.2.3, an expectation value of $\hat{L} = \sum_m (-1)^m \hat{n}_m$ with respect to arbitrary energy eigenstate is exactly zero, and thus the condition (C.15) is satisfied. Using this condition, we can rewrite $\langle \hat{L}^\dagger \hat{L} \rangle$ as

$$\langle \hat{L}^\dagger \hat{L} \rangle = \langle \sum_{z,w} \hat{l}_z^\dagger \hat{l}_w \rangle \simeq \sum_{z,w} (\langle \hat{l}_z^\dagger \hat{l}_w \rangle - \langle \hat{l}_z^\dagger \rangle \langle \hat{l}_w \rangle) \simeq \sum_z \left(\langle \hat{l}_z^\dagger \hat{l}_z \rangle - \langle \hat{l}_z^\dagger \rangle \langle \hat{l}_z \rangle \right) \propto O(V). \quad (C.16)$$

Here, in the first approximate equality we add the $o(V)$ contribution in Eq. (C.15), and in the second approximate equality we use the CDP of $\hat{\rho}$ and the scaling (C.15). The leading contribution in Δ_{xy} of Eq. (C.12) is obtained as

$$\Delta_{xy} \simeq \frac{\langle \hat{l}_x^\dagger \hat{h}_x \rangle \langle \hat{h}_y \hat{l}_y \rangle + \text{c.c.}}{\sum_z \left(\langle \hat{l}_z^\dagger \hat{l}_z \rangle - \langle \hat{l}_z^\dagger \rangle \langle \hat{l}_z \rangle \right)} \propto O\left(\frac{1}{V}\right), \quad (C.17)$$

again leading to the subextensive energy fluctuation:

$$\Delta E = \sqrt{\sum_x \Delta_{xx} + \sum_{x\neq y} \Delta_{xy}} \propto O(\sqrt{V}). \quad (C.18)$$

Appendix D: Minimally Destructive Spatial Observation of Indistinguishable Particles

We here provide details about taking the limit of strong atom-light coupling limit and weak-spatial resolution of Eq. (4.50), resulting in the stochastic differential equation (4.57) describing the random time evolution of indistinguishable atoms under minimally destructive spatial observation. From Eqs. (4.50) and (4.51), we obtain the following master equation:

4.5 Conclusions and Outlook

$$d\hat{\rho} = -\frac{i}{\hbar}\left[\hat{H}, \hat{\rho}\right] + \mathcal{D}[\hat{\rho}]dt + \int dX \left(\frac{\hat{M}(X)\hat{\rho}\hat{M}^\dagger(X)}{\langle\hat{M}^\dagger(X)\hat{M}(X)\rangle} - \hat{\rho}\right)\left(dN(X;t) - \langle\hat{M}^\dagger(X)\hat{M}(X)\rangle\right), \quad \text{(D.1)}$$

where we define the dissipator by

$$\mathcal{D}[\hat{\rho}] = \int dX \left(\hat{M}(X)\hat{\rho}\hat{M}^\dagger(X) - \frac{1}{2}\left\{\hat{M}^\dagger(X)\hat{M}(X), \hat{\rho}\right\}\right). \quad \text{(D.2)}$$

To be specific, we assume the measurement operator

$$\hat{M}(X) = \sqrt{\frac{\gamma}{\sqrt{\pi\sigma^2}}} \sum_m \exp\left[-\frac{(X-md)^2}{2\sigma^2}\right]\hat{n}_m, \quad \text{(D.3)}$$

where the point-spread function is chosen to be a Gaussian function. Then, the dissipator $\mathcal{D}[\hat{\rho}]$ in Eq. (D.2) can be approximated as

$$\mathcal{D}[\hat{\rho}] \simeq -\frac{\gamma}{2}\sum_{m,l}\left(1 - \frac{(m-l)^2 d^2}{4\sigma^2}\left[\hat{n}_l, [\hat{n}_m, \hat{\rho}]\right]\right) = -\frac{N^2\gamma d^2}{4\sigma^2}\left[\hat{X}_{\text{c.m.}}, \left[\hat{X}_{\text{c.m.}}, \hat{\rho}\right]\right], \quad \text{(D.4)}$$

where $\hat{X}_{\text{c.m.}} = \sum_m \hat{n}_m/N$ is the center-of-mass operator with N being the total number of atoms and we take the limit of weak-spatial resolution, i.e., we assume that interference peaks of particles cannot be spatially resolved. In other words, this condition is equivalent to the requirement for matrix elements of $\hat{n}_m \hat{n}_l \hat{\rho} + \hat{\rho}\hat{n}_l \hat{n}_m - 2\hat{n}_m \hat{\rho}\hat{n}_l$ to rapidly vanish in $|m - l| > \sigma/d$. Also, we employ the conservation of the particle number $\sum_m \hat{n}_m = N\hat{I}$ to derive the last equality. Equation (D.4) gives the first line in the time-evolution Eq. (4.57).

We next take the same limit of the last term in Eq. (D.1). To derive Eq. (4.57), we have to show that, in this limit, the following contribution goes to zero:

$$\int dX\, R(X;t)\frac{dN(X;t) - \langle\hat{M}^\dagger(X)\hat{M}(X)\rangle dt}{\sqrt{\langle\hat{M}^\dagger(X)\hat{M}(X)\rangle}} \simeq \int dX\, R(X;t)dW(X;t). \quad \text{(D.5)}$$

Here we employ (4.55) and define

$$R(X;t) = \sqrt{\langle\hat{M}^\dagger(X)\hat{M}(X)\rangle}\left(\frac{\hat{M}(X)\hat{\rho}\hat{M}^\dagger(X)}{\langle\hat{M}^\dagger(X)\hat{M}(X)\rangle} - \hat{\rho} - \frac{d}{\sqrt{2}\sigma}\left\{\hat{X}_{\text{c.m.}} - \langle\hat{X}_{\text{c.m.}}\rangle, \hat{\rho}\right\}\right). \quad \text{(D.6)}$$

The Gaussian point-spread function in $\hat{M}(X)$ can be expanded as

$$\sum_m e^{-\frac{(X-md)^2}{2\sigma^2}}\hat{n}_m \simeq e^{-\frac{X^2}{2\sigma^2}}\sum_m\left(1 + \frac{mXd}{\sigma^2}\right)\hat{n}_m = e^{-\frac{X^2}{2\sigma^2}}N\left(\hat{I} + \frac{Xd}{\sigma^2}\hat{X}_{\text{c.m.}}\right). \quad \text{(D.7)}$$

We can justify the use of the approximation (D.7) in Eq. (D.5) as follows. First, in the weak-spatial resolution limit, particles are assumed to be positioned around the site $m = 0$ without loss of generality. Since a cluster size of particles is smaller than σ, matrix elements of $\hat{n}_m \hat{\rho} \hat{n}_l$ in the Fock basis are vanishingly small if $m > \sigma/d$ or $l > \sigma/d$. Thus, higher-order terms of m in Eq. (D.7) can be neglected. Second, we note that higher-order terms of X can be also neglected. The reason is that, after performing the integration over X, they lead to higher-order terms of d/σ that can be neglected in the weak-spatial resolution limit $d/\sigma \to 0$. We then approximate the first two terms in Eq. (D.6) as

$$\int_{-\infty}^{\infty} dX \left(\frac{\hat{M}(X)\hat{\rho}\hat{M}^\dagger(X)}{\sqrt{\langle \hat{M}^\dagger(X)\hat{M}(X)\rangle}} - \sqrt{\langle \hat{M}^\dagger(X)\hat{M}(X)\rangle} \hat{\rho} \right) dW(X;t)$$

$$\simeq \sqrt{\frac{N^2 \gamma d^2}{\sqrt{\pi}\sigma^5}} \int_{-\infty}^{\infty} dX dW(X;t) X e^{-\frac{X^2}{2\sigma^2}} \left\{ \hat{X}_{\text{c.m.}} - \langle \hat{X}_{\text{c.m.}}\rangle, \hat{\rho} \right\}$$

$$= \sqrt{\frac{N^2 \gamma d^2}{2\sigma^2}} \left\{ \hat{X}_{\text{c.m.}} - \langle \hat{X}_{\text{c.m.}}\rangle, \hat{\rho} \right\} dW(t). \tag{D.8}$$

Here we note that a linear superposition of Wiener stochastic processes is a Wiener process:

$$\int_{-\infty}^{\infty} dX dW(X;t) X e^{-\frac{X^2}{2\sigma^2}} = \sqrt{\frac{\sigma^3 \sqrt{\pi}}{2}} dW(t). \tag{D.9}$$

Since Eq. (D.8) cancels out the last term in Eq. (D.6), we show that Eq. (D.5) goes to zero in the weak-spatial resolution limit. The resulting time-evolution equation is Eq. (4.57). In the case of a single particle, Eq. (4.57) reproduces the previous results. The derivation presented here is also applicable to a general point spread function if one replaces σ by the effective resolution introduced in Eq. (4.49).

Appendix E: Minimally Destructive Spatial Observation of Distinguishable Particles

We provide the derivation of Eq. (4.62) describing the motions of distinguishable particles under the spatial observation. Starting from Eq. (4.60), we use (4.61) to obtain the master equation:

$$d\hat{\rho} = -\frac{i}{\hbar}\left[\hat{H}, \hat{\rho}\right] + \sum_{i=1}^{N} \left[\mathcal{D}_i[\hat{\rho}]dt + \int dX_i \left(\frac{\hat{M}_i(X_i)\hat{\rho}\hat{M}_i^\dagger(X_i)}{\langle \hat{M}_i^\dagger(X_i)\hat{M}_i(X_i)\rangle} - \hat{\rho} \right) \left(dN_i(X_i;t) - \langle \hat{M}_i^\dagger(X_i)\hat{M}_i(X_i)\rangle \right) \right]. \tag{E.1}$$

4.5 Conclusions and Outlook

where we introduce the dissipator as

$$\mathcal{D}_i[\hat{\rho}] = \int dX \left(\hat{M}_i(X) \hat{\rho} \hat{M}_i^\dagger(X) - \frac{1}{2} \left\{ \hat{M}_i^\dagger(X) \hat{M}_i(X), \hat{\rho} \right\} \right). \tag{E.2}$$

In the same manner as in Appendix 4.5, the dissipator $\mathcal{D}_i[\hat{\rho}]$ in the weak-spatial resolution limit can be obtained as

$$\mathcal{D}_i[\hat{\rho}] \simeq -\frac{\gamma_i d^2}{4\sigma^2} \left[\hat{x}_i, [\hat{x}_i, \hat{\rho}] \right], \tag{E.3}$$

where $\hat{x}_i = \sum_m m |m\rangle_{ii}\langle m|$ is the position operator of each particle. This contribution provides the first line in Eq. (4.62).

To derive Eq. (4.62), we have to show that, in the weak resolution limit, the following contribution goes to zero:

$$\int dX \, R_i(X;t) \frac{dN_i(X;t) - \langle \hat{M}_i^\dagger(X) \hat{M}_i(X) \rangle dt}{\sqrt{\langle \hat{M}_i^\dagger(X) \hat{M}_i(X) \rangle}} \simeq \int dX \, R_i(X;t) dW_i(X;t), \tag{E.4}$$

where we introduce

$$R_i(X;t) = \sqrt{\langle \hat{M}_i^\dagger(X) \hat{M}_i(X) \rangle} \left(\frac{\hat{M}_i(X) \hat{\rho} \hat{M}_i^\dagger(X)}{\langle \hat{M}_i^\dagger(X) \hat{M}_i(X) \rangle} - \hat{\rho} - \frac{d}{\sqrt{2}\sigma} \left\{ \hat{x}_i - \langle \hat{x}_i \rangle, \hat{\rho} \right\} \right). \tag{E.5}$$

We can decompose the total density matrix $\hat{\rho}$ into the product of $\hat{\rho}_i$ for each particle i, as the particles are assumed to be noninteracting and not entangled in the initial state (see Sect. 4.3.4). The convergence of Eq. (E.4) thus follows from applying the derivation (D.8) in the previous Appendix to the case of a single particle.

Appendix F: Diffusion Constant for Distinguishable Particles

We provide the derivation of the diffusion constant (4.66) for distinguishable particles. To begin with, let us first study a single-particle case. Denoting $\langle n|\hat{\rho}|m\rangle = \rho_{n,m}$ with $|n\rangle$ being a particle localized at site n, the master equation can be written as

$$\dot{\rho}_{n,m} = iJ \left(\rho_{n+1,m} + \rho_{n-1,m} - \rho_{n,m+1} - \rho_{n,m-1} \right) - \frac{\Gamma}{4}(n-m)^2 \rho_{n,m}. \tag{F.1}$$

In the diffusive regime, we can neglect the off-diagonal elements between remote lattice sites. It thus suffices to take into account the off-diagonal elements with neighboring sites $\rho_{n,n\pm1}$. Using the stationary condition $\dot{\rho}_{n,n\pm1} \simeq 0$, we obtain

$$\rho_{n,n\pm 1} \simeq \frac{4iJ}{\Gamma}\left(\rho_{n\pm 1,n\pm 1} - \rho_{n,n}\right),$$

$$\rho_{n\pm 1,n} \simeq -\frac{4iJ}{\Gamma}\left(\rho_{n\pm 1,n\pm 1} - \rho_{n,n}\right). \tag{F.2}$$

Using these relations in the master equation for the diagonal elements $\rho_{n,n}$, we get

$$\dot\rho_{n,n} = iJ\left(\rho_{n+1,n} + \rho_{n-1,n} - \rho_{n,n+1} - \rho_{n,n-1}\right)$$
$$\simeq \frac{8J^2}{\Gamma}\left(\rho_{n+1,n+1} - 2\rho_{n,n} + \rho_{n-1,n-1}\right). \tag{F.3}$$

This is nothing but the diffusion equation with the diffusion constant $8J^2/\Gamma$.

In the case of two distinguishable particles, the variance of the relative distance is $\sigma_r^2 = \langle(\hat{x}_1 - \hat{x}_2)^2\rangle = \langle\hat{x}_1^2\rangle + \langle\hat{x}_2^2\rangle - 2\langle\hat{x}_1\rangle\langle\hat{x}_2\rangle$ with $\hat{x}_{1(2)}$ being the position of particle 1 (2). In the numerical simulations presented in this chapter, we assume the localized initial state and thus the expectation values of $\langle\hat{x}_{1,2}\rangle$ remain to be their initial values: $\langle\hat{x}_1\rangle_0 = 0$ and $\langle\hat{x}_2\rangle_0 = 1$. In the long-time limit, we then get

$$\sigma_r^2 = \frac{32J^2 t}{\Gamma} \equiv 2D_c t, \tag{F.4}$$

which completes the derivation of the expression (4.66) of the diffusion constant D_c.

References

1. Levitov LS, Lee H, Lesovik GB (1996) Electron counting statistics and coherent states of electric current. J Math Phys 37:4845–4866
2. Belzig W, Nazarov YV (2001) Full counting statistics of electron transfer between superconductors. Phys Rev Lett 87:197006
3. Saito K, Utsumi Y (2008) Symmetry in full counting statistics, fluctuation theorem, and relations among nonlinear transport coefficients in the presence of a magnetic field. Phys Rev B 78:115429
4. Esposito M, Harbola U, Mukamel S (2009) Nonequilibrium fluctuations, fluctuation theorems, and counting statistics in quantum systems. Rev Mod Phys 81:1665–1702
5. Lu W, Ji Z, Pfeiffer L, West KW, Rimberg AJ (2003) Real-time detection of electron tunneling in a quantum dot. Nature 423:422–425
6. Fujisawa T, Hayashi T, Hirayama Y, Cheong HD, Jeong YH (2004) Electron counting of single-electron tunneling current. Appl Phys Lett 84:2343–2345
7. Bylander J, Duty T, Delsing P (2005) Current measurement by real-time counting of single electrons. Nature 434:361–364
8. Gustavsson S, Leturcq R, Simovič B, Schleser R, Ihn T, Studerus P, Ensslin K, Driscoll DC, Gossard AC (2006) Counting statistics of single electron transport in a quantum dot. Phys Rev Lett 96:076605
9. Eisaman MD, Fan J, Migdall A, Polyakov SV (2011) Invited review article: Single-photon sources and detectors. Rev Sci Instrum 82:071101
10. Altman E, Demler E, Lukin MD (2004) Probing many-body states of ultracold atoms via noise correlations. Phys Rev A 70:013603

References

11. Polkovnikov A, Altman E, Demler E (2006) Interference between independent fluctuating condensates. Proc Natl Acad Sci USA 103:6125–6129
12. Hofferberth S, Lesanovsky I, Schumm T, Gritsev V, Demler E, Schmiedmayer J (2008) Probing quantum and thermal noise in an interacting many-body system. Nat Phys 4:489–495
13. Klich I, Levitov L (2009) Quantum noise as an entanglement meter. Phys Rev Lett 102:100502
14. Lieb EH, Robinson DW (1972) The finite group velocity of quantum spin systems. Commun Math Phys 28:251–257
15. Bender CM, Boettcher S (1998) Real spectra in non-Hermitian hamiltonians having PT symmetry. Phys Rev Lett 80:5243–5246
16. Dum R, Zoller P, Ritsch H (1992) Monte Carlo simulation of the atomic master equation for spontaneous emission. Phys Rev A 45:4879–4887
17. Dalibard J, Castin Y, Mølmer K (1992) Wave-function approach to dissipative processes in quantum optics. Phys Rev Lett 68:580–583
18. Carmichael H (1993) An open system approach to quantum optics. Springer, Berlin
19. Daley AJ (2014) Quantum trajectories and open many-body quantum systems. Adv Phys 63:77–149
20. Endres M, Cheneau M, Fukuhara T, Weitenberg C, Schauß P, Gross C, Mazza L, Bañuls MC, Pollet L, Bloch I, Kuhr S (2011) Observation of correlated particle-hole pairs and string order in low-dimensional Mott insulators. Science 334:200–203
21. Fukuhara T, Hild S, Zeiher J, Schauß P, Bloch I, Endres M, Gross C (2015) Spatially resolved detection of a spin-entanglement wave in a Bose-Hubbard chain. Phys Rev Lett 115:035302
22. Bloch I, Dalibard J, Zwerger W (2008) Many-body physics with ultracold gases. Rev Mod Phys 80:885–964
23. Islam R, Ma R, Preiss PM, Tai ME, Lukin A, Rispoli M, Greiner M (2015) Measuring entanglement entropy in a quantum many-body system. Nature 528:77–83
24. Poulin D (2010) Lieb-Robinson bound and locality for general Markovian quantum dynamics. Phys Rev Lett 104:190401
25. Kliesch M, Gogolin C, Eisert J (2014) Lieb-Robinson bounds and the simulation of time-evolution of local observables in lattice systems. In: Many-electron approaches in physics, chemistry and mathematics. Springer, Cham, pp 301–318
26. Calabrese P, Cardy J (2005) Evolution of entanglement entropy in one-dimensional systems. J Stat Mech P04010
27. Calabrese P, Cardy J (2006) Time dependence of correlation functions following a quantum quench. Phys Rev Lett 96:136801
28. Kollath C, Läuchli AM, Altman E (2007) Quench dynamics and nonequilibrium phase diagram of the Bose-Hubbard model. Phys Rev Lett 98:180601
29. Collura M, Calabrese P, Essler FHL (2015) Quantum quench within the gapless phase of the spin-$\frac{1}{2}$ Heisenberg XXZ spin chain. Phys Rev B 92:125131
30. Kim H, Huse DA (2013) Ballistic spreading of entanglement in a diffusive nonintegrable system. Phys Rev Lett 111:127205
31. Kormos M, Collura M, Takács G, Calabrese P (2017) Real-time confinement following a quantum quench to a non-integrable model. Nat Phys 13:246–249
32. Fagotti M, Calabrese P (2008) Evolution of entanglement entropy following a quantum quench: analytic results for the XY chain in a transverse magnetic field. Phys Rev A 78:010306
33. Eisler V, Peschel I (2008) Entanglement in a periodic quench. Ann Phys 17:410–423
34. Geiger R, Langen T, Mazets IE, Schmiedmayer J (2014) Local relaxation and light-cone-like propagation of correlations in a trapped one-dimensional Bose gas. New J Phys 16:053034
35. Ashida Y, Ueda M (2018) Full-counting many-particle dynamics: Nonlocal and chiral propagation of correlations. Phys Rev Lett 120:185301
36. Turlapov A, Tonyushkin A, Sleator T (2003) Optical mask for laser-cooled atoms. Phys Rev A 68:023408
37. Kato T (1966) Perturbation Theory for Linear Operators. Springer, New York
38. Bravyi S, Hastings MB, Verstraete F (2006) Lieb-Robinson bounds and the generation of correlations and topological quantum order. Phys Rev Lett 97:050401

39. Bernier J-S, Tan R, Bonnes L, Guo C, Poletti D, Kollath C (2018) Light-cone and diffusive propagation of correlations in a many-body dissipative system. Phys Rev Lett 120:020401
40. Gisin N (1981) A simple nonlinear dissipative quantum evolution equation. J Phys A 14:2259
41. Brody DC, Graefe E-M (2012) Mixed-state evolution in the presence of gain and loss. Phys Rev Lett 109:230405
42. Makris KG, El-Ganainy R, Christodoulides DN, Musslimani ZH (2008) Beam dynamics in $\mathcal{PT}$ symmetric optical lattices. Phys Rev Lett 100:103904
43. Ruter CE, Makris KG, El-Ganainy R, Christodoulides DN, Segev M, Kip D (2010) Observation of parity-time symmetry in optics. Nat Phys 6:192–195
44. Regensburger A, Bersch C, Miri M-A, Onishchukov G, Christodoulides DN, Peschel U (2012) Parity-time synthetic photonic lattices. Nature 488:167–171
45. Barontini G, Labouvie R, Stubenrauch F, Vogler A, Guarrera V, Ott H (2013) Controlling the dynamics of an open many-body quantum system with localized dissipation. Phys Rev Lett 110:035302
46. Peng P, Cao W, Shen C, Qu W, Jianming W, Jiang L, Xiao Y (2016) Anti-parity-time symmetry with flying atoms. Nat Phys 12:1139–1145
47. Gao T, Estrecho E, Bliokh KY, Liew TCH, Fraser MD, Brodbeck S, Kamp M, Schneider C, Hofling S, Yamamoto Y, Nori F, Kivshar YS, Truscott AG, Dall RG, Ostrovskaya EA (2015) Observation of non-Hermitian degeneracies in a chaotic exciton-polariton billiard. Nature 526:554–558
48. Konotop VV, Yang J, Zezyulin DA (2016) Nonlinear waves in $\mathcal{PT}$-symmetric systems. Rev Mod Phys 88:035002
49. Heiss W, Harney H (2001) The chirality of exceptional points. Eur Phys J D 17:149–151
50. Longhi S (2009) Bloch oscillations in complex crystals with $\mathcal{PT}$ symmetry. Phys Rev Lett 103:123601
51. Longhi S (2010) Spectral singularities and bragg scattering in complex crystals. Phys Rev A 81:022102
52. Lin Z, Ramezani H, Eichelkraut T, Kottos T, Cao H, Christodoulides DN (2011) Unidirectional invisibility induced by $\mathcal{PT}$-symmetric periodic structures. Phys Rev Lett 106:213901
53. Scott DD, Joglekar YN (2012) $\mathcal{PT}$-symmetry breaking and ubiquitous maximal chirality in a $\mathcal{PT}$-symmetric ring. Phys Rev A 85:062105
54. Feng L, Xu Y-L, Fegadolli WS, Lu M-H, Oliveira JEB, Almeida VR, Chen Y-F, Scherer A (2013) Experimental demonstration of a unidirectional reflectionless parity-time metamaterial at optical frequencies. Nat Mater 12:108–113
55. Wiersig J (2014) Chiral and nonorthogonal eigenstate pairs in open quantum systems with weak backscattering between counterpropagating traveling waves. Phys Rev A 89:012119
56. Peschel I (2003) Calculation of reduced density matrices from correlation functions. J Phys A 36:L205
57. Popescu S, Short AJ, Winter A (2006) Entanglement and the foundations of statistical mechanics. Nat Phys 2:754–758
58. Goldstein S, Lebowitz JL, Tumulka R, Zanghì N (2006) Canonical typicality. Phys Rev Lett 96:050403
59. Manmana SR, Wessel S, Noack RM, Muramatsu A (2007) Strongly correlated fermions after a quantum quench. Phys Rev Lett 98:210405
60. Reimann P (2008) Foundation of statistical mechanics under experimentally realistic conditions. Phys Rev Lett 101:190403
61. Cramer M, Flesch A, McCulloch IP, Schollwöck U, Eisert J (2008) Exploring local quantum many-body relaxation by atoms in optical superlattices. Phys Rev Lett 101:063001
62. Bañuls MC, Cirac JI, Hastings MB (2011) Strong and weak thermalization of infinite nonintegrable quantum systems. Phys Rev Lett 106:050405
63. Short AJ, Farrelly TC (2012) Quantum equilibration in finite time. New J Phys 14:013063
64. Gogolin C, Eisert J (2016) Equilibration, thermalisation, and the emergence of statistical mechanics in closed quantum systems. Rep Prog Phys 79:056001

65. Mori T, Ikeda TN, Kaminishi E, Ueda M (2018) Thermalization and prethermalization in isolated quantum systems: a theoretical overview. J Phys B 51:112001
66. Abanin DA, De Roeck W, Huveneers F (2015) Exponentially slow heating in periodically driven many-body systems. Phys Rev Lett 115:256803
67. Kuwahara T, Mori T, Saito K (2016) Floquet-Magnus theory and generic transient dynamics in periodically driven many-body quantum systems. Ann Phys 367:96–124
68. Rigol M, Dunjko V, Olshanii M (2008) Thermalization and its mechanism for generic isolated quantum systems. Nature 452:854–858
69. Rigol M (2009) Breakdown of thermalization in finite one-dimensional systems. Phys Rev Lett 103:100403
70. Biroli G, Kollath C, Läuchli AM (2010) Effect of rare fluctuations on the thermalization of isolated quantum systems. Phys Rev Lett 105:250401
71. Steinigeweg R, Herbrych J, Prelovšek P (2013) Eigenstate thermalization within isolated spin-chain systems. Phys Rev E 87:012118
72. Steinigeweg R, Khodja A, Niemeyer H, Gogolin C, Gemmer J (2014) Pushing the limits of the eigenstate thermalization hypothesis towards mesoscopic quantum systems. Phys Rev Lett 112:130403
73. Kim H, Ikeda TN, Huse DA (2014) Testing whether all eigenstates obey the eigenstate thermalization hypothesis. Phys Rev E 90:052105
74. Beugeling W, Moessner R, Haque M (2014) Finite-size scaling of eigenstate thermalization. Phys Rev E 89:042112
75. Khodja A, Steinigeweg R, Gemmer J (2015) Relevance of the eigenstate thermalization hypothesis for thermal relaxation. Phys Rev E 91:012120
76. D'Alessio L, Kafri Y, Polkovnikov A, Rigol M (2016) From quantum chaos and eigenstate thermalization to statistical mechanics and thermodynamics. Adv Phys 65:239–362
77. Garrison JR, Grover T (2018) Does a single eigenstate encode the full hamiltonian? Phys Rev X 8:021026
78. Yoshizawa T, Iyoda E, Sagawa T (2018) Numerical large deviation analysis of the eigenstate thermalization hypothesis. Phys Rev Lett 120:200604
79. Neumann, JV (1929) Z Phys 57:30–70
80. Jaynes ET (1957) Information theory and statistical mechanics. Phys Rev 106:620–630
81. Berry MV (1977) Regular and irregular semiclassical wavefunctions. J Phys A 10:2083
82. Peres A (1984) Ergodicity and mixing in quantum theory. I Phys Rev A 30:504–508
83. Jensen RV, Shankar R (1985) Statistical behavior in deterministic quantum systems with few degrees of freedom. Phys Rev Lett 54:1879–1882
84. Srednicki M (1994) Chaos and quantum thermalization. Phys Rev E 50:888–901
85. Deutsch JM (1991) Quantum statistical mechanics in a closed system. Phys Rev A 43:2046–2049
86. Tasaki H (1998) From quantum dynamics to the canonical distribution: general picture and a rigorous example. Phys Rev Lett 80:1373–1376
87. Davies EB (1974) Markovian master equations. Commun Math Phys 39:91–110
88. Spohn H, Lebowitz JL (1978) Irreversible thermodynamics for quantum systems weakly coupled to thermal reservoirs. Adv Chem Phys 109–142
89. Alicki R (1979) The quantum open system as a model of the heat engine. J Phys A 12:L103
90. Caldeira AO, Leggett AJ (1983) Path integral approach to quantum Brownian motion. Phys A 121:587–616
91. Weiss U (1999) Quantum dissipative systems. World Scientific, Singapore
92. Breuer H-P, Petruccione F (2002) The theory of open quantum systems. Oxford University Press, Oxford, England
93. Kinoshita T, Wenger T, Weiss DS (2006) A quantum Newton's cradle. Nature 440:900–903
94. Trotzky S, Chen Y-A, Flesch A, McCulloch IP, Schollwöck U, Eisert J, Bloch I (2012) Probing the relaxation towards equilibrium in an isolated strongly correlated one-dimensional Bose gas. Nat Phys 8:325–330

95. Gring M, Kuhnert M, Langen T, Kitagawa T, Rauer B, Schreitl M, Mazets I, Smith DA, Demler E, Schmiedmayer J (2012) Relaxation and prethermalization in an isolated quantum system. Science 337:1318–1322
96. Kaufman AM, Tai ME, Lukin A, Rispoli M, Schittko R, Preiss PM, Greiner M (2016) Quantum thermalization through entanglement in an isolated many-body system. Science 353:794–800
97. Tang Y, Kao W, Li K-Y, Seo S, Mallayya K, Rigol M, Gopalakrishnan S, Lev BL (2018) Thermalization near integrability in a dipolar quantum Newton's cradle. Phys Rev X 8:021030
98. Rigol M, Dunjko V, Yurovsky V, Olshanii M (2007) Relaxation in a completely integrable many-body quantum system: an ab initio study of the dynamics of the highly excited states of 1d lattice hard-core bosons. Phys Rev Lett 98:050405
99. Calabrese P, Essler FHL, Fagotti M (2011) Quantum quench in the transverse-field Ising chain. Phys Rev Lett 106:227203
100. Fagotti M, Essler FHL (2013) Reduced density matrix after a quantum quench. Phys Rev B 87:245107
101. Sotiriadis S, Calabrese P (2014) Validity of the gge for quantum quenches from interacting to noninteracting models. J Stat Mech 2014:P07024
102. Essler FHL, Fagotti M (2016) Quench dynamics and relaxation in isolated integrable quantum spin chains. J Stat Mech 2016:064002
103. Vidmar L, Rigol M (2016) Generalized Gibbs ensemble in integrable lattice models. J Stat Mech 2016:064007
104. Lange F, Lenarcic Z, Rosch A (2017) Pumping approximately integrable systems. Nat Commun 8:15767
105. Nandkishore R, Huse DA (2015) Many-body localization and thermalization in quantum statistical mechanics. Ann Rev Cond Matt Phys 6:15–38
106. Vasseur R, Moore JE (2016) Nonequilibrium quantum dynamics and transport: from integrability to many-body localization. J Stat Mech 2016:064010
107. Pichler H, Daley AJ, Zoller P (2010) Nonequilibrium dynamics of bosonic atoms in optical lattices: Decoherence of many-body states due to spontaneous emission. Phys Rev A 82:063605
108. Garrahan JP, Lesanovsky I (2010) Thermodynamics of quantum jump trajectories. Phys Rev Lett 104:160601
109. Lee TE, Reiter F, Moiseyev N (2014) Entanglement and spin squeezing in non-Hermitian phase transitions. Phys Rev Lett 113:250401
110. Lee TE, Chan C-K (2014) Heralded magnetism in non-Hermitian atomic systems. Phys Rev X 4:041001
111. Lee MD, Ruostekoski J (2014) Classical stochastic measurement trajectories: Bosonic atomic gases in an optical cavity and quantum measurement backaction. Phys Rev A 90:023628
112. Pedersen MK, Sørensen JJWH, Tichy MC, Sherson JF (2014) Many-body state engineering using measurements and fixed unitary dynamics. New J Phys 16:113038
113. Elliott TJ, Kozlowski W, Caballero-Benitez SF, Mekhov IB (2015) Multipartite entangled spatial modes of ultracold atoms generated and controlled by quantum measurement. Phys Rev Lett 114:113604
114. Wade ACJ, Sherson JF, Mølmer K (2015) Squeezing and entanglement of density oscillations in a Bose-Einstein condensate. Phys Rev Lett 115:060401
115. Dhar S, Dasgupta S, Dhar A, Sen D (2015) Detection of a quantum particle on a lattice under repeated projective measurements. Phys Rev A 91:062115
116. Ashida Y, Ueda M (2016) Precise multi-emitter localization method for fast super-resolution imaging. Opt Lett 41:72–75
117. Dhar S, Dasgupta S (2016) Measurement-induced phase transition in a quantum spin system. Phys Rev A 93:050103
118. Mazzucchi G, Kozlowski W, Caballero-Benitez SF, Elliott TJ, Mekhov IB (2016) Quantum measurement-induced dynamics of many-body ultracold bosonic and fermionic systems in optical lattices. Phys Rev A 93:023632

119. Ashida Y, Ueda M (2017) Multiparticle quantum dynamics under real-time observation. Phys Rev A 95:022124
120. Schemmer M, Johnson A, Photopoulos R, Bouchoule I (2017) Monte Carlo wave-function description of losses in a one-dimensional Bose gas and cooling to the ground state by quantum feedback. Phys Rev A 95:043641
121. Ashida Y, Furukawa S, Ueda M (2017) Parity-time-symmetric quantum critical phenomena. Nat Commun 8:15791
122. Mehboudi M, Lampo A, Charalambous C, Correa LA, Garcia-March MA, Lewenstein M (2018) Using polarons for sub-nK quantum non-demolition thermometry in a Bose-Einstein condensate. arXiv:1806.07198
123. Buffoni L, Solfanelli A, Verrucchi P, Cuccoli A, Campisi M (2018) Quantum measurement cooling. arXiv:1806.07814
124. Soerensen JJ, Dalgaard M, Kiilerich AH, Moelmer K, Sherson J (2018) Quantum control with measurements and quantum Zeno dynamics. arXiv:1806.07793
125. Beige A, Braun D, Tregenna B, Knight PL (2000) Quantum computing using dissipation to remain in a decoherence-free subspace. Phys Rev Lett 85:1762–1765
126. Kraus B, Büchler HP, Diehl S, Kantian A, Micheli A, Zoller P (2008) Preparation of entangled states by quantum Markov processes. Phys Rev A 78:042307
127. Diehl S, Micheli A, Kantian A, Kraus B, Büchler HP, Zoller P (2008) Quantum states and phases in driven open quantum systems with cold atoms. Nat Phys 4:878–883
128. Verstraete F, Wolf MM, Cirac JI (2009) Quantum computation and quantum-state engineering driven by dissipation. Nat Phys 5:633–636
129. Yi W, Diehl S, Daley AJ, Zoller P (2012) Driven-dissipative many-body pairing states for cold fermionic atoms in an optical lattice. New J Phys 14:055002
130. Poletti D, Bernier J-S, Georges A, Kollath C (2012) Interaction-induced impeding of decoherence and anomalous diffusion. Phys Rev Lett 109:045302
131. Poletti D, Barmettler P, Georges A, Kollath C (2013) Emergence of glasslike dynamics for dissipative and strongly interacting bosons. Phys Rev Lett 111:195301
132. Joshi C, Larson J, Spiller TP (2016) Quantum state engineering in hybrid open quantum systems. Phys Rev A 93:043818
133. Gong Z, Ashida Y, Kawabata K, Takasan K, Higashikawa S, Ueda M (2018) Topological phases of non-Hermitian systems. Phys Rev X 8:031079
134. Patil YS, Chakram S, Vengalattore M (2015) Measurement-induced localization of an ultracold lattice gas. Phys Rev Lett 115:140402
135. Labouvie R, Santra B, Heun S, Ott H (2016) Bistability in a driven-dissipative superfluid. Phys Rev Lett 116:235302
136. Rauer B, Grišins P, Mazets IE, Schweigler T, Rohringer W, Geiger R, Langen T, Schmiedmayer J (2016) Cooling of a one-dimensional Bose gas. Phys Rev Lett 116:030402
137. Lüschen HP, Bordia P, Hodgman SS, Schreiber M, Sarkar S, Daley AJ, Fischer MH, Altman E, Bloch I, Schneider U (2017) Signatures of many-body localization in a controlled open quantum system. Phys Rev X 7:011034
138. Zhu B, Gadway B, Foss-Feig M, Schachenmayer J, Wall ML, Hazzard KRA, Yan B, Moses SA, Covey JP, Jin DS, Ye J, Holland M, Rey AM (2014) Suppressing the loss of ultracold molecules via the continuous quantum Zeno effect. Phys Rev Lett 112:070404
139. Schachenmayer J, Pollet L, Troyer M, Daley AJ (2014) Spontaneous emission and thermalization of cold bosons in optical lattices. Phys Rev A 89:011601
140. Ashida Y, Saito K, Ueda M (2018) Thermalization and heating dynamics in open generic many-body systems. Phys Rev Lett 121:170402
141. Marino J, Silva A (2012) Relaxation, prethermalization, and diffusion in a noisy quantum Ising chain. Phys Rev B 86:060408
142. Pichler H, Schachenmayer J, Daley AJ, Zoller P (2013) Heating dynamics of bosonic atoms in a noisy optical lattice. Phys Rev A 87:033606
143. Chenu A, Beau M, Cao J, del Campo A (2017) Quantum simulation of generic many-body open system dynamics using classical noise. Phys Rev Lett 118:140403

144. Banchi L, Burgarth D, Kastoryano MJ (2017) Driven quantum dynamics: will it blend? Phys Rev X 7:041015
145. Hayden P, Nezami S, Qi X-L, Thomas N, Walter M, Yang Z (2016) Holographic duality from random tensor networks. J High Energy Phys 2016:9
146. Nahum A, Ruhman J, Vijay S, Haah J (2017) Quantum entanglement growth under random unitary dynamics. Phys Rev X 7:031016
147. Nahum A, Vijay S, Haah J (2018) Operator spreading in random unitary circuits. Phys Rev X 8:021014
148. von Keyserlingk CW, Rakovszky T, Pollmann F, Sondhi SL (2018) Operator hydrodynamics, otocs, and entanglement growth in systems without conservation laws. Phys Rev X 8:021013
149. Sünderhauf C, Pérez-Garcia D, Huse DA, Schuch N, Cirac JI (2018) Localisation with random time-periodic quantum circuits. arXiv:1805.08487
150. Knap M (2018) Entanglement production and information scrambling in a noisy spin system. arXiv:1806.04686
151. Hernández-Santana S, Riera A, Hovhannisyan KV, Perarnau-Llobet M, Tagliacozzo L, Acin A (2015) Locality of temperature in spin chains. New J Phys 17:085007
152. Kliesch M, Gogolin C, Kastoryano MJ, Riera A, Eisert J (2014) Locality of temperature. Phys Rev X 4:031019
153. Santos LF, Polkovnikov A, Rigol M (2011) Entropy of isolated quantum systems after a quench. Phys Rev Lett 107:040601
154. Polkovnikov A (2011) Microscopic diagonal entropy and its connection to basic thermodynamic relations. Ann Phys 326:486–499
155. Ikeda TN, Sakumichi N, Polkovnikov A, Ueda M (2015) The second law of thermodynamics under unitary evolution and external operations. Ann Phys 354:338–352
156. Thirring W (2002) Quantum mathematical physics. Springer, Berlin
157. Neuenhahn C, Marquardt F (2012) Thermalization of interacting fermions and delocalization in fock space. Phys Rev E 85:060101
158. Mori T (2016) Weak eigenstate thermalization with large deviation bound. arXiv:1609.09776
159. Gemelke N, Zhang X, Hung C-L, Chin C (2009) In situ observation of incompressible Mott-insulating domains in ultracold atomic gases. Nature 460:995–998
160. Patil YS, Chakram S, Aycock LM, Vengalattore M (2014) Nondestructive imaging of an ultracold lattice gas. Phys Rev A 90:033422
161. Preiss PM, Ma R, Tai ME, Simon J, Greiner M (2015) Quantum gas microscopy with spin, atom-number, and multilayer readout. Phys Rev A 91:041602
162. Ashida Y, Ueda M (2015) Diffraction-unlimited position measurement of ultracold atoms in an optical lattice. Phys Rev Lett 115:095301
163. Alberti A, Robens C, Alt W, Brakhane S, Karski M, Reimann R, Widera A, Meschede D (2016) Super-resolution microscopy of single atoms in optical lattices. New J Phys 18:053010
164. Wigley PB, Everitt PJ, Hardman KS, Hush MR, Wei CH, Sooriyabandara MA, Manju P, Close JD, Robins NP, Kuhn CCN (2016) Non-destructive shadowgraph imaging of ultra-cold atoms. Opt Lett 41:4795–4798
165. Wiseman HM, Milburn GJ (1993) Quantum theory of optical feedback via homodyne detection. Phys Rev Lett 70:548–551
166. Sayrin C, Dotsenko I, Zhou X, Peaudecerf B, Rybarczyk T, Gleyzes S, Rouchon P, Mirrahimi M, Amini H, Brune M, Raimond J-M, Haroche S (2011) Real-time quantum feedback prepares and stabilizes photon number states. Nature 477:73–77
167. Bakr WS, Gillen JI, Peng A, Fölling S, Greiner M (2009) A quantum gas microscope for detecting single atoms in a Hubbard-regime optical lattice. Nature 462:74–77
168. Bloch I, Dalibard J, Nascimbène S (2012) Quantum simulations with ultracold quantum gases. Nat Phys 8:267–276
169. Caves CM, Milburn GJ (1987) Quantum-mechanical model for continuous position measurements. Phys Rev A 36:5543–5555
170. Diósi L (1988) Continuous quantum measurement and its formalism. Phys Lett A 129:419–423

171. Diósi L (1988) Localized solution of a simple nonlinear quantum langevin equation. Phys Lett A 132:233–236
172. Belavkin VP, Staszewski P (1992) Nondemolition observation of a free quantum particle. Phys Rev A 45:1347–1356
173. Gagen MJ, Wiseman HM, Milburn GJ (1993) Continuous position measurements and the quantum Zeno effect. Phys Rev A 48:132–142
174. Bassi A, Lochan K, Satin S, Singh TP, Ulbricht H (2013) Models of wave-function collapse, underlying theories, and experimental tests. Rev Mod Phys 85:471–527
175. Keßler S, Holzner A, McCulloch IP, von Delft J, Marquardt F (2012) Stroboscopic observation of quantum many-body dynamics. Phys Rev A 85:011605
176. Yanay Y, Mueller EJ (2014) Heating from continuous number density measurements in optical lattices. Phys Rev A 90:023611
177. Barchielli A, Gregoratti M (2012) Quantum measurements in continuous time, non-Markovian evolutions and feedback. Phil Trans R Soc A 370:5364–5385
178. Gullans M, Tiecke TG, Chang DE, Feist J, Thompson JD, Cirac JI, Zoller P, Lukin MD (2012) Nanoplasmonic lattices for ultracold atoms. Phys Rev Lett 109:235309
179. Romero-Isart O, Navau C, Sanchez A, Zoller P, Cirac JI (2013) Superconducting vortex lattices for ultracold atoms. Phys Rev Lett 111:145304
180. González-Tudela A, Hung C-L, Chang DE, Cirac IJ, Kimble JH (2015) Subwavelength vacuum lattices and atom-atom interactions in two-dimensional photonic crystals. Nat Photon 9:320–325
181. Nascimbene S, Goldman N, Cooper NR, Dalibard J (2015) Dynamic optical lattices of subwavelength spacing for ultracold atoms. Phys Rev Lett 115:140401
182. Wiseman H, Milburn G (2010) Quantum measurement and control. Cambridge University Press, Cambridge, England
183. Diósi L (1989) Models for universal reduction of macroscopic quantum fluctuations. Phys Rev A 40:1165–1174
184. Weitenberg C, Endres M, Sherson JF, Cheneau M, Schauß P, Fukuhara T, Bloch I, Kuhr S (2011) Single-spin addressing in an atomic Mott insulator. Nature 471:319–324
185. Fukuhara T, Kantian A, Endres M, Cheneau M, Schauß P, Hild S, Bellem D, Schollwöck U, Giamarchi T, Gross C, Bloch I, Kuhr S (2013) Quantum dynamics of a mobile spin impurity. Nat Phys 9:235–241
186. Fukuhara T, Schausz P, Endres M, Hild S, Cheneau M, Bloch I, Gross C (2013) Microscopic observation of magnon bound states and their dynamics. Nature 502:76–79
187. Schauß P, Zeiher J, Fukuhara T, Hild S, Cheneau M, Macri T, Pohl T, Bloch I, Gross C (2015) Crystallization in Ising quantum magnets. Science 347:1455–1458
188. Ghirardi GC, Rimini A, Weber T (1986) Unified dynamics for microscopic and macroscopic systems. Phys Rev D 34:470–491
189. Ghirardi GC, Pearle P, Rimini A (1990) Markov processes in hilbert space and continuous spontaneous localization of systems of identical particles. Phys Rev A 42:78–89
190. Schwarzer E, Haken H (1972) The moments of the coupled coherent and incoherent motion of excitons. Phys Lett A 42:317–318
191. Li J, Harter AK, Liu J, de Melo L, Joglekar YN, Luo L (2016) Observation of parity-time symmetry breaking transitions in a dissipative Floquet system of ultracold atoms. arXiv:1608.05061
192. Seiberg N (1990) Notes on quantum Liouville theory and quantum gravity. Prog Theor Phys Supp 102:319–349

Chapter 5
Quantum Spin in an Environment

Abstract A quantum system that strongly correlates with an external world is ubiquitous in nature. There, one has to deal with the strong entanglement between the system and the environment. Ideally, this can be achieved by explicitly taking into account all the degrees of freedom of the environment rather than eliminating them as done in the master-equation approach. In this chapter, we move on to studies of in- and out-of-equilibrium physics in such a strongly correlated open quantum system by focusing on its most fundamental paradigm, that is, a quantum impurity. We develop a versatile and efficient theoretical approach to study ground-state and out-of-equilibrium properties of generic quantum spin-impurity systems. In particular, we introduce a new canonical transformation that can completely disentangle the localized spin and the environmental degrees of freedom. After introducing our general variational formalism for a fermionic environment, we benchmark our approach by comparing it with other numerical and analytical results in both in- and out-of-equilibrium regimes. We also reveal new types of nonequilibrium dynamics such as long-time crossover dynamics mimicking nonmonotonic renormalization group flows, which has been difficult to study in other methods. We propose a possible experiment to test the predicted dynamics by using quantum gas microscopy. We also generalize our approach to a bosonic environment and apply it to study a novel type of strongly correlated systems realized in the state-of-the-art experiments of Rydberg molecules, which have been otherwise challenging to analyze in previous theoretical approaches.

Keywords Open quantum systems · Quantum impurity · Strongly correlated systems · Kondo effect · Rydberg atoms

5.1 Introduction

Understanding physics of quantum systems open to an external world has now become one of the most important problems in physics from both fundamental and application-oriented points of views. Open quantum systems can be studied based on the master equation or the quantum trajectory approach if the correlation between the system and an external observer or environment is vanishingly small and thus the information flow is unidirectional (i.e., the Born-Markov approximation is valid), as typically realized in quantum-optical setups. Yet, in a variety of physical systems ranging from solid-state materials to artificial quantum simulators, it is ubiquitous that a quantum system can be strongly correlated with an external world. In such a case, the strong system-environment entanglement invalidates the Markovian description and poses a theoretical challenge. A quantum impurity is the most fundamental class of such strongly correlated open quantum systems.

A variety of problems corresponding to quantum impurities have been at the forefront of condensed matter physics, atomic, molecular and optical (AMO) physics, and quantum information. For instance, they have played central roles to understand decoherence in quantum nanodevices [1–3], strongly correlated physics in heavy fermion materials [4–8], transport phenomena in mesoscopic systems [9–20], and laid the basis of dynamical mean-field theory [21] (DMFT). To understand physics of these vastly different systems, it is essential to reveal the role of entanglement between the impurity and the environment; this is best illustrated by the formation of the Kondo-singlet state [22], which is a many-body bound state between a localized spin-1/2 impurity and itinerant fermions in the environment.

The equilibrium properties of quantum impurities, especially for the Kondo problem, are now theoretically well understood from the perturbative renormalization group (RG) [23], Wilson's numerical renormalization group (NRG) [24–30] and the exact solution via the Bethe ansatz [31–33]. However, its nonequilibrium property is still a challenging and active area of research in both experiments [34–39] and theory [40–80]. Previous theoretical works include the real-time Monte Carlo calculations [40–44], the perturbative RG analyses [45–49], the Hamiltonian RG approach [50–52], the coherent-state expansion [53–55], the density-matrix renormalization group (DMRG) [56–63], the time-dependent NRG (TD-NRG) [64–70], the time evolving decimation (TEBD) [71, 72], and analytical solutions [73–80]. Despite the rich variety of the methods, revealing the long-time many-body dynamics remains a major challenge. Due to high computational cost, the previous theoretical approaches become increasingly challenging at longer times. For instance, numerical methods based on the matrix-product states (MPS) become very challenging in long-time regimes due to the requirement of an exponentially large bond dimension [81]. Also, in TD-NRG, the logarithmic discretization has been argued to cause artifacts in long-time regimes [82]. Another challenge is to reveal many-body spatiotemporal dynamics of the environmental degrees of freedom, as they are often replaced by a simplified effective bath or integrated out so that their microscopic details are lost. Furthermore, the previous methods are restricted to only a particular class of quantum impurity problems, where a type of interactions between the impurity and surrounding particles is specified and bath particles are often assumed to move ballistically. It is not obvious how one could extend the previous techniques to more generic situations including strong disorder and nonlocal couplings to the impurity, which are relevant to state-of-the-art experimental systems realized in AMO physics, condensed matter physics and quantum information. These challenges call for a new theoretical approach to solving quantum impurity problems.

5.1 Introduction

A variational approach is one of the most powerful and successful methods to solve quantum many-body problems. Its central guiding principle is to design quantum states that can capture essential physics behind the problems in an efficient way with avoiding the exponential complexity of the exact wavefunction. For example, a Bose-Einstein condensate (BEC) in both in- and out-of-equilibrium regimes can be well described by a coherent state [83], and low-temperature equilibrium physics of one-dimensional local Hamiltonians can be efficiently described by the MPS [84]. The essential physical feature in the former (latter) is the presence of the off-diagonal long-range order (the small amount of entanglement), which allows one to develop an efficient description avoiding the exponential complexity.

Over the last several decades, a variational approach has also proven useful to understand quantum impurity systems. Studies in this direction date back to Tomonaga's meson-nucleon theory in 1947 [86]. Lee, Low and Pines (LLP) then applied it to analyzing a polaron, which is a mobile spinless impurity dressed by phonon excitations [87]. The essential feature in quantum impurity problems is the presence of the entanglement between the impurity and the environment. The key idea by LLP is to take into account this entanglement by introducing the unitary transformation (now known as the LLP transformation)

$$\hat{U}_{\text{LLP}} = e^{-i\hat{\mathbf{x}} \cdot \hat{\mathbf{P}}_{\text{bath}}}, \tag{5.1}$$

where $\hat{\mathbf{x}}$ is the position operator of the impurity and $\hat{\mathbf{P}}_{\text{bath}}$ is the total momentum operator of phonons. In the laboratory frame, the total momentum $\hat{\mathbf{P}}_{\text{tot}} = \hat{\mathbf{p}} + \hat{\mathbf{P}}_{\text{bath}}$ of the impurity and the environment is conserved. Yet, via the LLP transformation, the conserved quantity turns out to be the momentum operator $\hat{\mathbf{p}}$ of the impurity:

$$\hat{U}_{\text{LLP}}^{\dagger} \hat{\mathbf{P}}_{\text{tot}} \hat{U}_{\text{LLP}} = \hat{\mathbf{p}}. \tag{5.2}$$

This indicates that, after the transformation, one can take $\hat{\mathbf{p}}$ as a classical variable, i.e, the impurity dynamics is completely frozen. The LLP transformation then leads to the following variational states

$$|\Psi_{\text{tot}}(\xi)\rangle = \hat{U}_{\text{LLP}} |\mathbf{p}\rangle |\Psi_{\text{bath}}(\xi)\rangle, \tag{5.3}$$

where $|\mathbf{p}\rangle$ is a momentum eigenstate of the impurity, and $|\Psi_{\text{bath}}(\xi)\rangle$ is a bath wavefunction with variational parameters ξ. For instance, one can choose a bath state $|\Psi_{\text{bath}}\rangle$ to be the product of coherent states as in the original papers [86, 87]. While such an efficiently parametrizable environment wavefunction $|\Psi_{\text{bath}}(\xi)\rangle$ is factorizable and thus does not incorporate the entanglement, the variational states (5.3) can efficiently represent the impurity-environment entanglement via the transformation (5.1). This class of variational states has proven very successful to reveal

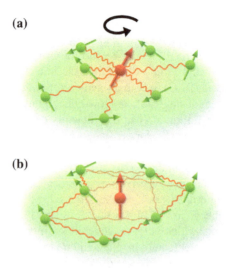

Fig. 5.1 Disentangling canonical transformation in quantum impurity problems. **a** The impurity spin interacts with environmental particles in an arbitrary way, leading to a strong impurity-environment entanglement. **b** The disentangling canonical transformation allows one to move on to the "corotating" frame of the impurity, where the impurity dynamics is frozen at the expense of generating an effective interaction among bath particles. Reproduced from Fig. 1 of Ref. [85]. Copyright © 2018 by the American Physical Society

ground-state properties and nonequilibrium dynamics of mobile impurities and laid the foundation for successive studies in polaron physics[1] [88–107].

The aim of this chapter is to generalize this variational approach to a localized spin-impurity model (SIM), which is a fundamental paradigm of strongly correlated open quantum systems (see Fig. 5.1a). Introducing a new disentangling transformation, we present a versatile and efficient variational approach to studying in- and out-of-equilibrium properties of generic SIM. In Sect. 5.2, we present the disentangling transformation that completely decouples the impurity and the environment (see Fig. 5.1b for an illustration). Different from the LLP transformation (5.1), constructing the canonical transformation for SIM is not obvious due to the noncommutativity of the spin operators. We explain a general strategy to obtain the disentangling transformations and apply it to single-impurity systems, two-impurity systems and the Anderson model. In Sect. 5.3, combining the canonical transformation with fermionic Gaussian states, we introduce a family of variational states that can efficiently capture nontrivial impurity-environment entanglement. We present a set of nonlinear equations of motions to reveal ground-state properties and nonequilibrium dynamics of generic SIM. In Sect. 5.4, applying the present theoretical approach to the anisotropic Kondo model, we benchmark

---

[1] In the next chapter, we will apply this approach to discuss new types of nonequilibrium dynamics of magnetic polarons.

5.1 Introduction

it with the MPS, the Bethe-ansatz solution and the known nonperturbative scalings. Besides the ability to efficiently represent nontrivial impurity-environment correlations, our approach reveals new types of long-time nonequilibrium dynamics that have been difficult to obtain in the previous approaches. In Sect. 5.5, applying our approach to the two-lead Kondo model, we analyze its transport properties and nonequilibrium spatiotemporal dynamics and demonstrate that the obtained results are consistent with the previous studies [58, 69, 80, 108]. In Sect. 5.6, we propose a possible experiment to realize the anisotropic Kondo model in ultracold gases. In particular, we argue that the long-time spatiotemporal dynamics predicted in our theory can readily be tested by the current techniques of quantum gas microscopy. In Sect. 5.7, we generalize our theoretical approach to a bosonic environment. This can be done by combining the canonical transformation introduced in Sect. 5.2 with the bosonic Gaussian states. We apply the variational approach to a novel type of quantum impurity problems, namely, the Rydberg central spin problem [109–111], in which the impurity spin interacts with surrounding bosons via long-range couplings. We present numerical results on the absorption spectrum and the magnetization dynamics, which can be directly tested by the state-of-the-art experiments in ultracold gases. Finally, we conclude this chapter with an outlook in Sect. 5.8.

5.2 Disentangling Canonical Transformation

We here introduce a disentangling canonical transformation of SIM. Let us first outline the general idea here. The central point is that a spin-1/2 operator can completely be decoupled from the environmental degrees of freedom if a generic Hamiltonian $\hat{H}$ has some parity symmetry $\hat{\mathbb{P}}$, i.e., $[\hat{H}, \hat{\mathbb{P}}] = 0$ with $\hat{\mathbb{P}}^2 = 1$. The reason is that, as both the spin-1/2 operator and the parity operator have the same set of eigenvalues $\{1, -1\}$, we can always construct a unitary transformation $\hat{U}$ that transforms the conserved parity $\hat{\mathbb{P}}$ to a spin-1/2 operator via (compare Eq. (5.4) with the LLP transformation (5.2))

$$\hat{U}^\dagger \hat{\mathbb{P}} \hat{U} = \mathbf{n} \cdot \hat{\boldsymbol{\sigma}}, \qquad (5.4)$$

where $\hat{\boldsymbol{\sigma}} = (\hat{\sigma}^x, \hat{\sigma}^y, \hat{\sigma}^z)^\mathrm{T}$ is a vector of the Pauli matrices and $\mathbf{n}$ is a three-dimensional real unit vector. As a result, the transformed Hamiltonian commutes with the spin-1/2 operator $[\hat{U}^\dagger \hat{H} \hat{U}, \mathbf{n} \cdot \hat{\boldsymbol{\sigma}}] = 0$, meaning that the spin dynamics is frozen, or said differently, the transformed Hamiltonian conditioned on a classical number $\mathbf{n} \cdot \hat{\boldsymbol{\sigma}} = \pm 1$ only contains the environmental degrees of freedom. While a construction of such a unitary transformation $\hat{U}$ is not obvious due to the noncommutativity of the spin operators, we find its explicit and simple form (see Eq. (5.13) below). We can then combine this disentangling transformation with an efficiently parametrizable bath wavefunction and introduce a family of variational states to efficiently encode the nontrivial entanglement between the localized spin-1/2 impurity and the environment.

5.2.1 Disentangling a Single Spin-1/2 Impurity and an Environment

We first apply the disentangling transformation to a localized spin-1/2 impurity coupled to a fermionic environment:

$$\hat{H} = \hat{H}_{\text{bath}} + \hat{H}_{\text{imp}} + \hat{H}_{\text{int}}, \tag{5.5}$$

where

$$\hat{H}_{\text{bath}} = \sum_{lm\alpha} \hat{\Psi}^\dagger_{l\alpha} h_{lm} \hat{\Psi}_{m\alpha} \tag{5.6}$$

is a noninteracting fermionic bath Hamiltonian, $\hat{\Psi}^\dagger_{l\alpha}(\hat{\Psi}_{l\alpha})$ creates (annihilates) a fermion with an energy level $l = 1, 2, \ldots, N_f$ and a spin-z component $\alpha = \uparrow, \downarrow$, and h_{lm} is the (l, m) component of an arbitrary $N_f \times N_f$ Hermitian matrix corresponding to a single-particle bath Hamiltonian. The impurity Hamiltonian

$$\hat{H}_{\text{imp}} = -h_z \hat{s}^z_{\text{imp}} \tag{5.7}$$

describes a magnetic field h_z acting on the spin-1/2 impurity operator $\hat{s}^\gamma_{\text{imp}} = \hat{\sigma}^\gamma_{\text{imp}}/2$ with $\gamma = x, y, z$ and

$$\hat{H}_{\text{int}} = \hat{\mathbf{s}}_{\text{imp}} \cdot \hat{\boldsymbol{\Sigma}} \tag{5.8}$$

represents a generic impurity-bath interaction, where $\hat{\boldsymbol{\Sigma}}$ is the bath-spin density operator including the interaction couplings

$$\hat{\Sigma}^\gamma = \frac{1}{2} \sum_{lm\alpha\beta} g^\gamma_{lm} \hat{\Psi}^\dagger_{l\alpha} \sigma^\gamma_{\alpha\beta} \hat{\Psi}_{m\beta}, \tag{5.9}$$

where the impurity-bath couplings g^γ_{lm} labeled by $\gamma = x, y, z$ are determined by arbitrary $N_f \times N_f$ Hermitian matrices; they can be long-range and anisotropic. Paradigmatic models with the interaction Hamiltonian that can be written as Eq. (5.8) include, e.g., the central spin model [112] and the Kondo-type Hamiltonians [22].

The canonical transformation can be constructed on the basis of the parity symmetry hidden in the total Hamiltonian (5.5):

$$\hat{\mathbb{P}} = \hat{\sigma}^z_{\text{imp}} \hat{\mathbb{P}}_{\text{bath}}, \tag{5.10}$$

where $\hat{\mathbb{P}}_{\text{bath}}$ is a bath parity defined by

$$\hat{\mathbb{P}}_{\text{bath}} = e^{(i\pi/2)(\sum_l \hat{\sigma}^z_l + \hat{N})} = e^{i\pi \hat{N}_\uparrow}. \tag{5.11}$$

5.2 Disentangling Canonical Transformation

Here, $\hat{\sigma}_l^\gamma = \sum_{\alpha\beta} \hat{\Psi}_{l\alpha}^\dagger \sigma_{\alpha\beta}^\gamma \hat{\Psi}_{l\beta}$ with $\gamma = x, y, z$ is a spin density for a bath mode l, $\hat{N}$ is the total bath particle number, and $\hat{N}_\uparrow$ is the number of spin-up fermions. It is easy to check $\hat{\mathbb{P}}_{\text{bath}}^2 = \hat{\mathbb{P}}^2 = 1$. The parity symmetry $[\hat{H}, \hat{\mathbb{P}}] = 0$ follows from the fact that $\hat{H}$ is invariant with respect to the rotation around the z axis by π, transforming impurity and bath spins as $\hat{\mathbb{P}}^{-1} \hat{\sigma}^{x,y} \hat{\mathbb{P}} = -\hat{\sigma}^{x,y}$ while keeping $\hat{\mathbb{P}}^{-1} \hat{\sigma}^z \hat{\mathbb{P}} = \hat{\sigma}^z$. We use this conserved parity to introduce the disentangling transformation $\hat{U}$ obeying

$$\hat{U}^\dagger \hat{\mathbb{P}} \hat{U} = \hat{U}^\dagger \left(\hat{\sigma}_{\text{imp}}^z \hat{\mathbb{P}}_{\text{bath}} \right) \hat{U} = \hat{\sigma}_{\text{imp}}^x. \tag{5.12}$$

Then, in the transformed frame, the impurity spin turns into a conserved quantity.[2] We find an explicit and simple form of a unitary transformation $\hat{U}$ satisfying the condition (5.12) as

$$\hat{U} = \exp\left[\frac{i\pi}{4} \hat{\sigma}_{\text{imp}}^y \hat{\mathbb{P}}_{\text{bath}} \right] = \frac{1}{\sqrt{2}} \left(1 + i\hat{\sigma}_{\text{imp}}^y \hat{\mathbb{P}}_{\text{bath}} \right). \tag{5.13}$$

Applying this unitary transformation to the Hamiltonian, we obtain the transformed Hamiltonian $\hat{\tilde{H}} = \hat{U}^\dagger \hat{H} \hat{U}$ as

$$\hat{\tilde{H}} = \hat{H}_{\text{bath}} - h_z \hat{s}_{\text{imp}}^x \hat{\mathbb{P}}_{\text{bath}} + \hat{s}_{\text{imp}}^x \hat{\Sigma}^x + \hat{\mathbb{P}}_{\text{bath}} \left(-\frac{i\hat{\Sigma}^y}{2} + \hat{s}_{\text{imp}}^x \hat{\Sigma}^z \right). \tag{5.14}$$

It obeys $[\hat{\tilde{H}}, \hat{\sigma}_{\text{imp}}^x] = 0$ by construction. The impurity can be taken as a conserved classical number $\sigma_{\text{imp}}^x = \pm 1$. We remark that the disentangling transformation is also applicable to a bosonic bath (see Sect. 5.7) and to an interacting bath having an arbitrary bath spin. For the sake of simplicity, we omit the magnetic-field term acting on the bath.

The decoupling of the impurity came at the expense of generating interactions among environmental particles. This can be inferred from the second and the last terms on the right-hand side of Eq. (5.14). Physically, the emergent bath interactions indicate effective spin-exchange interactions between bath fermions mediated by the impurity spin. This situation is analogous to the generation of a nonlocal phonon coupling after the LLP transformation in the mobile spinless impurity [87]. We stress that the decoupling of the spin-1/2 operator has a broad applicability, as it is based on the elemental parity symmetry in the Hamiltonian. For example, we will demonstrate its extension to two-impurity systems and the single-impurity Anderson model later.

We remark that, at zero field $h_z = 0$ with an even particle number N, there are two degenerate energy sectors associated with the conserved quantity $\sigma_{\text{imp}}^x = \pm 1$. This can be shown by noticing the fact that the two sectors in Eq. (5.14) relate to each other by $\hat{U}_{\text{bath}}^y = e^{(i\pi/2) \sum_l \hat{\sigma}_l^y}$, which transforms the bath spins via $\hat{\sigma}_l^{x,z} \to -\hat{\sigma}_l^{x,z}$. To

---

[2] While we here specify $\mathbf{n} = (1, 0, 0)^T$ in Eq. (5.4), other choices result in the same class of variational states.

the best of our knowledge, this exact degeneracy in a generic case of SIM has not previously been discussed although there have been works on some exactly solvable cases [31–33, 113].

Employing the unitary transformation, we will introduce the following family of variational states (compare it with Eq. (5.3) for the LLP transformation)

$$|\Psi\rangle = \hat{U}|\pm_x\rangle_{\text{imp}}|\Psi_{\text{bath}}\rangle$$
$$= |\uparrow\rangle_{\text{imp}}\hat{\mathbb{P}}_{\pm}|\Psi_{\text{bath}}\rangle \pm |\downarrow\rangle_{\text{imp}}\hat{\mathbb{P}}_{\mp}|\Psi_{\text{bath}}\rangle, \quad (5.15)$$

where $|\Psi_{\text{bath}}\rangle$ is a variational bath wavefunction, $|\pm_x\rangle_{\text{imp}}$ is the eigenstate of the impurity spin with $\hat{\sigma}^x_{\text{imp}} = \pm 1$, and $\hat{\mathbb{P}}_{\pm} = (1 \pm \hat{\mathbb{P}}_{\text{bath}})/2$ is the projection onto the subspace with even or odd bath parity. As we detail later, the variational states can capture the strong impurity-environment entanglement in a very efficient manner. It is worthwhile to make a remark on the unbiasedness of the variational states; we use no a priori knowledge about physical properties of the underlying impurity model (e.g., the formation of the Kondo-singlet state). The unitary transformation $\hat{U}$ contains neither values of parameters in the Hamiltonian nor variational parameters. Although we will specify $|\Psi_{\text{bath}}\rangle$ as the Gaussian states later, a family of variational states (5.15) is still unbiased because the Gaussian states include all the possible two-particle excitations in an unbiased way.

5.2.2 Disentangling Two Spin-1/2 Impurities and an Environment

It is possible to generalize the disentangling canonical transformation to two spin-1/2 impurities if the system possesses two independent parity symmetries. The main point is that we can then apply the disentangling transformation twice to change the two conserved parity operators into the two spin-1/2 operators. To be concrete, we consider the following Hamiltonian:

$$\hat{H}_{\text{two}} = \hat{H}_{\text{bath}} + \hat{H}_{\text{direct}} + \hat{H}_{\text{imp-bath}}, \quad (5.16)$$

where the bath Hamiltonian $\hat{H}_{\text{bath}}$ and the direct-spin interaction $\hat{H}_{\text{direct}}$ are given by

$$\hat{H}_{\text{bath}} = \sum_{lm\alpha} \hat{\Psi}^\dagger_{l\alpha} h_{lm} \hat{\Psi}_{m\alpha}, \quad \hat{H}_{\text{direct}} = \sum_\gamma K^\gamma \hat{s}^\gamma_{\text{imp},1} \hat{s}^\gamma_{\text{imp},2}, \quad (5.17)$$

where K^γ are couplings between the two impurity spins $\hat{s}^\gamma_{\text{imp},i} = \hat{\sigma}^\gamma_{\text{imp},i}/2$ with $i = 1, 2$ and $\gamma = x, y, z$. The Hamiltonian $\hat{H}_{\text{imp-bath}}$ represents a general interaction between the two impurities and the environment:

5.2 Disentangling Canonical Transformation

$$\hat{H}_{\text{imp-bath}} = \sum_i \hat{\mathbf{s}}_{\text{imp},i} \cdot \hat{\boldsymbol{\Sigma}}_i, \tag{5.18}$$

where we introduce the bath-spin density operator including the impurity-bath couplings $g^{\gamma}_{lm,i}$ for each impurity spin $i = 1, 2$ as

$$\hat{\Sigma}^{\gamma}_i = \frac{1}{2} \sum_{lm\alpha\beta} g^{\gamma}_{lm,i} \hat{\Psi}^{\dagger}_{l\alpha} \sigma^{\gamma}_{\alpha\beta} \hat{\Psi}_{m\beta}. \tag{5.19}$$

To decouple the first impurity, we use the unitary transformation

$$\hat{U}^{\text{two}}_1 = \frac{1}{\sqrt{2}} \left(1 + i\hat{\sigma}^{y}_{\text{imp},1} \hat{\sigma}^{z}_{\text{imp},2} \hat{\mathbb{P}}_z \right), \tag{5.20}$$

where we denote the parity operator acting on the bath as

$$\hat{\mathbb{P}}_{\gamma} = e^{(i\pi/2)(\sum_l \hat{\sigma}^{\gamma}_l + \hat{N})} \quad (\gamma = x, y, z). \tag{5.21}$$

We observe that the original Hamiltonian is invariant under the π rotation about the z axis, leading to the parity conservation

$$[\hat{H}_{\text{two}}, \hat{\mathbb{P}}_1] = 0, \quad \hat{\mathbb{P}}_1 = \hat{\sigma}^{z}_{\text{imp},1} \hat{\sigma}^{z}_{\text{imp},2} \hat{\mathbb{P}}_z. \tag{5.22}$$

The unitary transformation $\hat{U}^{\text{two}}_1$ can map this conserved parity (5.22) onto the first impurity operator:

$$\hat{U}^{\dagger\text{two}}_1 \hat{\mathbb{P}}_1 \hat{U}^{\text{two}}_1 = \hat{\sigma}^{x}_{\text{imp},1}. \tag{5.23}$$

Thus, the first impurity is disentangled after the first unitary transformation. We can explicitly obtain the transformed Hamiltonian $\hat{\tilde{H}}_{1,\text{two}} = \hat{U}^{\dagger\text{two}}_1 \hat{H}_{\text{two}} \hat{U}^{\text{two}}_1$ as

$$\hat{\tilde{H}}_{1,\text{two}} = \hat{H}_{\text{bath}} + K^x \hat{s}^x_{\text{imp},1} \hat{s}^x_{\text{imp},2} + \hat{\mathbb{P}}_z \left(-K^y \hat{s}^x_{\text{imp},2} + K^z \hat{s}^x_{\text{imp},1} \right)$$

$$+ \hat{s}^x_{\text{imp},1} \hat{\Sigma}^x_1 + \hat{s}^z_{\text{imp},2} \hat{\mathbb{P}}_z \left(-\frac{i\hat{\Sigma}^y_1}{2} + \hat{s}^x_{\text{imp},1} \hat{\Sigma}^z_1 \right) + \hat{\mathbf{s}}_{\text{imp},2} \cdot \hat{\boldsymbol{\Sigma}}_2, \tag{5.24}$$

which commutes with $\hat{\sigma}^x_{\text{imp},1}$ as expected from the construction. To decouple the second impurity, we use the following second unitary transformation:

$$\hat{U}^{\text{two}}_2 = \frac{1}{\sqrt{2}} \left(1 + i\hat{\sigma}^y_{\text{imp},2} \hat{\mathbb{P}}_x \right). \tag{5.25}$$

We observe that the Hamiltonian $\hat{\tilde{H}}_{1,\text{two}}$ in Eq. (5.24) is still invariant under the π rotation about the x axis, and this leads to another parity conservation

$$[\hat{\tilde{H}}_{1,\text{two}}, \hat{\mathbb{P}}_2] = 0, \quad \hat{\mathbb{P}}_2 = \hat{\sigma}^x_{\text{imp},2}\hat{\mathbb{P}}_x. \tag{5.26}$$

We can use the second unitary transformation $\hat{U}_2^{\text{two}}$ to map the second conserved parity $\hat{\mathbb{P}}_2$ onto the second impurity operator as

$$\hat{U}_2^{\dagger\text{two}}\hat{\mathbb{P}}_2\hat{U}_2^{\text{two}} = -\hat{\sigma}^z_{\text{imp},2}. \tag{5.27}$$

We note that $\hat{U}_2^{\text{two}}$ commutes with the first impurity operator and thus $\hat{\sigma}^x_{\text{imp},1}$ remains to be a conserved quantity. The final Hamiltonian $\hat{\tilde{H}}_{2,\text{two}} = \hat{U}_2^{\dagger\text{two}}\hat{\tilde{H}}_{1,\text{two}}\hat{U}_2^{\text{two}}$ after the decoupling of the two impurities is

$$\begin{aligned}\hat{\tilde{H}}_{2,\text{two}} &= \hat{H}_{\text{bath}} - K^x\hat{\mathbb{P}}_x\hat{s}^x_{\text{imp},1}\hat{s}^z_{\text{imp},2} + (-i)^N K^y\hat{\mathbb{P}}_y\hat{s}^z_{\text{imp},2} + K^z\hat{\mathbb{P}}_z\hat{s}^x_{\text{imp},1} \\ &+ \hat{s}^x_{\text{imp},1}\hat{\Sigma}^x_1 + \hat{s}^z_{\text{imp},2}\hat{\Sigma}^z_2 + \hat{s}^z_{\text{imp},2}\hat{\mathbb{P}}_z\left(-\frac{i\hat{\Sigma}^y_1}{2} + \hat{s}^x_{\text{imp},1}\hat{\Sigma}^z_1\right) \\ &+ \hat{\mathbb{P}}_x\left(-\frac{i\hat{\Sigma}^y_2}{2} - \hat{s}^z_{\text{imp},2}\hat{\Sigma}^x_2\right),\end{aligned} \tag{5.28}$$

which commutes with the two impurities $\hat{s}^x_{\text{imp},1}$ and $\hat{s}^z_{\text{imp},2}$. Here, we use the relation $\hat{\mathbb{P}}_z\hat{\mathbb{P}}_x = (-i)^N\hat{\mathbb{P}}_y$. To see the entanglement structure encoded by the decoupling transformation, let us specify the sector $\hat{\sigma}^x_{\text{imp},1} = +1$ and $\hat{\sigma}^z_{\text{imp},2} = +1$ for the sake of concreteness. The variational state is then represented by

$$\begin{aligned}|\Psi\rangle &= \hat{U}_2^{\text{two}}\hat{U}_1^{\text{two}}|+_x\rangle_1|\uparrow\rangle_2|\Psi_{\text{bath}}\rangle \\ &= \hat{U}_2^{\text{two}}\left[|\uparrow\rangle_1|\uparrow\rangle_2\frac{(1+\mathbb{P}_z)}{2}|\Psi_{\text{bath}}\rangle + |\downarrow\rangle_1|\uparrow\rangle_2\frac{(1-\mathbb{P}_z)}{2}|\Psi_{\text{bath}}\rangle\right] \\ &= \frac{1}{2}|+_x\rangle_1|\uparrow\rangle_2|\Psi_{\text{bath}}\rangle + \frac{1}{2}|-_x\rangle_1|\uparrow\rangle_2\hat{\mathbb{P}}_z|\Psi_{\text{bath}}\rangle - \frac{1}{2}|+_x\rangle_1|\downarrow\rangle_2\hat{\mathbb{P}}_x|\Psi_{\text{bath}}\rangle \\ &- \frac{(-i)^N}{2}|-_x\rangle_1|\downarrow\rangle_2\hat{\mathbb{P}}_y|\Psi_{\text{bath}}\rangle.\end{aligned} \tag{5.29}$$

We remark that this canonical transformation has already found applications in analyzing the string-breaking dynamics of quark-antiquark pairs in the prototypical lattice gauge theory [114].

5.2.3 Disentangling the Single-Impurity Anderson Model

We can also generalize the transformation to the single-impurity Anderson model (Ads) such that the impurity degrees of freedom can be partially disentangled from the environment. We start from the Hamiltonian

$$\hat{H}_{\text{Ads}} = \sum_{lm} h_{lm} \hat{\Psi}^\dagger_{l\sigma} \hat{\Psi}_{m\sigma} + V\left(\hat{\Psi}^\dagger_{0\sigma} \hat{\Psi}_{d\sigma} + \text{H.c.}\right) + \epsilon_d \sum_\sigma \hat{\Psi}^\dagger_{d\sigma} \hat{\Psi}_{d\sigma} + U \hat{\Psi}^\dagger_{d\uparrow} \hat{\Psi}^\dagger_{d\downarrow} \hat{\Psi}_{d\downarrow} \hat{\Psi}_{d\uparrow}, \quad (5.30)$$

where $\hat{\Psi}_{d\sigma}$ ($\hat{\Psi}^\dagger_{d\sigma}$) annihilates (creates) a fermion at the impurity orbital with spin σ, V is the hybridization term, ϵ_d is an energy of the impurity orbital, and U denotes the on-site Coulomb interaction. We represent the impurity in the following four-dimensional basis:

$$\{|0\rangle_d, |\uparrow\rangle_d, |\downarrow\rangle_d, |\uparrow\downarrow\rangle_d\}, \quad (5.31)$$

and use the spin representation of $\hat{\Psi}_{d\sigma}$ in this basis as

$$\hat{\Psi}_{d\uparrow} = e^{i\pi \hat{N}} I_2 \otimes \sigma^+, \quad \hat{\Psi}_{d\downarrow} = e^{i\pi \hat{N}} \sigma^+ \otimes \sigma^z, \quad (5.32)$$

where we introduce $e^{i\pi \hat{N}}$ with $\hat{N} = \Sigma_{l\sigma} \hat{\Psi}^\dagger_{l\sigma} \hat{\Psi}_{l\sigma}$ to maintain the anticommutation relation $\{\hat{\Psi}_{d\sigma}, \hat{\Psi}_{l\sigma'}\} = 0$. We observe that the Hamiltonian (5.30) has the parity symmetry

$$[\hat{H}_{\text{Ads}}, \hat{\mathbb{P}}_{\text{Ads}}] = 0, \quad \hat{\mathbb{P}}_{\text{Ads}} = \hat{S}^z \hat{\mathbb{P}}_z, \quad (5.33)$$

where $\hat{\mathbb{P}}_z = e^{(i\pi/2)(\Sigma_l \hat{\sigma}^z_l + \hat{N})} = e^{i\pi \Sigma_l \hat{\Psi}^\dagger_{l\uparrow} \hat{\Psi}_{l\uparrow}}$ and we define an operator $\hat{S}^z = -I_2 \otimes \sigma^z$ acting on the impurity. This symmetry corresponds to the invariance of $\hat{H}_{\text{Ads}}$ under the following transformation:

$$(\hat{\Psi}_{0\uparrow}, \hat{\Psi}_{d\uparrow}) \to (-\hat{\Psi}_{0\uparrow}, -\hat{\Psi}_{d\uparrow}), \quad (\hat{\Psi}_{0\downarrow}, \hat{\Psi}_{d\downarrow}) \to (\hat{\Psi}_{0\downarrow}, \hat{\Psi}_{d\downarrow}). \quad (5.34)$$

In the similar manner as in the transformation (5.13) for the localized single-impurity case, we introduce the unitary operator

$$\hat{U}_{\text{Ads}} = \frac{1}{\sqrt{2}}(1 + i \hat{S}^y \hat{\mathbb{P}}_z), \quad (5.35)$$

where we define $\hat{S}^y = -\sigma^x \otimes \sigma^y$. This transformation can map the conserved parity $\hat{\mathbb{P}}_{\text{Ads}}$ onto the impurity operator as follows:

$$\hat{U}^\dagger_{\text{Ads}} \hat{\mathbb{P}}_{\text{Ads}} \hat{U}_{\text{Ads}} = i \hat{S}^z \hat{S}^y = \hat{X}, \quad (5.36)$$

where we introduce the impurity operator $\hat{X} = \sigma^x \otimes \sigma^x$. The explicit form of the transformed Hamiltonian $\hat{\tilde{H}}_{\text{Ads}} = \hat{U}^\dagger_{\text{Ads}} \hat{H} \hat{U}_{\text{Ads}}$ can be given by

$$\hat{\tilde{H}}_{\text{Ads}} = \hat{H}_{\text{bath}} + \frac{V e^{i\pi \hat{N}}}{2} \left(\hat{\Psi}^\dagger_{0\uparrow} I_2 \otimes \sigma^x + i\hat{\Psi}^\dagger_{0\downarrow} \sigma^y \otimes \sigma^z - \hat{\mathbb{P}}_z \hat{\Psi}^\dagger_{0\uparrow} \sigma^x \otimes I_2 \right.$$
$$\left. - \hat{\mathbb{P}}_z \hat{\Psi}^\dagger_{0\downarrow} I_2 \otimes \sigma^x + \text{H.c.} \right)$$
$$+ \epsilon_d + \frac{U}{4} + \frac{U}{4}\hat{Z} + \frac{1}{4}(2\epsilon_d + U)(\hat{X} - \hat{Y})\hat{\mathbb{P}}_z, \qquad (5.37)$$

where we define $\hat{Y} = \sigma^y \otimes \sigma^y$ and $\hat{Z} = \sigma^z \otimes \sigma^z$. The conservation of the parity $[\hat{H}_{\text{Ads}}, \hat{\mathbb{P}}_{\text{Ads}}] = 0$ in the transformed frame now turns out to be the partial decoupling of the impurity $[\hat{\tilde{H}}_{\text{Ads}}, \hat{X}] = 0$. Since $\hat{X}$ has two degenerate eigenvalues ± 1, the transformed Hamiltonian $\hat{\tilde{H}}_{\text{Ads}}$ has two sectors $\hat{\tilde{H}}_P$ corresponding to the impurity subspace $\{|P_c\rangle, |P_s\rangle\}$ spanned by two eigenstates of $\hat{X}$ having the eigenvalues $P = \pm 1$:

$$|\pm_c\rangle = \frac{1}{\sqrt{2}}(|0\rangle_d \pm |\uparrow\downarrow\rangle_d), \quad |\pm_s\rangle = \frac{1}{\sqrt{2}}(|\uparrow\rangle_d \pm |\downarrow\rangle_d). \qquad (5.38)$$

The states $|\pm_c\rangle$ ($|\pm_s\rangle$) represent the charge (spin) sector of the impurity. Introducing the fermionic operator

$$\hat{f}_P = e^{i\pi \hat{N}} |P_s\rangle \langle P_c|, \qquad (5.39)$$

we can simplify the transformed Hamiltonian as follows:

$$\hat{\tilde{H}}_P = \hat{H}_{\text{bath}} + \frac{V}{2}\left[\hat{\Psi}^\dagger_{0\uparrow}(\hat{f}^\dagger_P + \hat{f}_P) - P\hat{\Psi}^\dagger_{0\downarrow}(\hat{f}_P - \hat{f}^\dagger_P) - P\hat{\mathbb{P}}_z \hat{\Psi}^\dagger_{0\uparrow}(\hat{f}^\dagger_P + \hat{f}_P) \right.$$
$$\left. - \hat{\mathbb{P}}_z \hat{\Psi}^\dagger_{0\downarrow}(\hat{f}^\dagger_P + \hat{f}_P) + \text{H.c.} \right]$$
$$+ \epsilon_d + \frac{U}{2}\hat{f}^\dagger_P \hat{f}_P + \left(\epsilon_d + \frac{U}{2}\right) P\hat{\mathbb{P}}_z \hat{f}^\dagger_P \hat{f}_P. \qquad (5.40)$$

The Hamiltonian $\hat{\tilde{H}}_P$ shows that in a large $U \gg V$ limit the small charge excitation can be adiabatically eliminated, resulting in the low-energy physics described by the spin fluctuation as realized in the Kondo model.

5.3 Efficient Variational Approach to Generic Spin-Impurity Systems

Here and henceforth, we focus on a class of single spin-impurity models (SIM) introduced in Sect. 5.2.1. To solve generic SIM efficiently, we need versatile variational states capturing essential physics behind impurity problems while they have to be simple enough such that calculations are still tractable. In this section, we achieve this aim by combining the disentangling transformation $\hat{U}$ with the fermionic Gaussian states [115–117] in which the number of parameters grows only quadratically with the system size. While the Gaussian states cannot represent nonfactorizable correlations, the transformation $\hat{U}$ can generate the entanglement between the impurity and the environment, making the variational states sufficiently flexible to capture essential physics of SIM.

5.3.1 Fermionic Gaussian States

The fermionic Gaussian states are defined by density operators represented by exponentials of a quadratic form of fermionic operators. For the sake of simplicity, we focus on the pure fermionic Gaussian states and use them to approximate the environmental wavefunction in the transformed frame, where the impurity dynamics is decoupled from the bath. It is convenient to introduce the Majorana operators

$$\hat{\psi}_{1,l\alpha} = \hat{\Psi}_{l\alpha}^{\dagger} + \hat{\Psi}_{l\alpha},$$
$$\hat{\psi}_{2,l\alpha} = i(\hat{\Psi}_{l\alpha}^{\dagger} - \hat{\Psi}_{l\alpha}), \qquad (5.41)$$

which satisfy the anticommutation relation

$$\{\hat{\psi}_{\xi,l\alpha}, \hat{\psi}_{\eta,m\beta}\} = 2\delta_{\xi\eta}\delta_{lm}\delta_{\alpha\beta}, \quad (\xi, \eta = 1, 2). \qquad (5.42)$$

One can fully characterize a pure fermionic Gaussian state $|\Psi_G\rangle$ by a $4N_f \times 4N_f$ real-antisymmetric matrix Γ, which gives the covariance matrix [115–117]:

$$\Gamma = \frac{i}{2}\left\langle [\hat{\boldsymbol{\psi}}, \hat{\boldsymbol{\psi}}^{\mathrm{T}}] \right\rangle_G, \qquad (5.43)$$

where $\langle \cdots \rangle_G$ is an expectation value with respect to $|\Psi_G\rangle$. A Gaussian state is pure if and only if its covariance matrix satisfies $\Gamma^2 = -\mathrm{I}_{4N_f}$, where I_d is the $d \times d$ identity matrix. We denote a vector of the Majorana operators by $\hat{\boldsymbol{\psi}} = (\hat{\boldsymbol{\psi}}_1, \hat{\boldsymbol{\psi}}_2)^{\mathrm{T}}$, in which a row vector $\hat{\boldsymbol{\psi}}_\xi$ with $\xi = 1, 2$ is ordered as

$$\hat{\boldsymbol{\psi}}_\xi = (\hat{\psi}_{\xi,1\uparrow}, \ldots, \hat{\psi}_{\xi,N_f\uparrow}, \hat{\psi}_{\xi,1\downarrow}, \ldots, \hat{\psi}_{\xi,N_f\downarrow}). \qquad (5.44)$$

The pure Gaussian state can be expressed as

$$|\Psi_G\rangle = e^{\frac{1}{4}\hat{\psi}^T X \hat{\psi}}|0\rangle \equiv \hat{U}_G|0\rangle, \tag{5.45}$$

where X is a $4N_f \times 4N_f$ real-antisymmetric matrix that is related to Γ as

$$\Gamma = -\Xi \sigma (\Xi)^T, \tag{5.46}$$

in which we denote $\Xi = e^X$ and $\sigma = i\sigma^y \otimes I_{2N_f}$. It will be also useful to define the $2N_f \times 2N_f$ correlation matrix:

$$\Gamma_f = \langle \hat{\Psi}^\dagger \hat{\Psi} \rangle_G \tag{5.47}$$

in terms of Dirac fermions

$$\hat{\Psi} = (\hat{\Psi}_{1\uparrow}, \ldots, \hat{\Psi}_{N_f\uparrow}, \hat{\Psi}_{1\downarrow}, \ldots, \hat{\Psi}_{N_f\downarrow}). \tag{5.48}$$

Since the Gaussian states are factorizable, using Wick's theorem, all the higher-order correlations of fermionic operators can be obtained from the covariance matrix Γ.

5.3.2 Variational Time Evolution of the Covariance Matrix

Since the impurity dynamics is decoupled from the bath in the transformed frame, only the environmental degrees of freedom, which are denoted as $|\Psi_{\text{bath}}\rangle$ in Eq. (5.15), evolve in time. Its exact evolution can be approximated by projecting it onto the manifold spanned by the family of variational states, which are chosen to be the fermionic Gaussian states here. The variational time-evolution equations can be derived based on the time-dependent variational principle [117–120]. Integrating the variational imaginary- and real-time evolutions, one can analyze ground-state properties and nonequilibrium dynamics of a general SIM, respectively.

We first derive a variational equation for the imaginary-time evolution. Its exact form is defined by

$$|\Psi_{\text{bath}}(\tau)\rangle = \frac{e^{-\hat{H}\tau}|\Psi_{\text{bath}}(0)\rangle}{\left\| e^{-\hat{H}\tau}|\Psi_{\text{bath}}(0)\rangle \right\|}. \tag{5.49}$$

If the overlap between the initial state $|\Psi_{\text{bath}}(0)\rangle$ and the ground state is nonzero, this evolution will provide the ground state in $\tau \to \infty$. A differential form of the equation of motion is given by

$$\frac{d}{d\tau}|\Psi_{\text{bath}}(\tau)\rangle = -(\hat{H} - E)|\Psi_{\text{bath}}(\tau)\rangle, \tag{5.50}$$

5.3 Efficient Variational Approach to Generic Spin-Impurity Systems

where we denote $E = \langle \Psi_{\text{bath}}(\tau)|\hat{\tilde{H}}|\Psi_{\text{bath}}(\tau)\rangle$. In our consideration, we approximate $|\Psi_{\text{bath}}\rangle$ by the Gaussian state $|\Psi_G\rangle$ whose covariance matrix Γ corresponds to the time-dependent variational parameter. Its imaginary-time evolution equation is given by minimizing the deviation ϵ from the exact imaginary-time evolution in the variational manifold:

$$\epsilon = \left\| \frac{d}{d\tau}|\Psi_G(\tau)\rangle + (\hat{\tilde{H}} - E_{\text{var}})|\Psi_G(\tau)\rangle \right\|^2, \tag{5.51}$$

where $E_{\text{var}} = \langle \Psi_G(\tau)|\hat{\tilde{H}}|\Psi_G(\tau)\rangle$. The minimization condition can formally be written as the following differential equation:

$$\frac{d}{d\tau}|\Psi_G(\tau)\rangle = -\hat{P}_{\partial\Gamma}(\hat{\tilde{H}} - E_{\text{var}})|\Psi_G(\tau)\rangle, \tag{5.52}$$

where $\hat{P}_{\partial\Gamma}$ projects quantum states onto the subspace spanned by tangent vectors of the variational manifold. The left-hand side of Eq. (5.52) leads to

$$\frac{d}{d\tau}|\Psi_G(\tau)\rangle = \hat{U}_G \left(\frac{1}{4} : \hat{\psi}^T \Xi^T \frac{d\Xi}{d\tau} \hat{\psi} : + \frac{i}{4}\text{Tr}\left[\Xi^T \frac{d\Xi}{d\tau} \Gamma \right] \right) |0\rangle, \tag{5.53}$$

where $: \;:$ denotes the normal order. The right-hand side of Eq. (5.52) is given by

$$-(\hat{\tilde{H}} - E_{\text{var}})|\Psi_G(\tau)\rangle = -\hat{U}_G \left(\frac{i}{4} : \hat{\psi}^T \Xi^T \mathcal{H} \Xi \hat{\psi} : + \delta\hat{O} \right) |0\rangle, \tag{5.54}$$

where $\delta\hat{O}$ represents the higher-order terms of $\hat{\psi}$ which are projected out by $\hat{P}_{\partial\Gamma}$ in Eq. (5.52), and

$$\mathcal{H} = 4\frac{\delta E_{\text{var}}}{\delta \Gamma} \tag{5.55}$$

is the functional derivative of the variational energy. From Eqs. (5.46), (5.53) and (5.54), the variational imaginary time-evolution equation is obtained as

$$\frac{d\Gamma}{d\tau} = -\mathcal{H} - \Gamma \mathcal{H} \Gamma. \tag{5.56}$$

The variational energy E_{var} monotonically decreases in this imaginary-time evolution and the variational ground state will be obtained in the limit of $\tau \to \infty$. In this limit, the deviation ϵ in Eq. (5.51) is equivalent to the energy variance, which can be used to assess the accuracy of the reached variational ground state.

Following the same procedure, the variational equation for the real-time evolution

$$|\Psi_{\text{bath}}(t)\rangle = e^{-i\hat{\tilde{H}}t}|\Psi_{\text{bath}}(0)\rangle \tag{5.57}$$

can also be obtained. The projection

$$\frac{d}{dt}|\Psi_G(t)\rangle = -i\hat{P}_{\partial\Gamma}\hat{\tilde{H}}|\Psi_G(t)\rangle \tag{5.58}$$

results in the following variational real-time evolution equation:

$$\frac{d\Gamma}{dt} = \mathcal{H}\Gamma - \Gamma\mathcal{H}. \tag{5.59}$$

5.3.3 General Expression of the Functional Derivative

To integrate the equations of motions in Eqs. (5.56) and (5.59), it is necessary to derive the analytical expression of the functional derivative $\mathcal{H} = 4\delta E_{\text{var}}/\delta\Gamma$. The variational energy $E_{\text{var}} = \langle\hat{\tilde{H}}\rangle_G$ in the transformed frame can be obtained from Eq. (5.14) as

$$E_{\text{var}} = \sum_{lm\alpha} h_{lm}(\Gamma_f)_{l\alpha,m\alpha} - \frac{h_z}{2}\sigma^x_{\text{imp}}\langle\hat{\mathbb{P}}_{\text{bath}}\rangle_G + \frac{1}{4}\sum_{lm\alpha\beta} g^x_{lm}\sigma^x_{\text{imp}}\sigma^x_{\alpha\beta}(\Gamma_f)_{l\alpha,m\beta}$$
$$+ \frac{1}{4}\sum_{lm\alpha\beta}\left(-ig^y_{lm}\sigma^y + g^z_{lm}\sigma^x_{\text{imp}}\sigma^z\right)_{\alpha\beta}(\Gamma^P_f)_{l\alpha,m\beta}, \tag{5.60}$$

where the impurity spin is conditioned on a classical number $\sigma^x_{\text{imp}} = \pm 1$ and we introduce the $2N_f \times 2N_f$ correlation matrix containing the parity operator as

$$\Gamma^P_f = \langle\hat{\mathbb{P}}_{\text{bath}}\hat{\Psi}^\dagger\hat{\Psi}\rangle_G. \tag{5.61}$$

We can calculate $\langle\hat{\mathbb{P}}_{\text{bath}}\rangle_G$ and Γ^P_f as

$$\langle\hat{\mathbb{P}}_{\text{bath}}\rangle_G = (-1)^{N_f}\text{Pf}\left[\frac{\Gamma_F}{2}\right], \tag{5.62}$$

$$\Gamma^P_f = \frac{1}{4}\langle\hat{\mathbb{P}}_{\text{bath}}\rangle_G \Sigma_z(I_{2N_f}, -iI_{2N_f})\Upsilon^{-1}(\Gamma\sigma - I_{4N_f})\begin{pmatrix}I_{2N_f}\\iI_{2N_f}\end{pmatrix}, \tag{5.63}$$

where Pf denotes the Pfaffian of a real antisymmetric matrix and we introduce the matrices

5.3 Efficient Variational Approach to Generic Spin-Impurity Systems

$$\Gamma_F = \sqrt{I_{4N_f} + \Lambda}\,\Gamma\sqrt{I_{4N_f} + \Lambda} - (I_{4N_f} - \Lambda)\sigma, \tag{5.64}$$

$$\Upsilon = I_{4N_f} + \frac{1}{2}\left(\Gamma\sigma - I_{4N_f}\right)\left(I_{4N_f} + \Lambda\right), \tag{5.65}$$

which are defined by $\Lambda = I_2 \otimes \Sigma_z$ and $\Sigma_z = \sigma^z \otimes I_{N_f}$. We remark that σ is introduced below Eq. (5.46). We can represent the quadratic part of the Hamiltonian (5.14) in the Majorana basis as $i\hat{\psi}^T \mathcal{H}_0 \hat{\psi}/4$ with

$$\mathcal{H}_0 = i\sigma^y \otimes [I_2 \otimes h_{lm} + (\sigma^x_{\mathrm{imp}}/4)\,\sigma^x \otimes g^x_{lm}], \tag{5.66}$$

where h_{lm} and g^x_{lm} are understood to be $N_f \times N_f$ real matrices. Let us also introduce the matrix

$$\mathcal{G} = -i\sigma^y \otimes g^y_{lm} + \sigma^x_{\mathrm{imp}}\sigma^z \otimes g^z_{lm}. \tag{5.67}$$

We can then simplify the expression of the mean energy (5.60) as

$$E_{\mathrm{var}} = \frac{1}{4}\mathrm{Tr}\left(\mathcal{H}_0^T \Gamma\right) - \frac{h_z}{2}\sigma^x_{\mathrm{imp}}\langle\hat{\mathbb{P}}_{\mathrm{bath}}\rangle_G + \frac{1}{4}\mathrm{Tr}\left(\mathcal{G}^T \Gamma_f^P\right). \tag{5.68}$$

Its functional derivative $\mathcal{H} = 4\delta E_{\mathrm{var}}/\delta\Gamma$ is given by

$$\mathcal{H} = \mathcal{H}_0 + \frac{\delta}{\delta\Gamma}\left[-2h_z\sigma^x_{\mathrm{imp}}\langle\hat{\mathbb{P}}_{\mathrm{bath}}\rangle_G + \mathrm{Tr}(\mathcal{G}^T\Gamma_f^P)\right]. \tag{5.69}$$

Doing some calculations to take the derivatives of $\langle\hat{\mathbb{P}}_{\mathrm{bath}}\rangle_G$ and Γ_f^P with respect to Γ (c.f. Ref. [121]), we can rewrite Eq. (5.69) to get the analytical form of $\mathcal{H}$:

$$\mathcal{H} = \mathcal{H}_0 + \left[h_z\sigma^x_{\mathrm{imp}}\langle\hat{\mathbb{P}}_{\mathrm{bath}}\rangle_G - \frac{1}{2}\mathrm{Tr}(\mathcal{G}^T\Gamma_f^P)\right]\mathcal{P} - \frac{i}{4}\langle\hat{\mathbb{P}}_{\mathrm{bath}}\rangle_G\,\mathcal{A}\left[\mathcal{V}\mathcal{G}^T\Sigma_z\mathcal{V}^\dagger\right], \tag{5.70}$$

where $\mathcal{A}[M] = (M - M^T)/2$ indicates the matrix antisymmetrization and we denote

$$\mathcal{P} = \sqrt{I_{4N_f} + \Lambda}\,\Gamma_F^{-1}\sqrt{I_{4N_f} + \Lambda}, \tag{5.71}$$

$$\mathcal{V} = (\Upsilon^T)^{-1}\begin{pmatrix} I_{2N_f} \\ iI_{2N_f} \end{pmatrix}. \tag{5.72}$$

Integrating the variational Eqs. (5.56) and (5.59) by using a general expression (5.70) of the functional derivative $\mathcal{H}$, one can study ground-state properties and nonequilibrium dynamics of SIM on demand.

5.4 Application to the Anisotropic Kondo Model

In this section, the variational approach is applied to the anisotropic Kondo model. We benchmark our approach with the MPS [84], the Bethe-ansatz solution [31–33, 122–124] and the analytically known nonperturbative scalings.

5.4.1 Kondo Problem

Among many spin-impurity systems, the Kondo model is one of the most important and well-understood problems. It describes physics of a localized spin impurity coupled to a fermionic environment:

$$\hat{H}_K = -t_h \sum_{l=-L}^{L} \left(\hat{c}_{l\alpha}^\dagger \hat{c}_{l+1\alpha} + \text{h.c.} \right) + \frac{1}{4} \sum_\gamma J_\gamma \hat{\sigma}_{\text{imp}}^\gamma \hat{c}_{0\alpha}^\dagger \sigma_{\alpha\beta}^\gamma \hat{c}_{0\beta}, \tag{5.73}$$

where $\hat{c}_{l\alpha}^\dagger$ and $\hat{c}_{l\alpha}$ are fermionic creation and annhilation operators with spin α and position l.[3] The localized spin-1/2 impurity $\hat{\sigma}_{\text{imp}}^\gamma$ interacts with fermions at site $l = 0$ via the anisotropic couplings $J_{x,y} = J_\perp$ and $J_z = J_\parallel$. Here and henceforth, we choose the unit $t_h = 1$ and tacitly assume that repeated indices over α, β are to be summed.

The ground-state physics of the anisotropic Kondo problem (5.73) can be understood from the perturbative renormalization group (RG) analysis, which is valid for small couplings $J_\perp$ and $J_\parallel$. The leading terms of the RG equations for the dimensionless couplings $j_{\parallel,\perp} = \rho_F J_{\parallel,\perp}$ normalized by the density of states at the Fermi energy $\rho_F = 1/(2\pi t_h)$ are [23]

$$\frac{dj_\parallel}{dl} = j_\perp^2, \tag{5.74}$$

$$\frac{dj_\perp}{dl} = j_\parallel j_\perp. \tag{5.75}$$

The corresponding RG flows are plotted in Fig. 5.2. We note that the RG phase diagram is equivalent to that of the (conventional) sine-Gordon model with the correspondence $j_\parallel \leftrightarrow 2 - K$ and $j_\perp \leftrightarrow 2g_r$. The model exhibits a quantum phase transition [1] between the antiferromagnetic (AFM) phase and the ferromagnetic (FM) phase. In the AFM phase, the ground state forms the singlet state between the impurity spin and bath fermions (known as the Kondo singlet state), resulting in zero magnetization $\langle \hat{\sigma}_{\text{imp}}^z \rangle = 0$. The FM phase exhibits the triplet formation leading to a nonzero impurity magnetization. In this phase, the parameter $j_\perp$ flows to zero in the

---

[3]Generalizations to higher dimensions can be straightforwardly done by modifying the hopping matrix.

5.4 Application to the Anisotropic Kondo Model

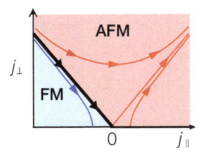

Fig. 5.2 Renormalization group flows in the anisotropic Kondo model. The impurity spin forms the triplet state in the ferromagnetic (FM) regime with $j_\parallel < 0$ and $|j_\parallel| > |j_\perp|$, where $j_\perp$ scales to zero and thus the impurity is decoupled in the low-energy limit. In the other regime, the spin forms the singlet state and develops the antiferromagnetic (AFM) correlation with the couplings flowing to large values. The structure of RG flows is equivalent to that of the Berezinskii-Kosterlitz-Thouless transition

low-energy limit and thus the impurity dynamics is eventually decoupled from the environment.

Let us apply our general variational approach to the anisotropic Kondo model. We note that, in Eq. (5.73), the impurity is coupled to only the symmetric bath modes. We can thus choose a basis of the fermionic modes $\hat{\Psi}$ used in the variational calculation as

$$\hat{\Psi}_{0\alpha} = \hat{c}_{0\alpha}, \quad \hat{\Psi}_{l\alpha} = \frac{1}{\sqrt{2}}\left(\hat{c}_{l\alpha} + \hat{c}_{-l\alpha}\right), \quad l = 1, 2, \ldots, L. \tag{5.76}$$

The hopping matrix h_{lm} is then given by the following $(L+1) \times (L+1)$ matrix:

$$h_1 = (-t_\mathrm{h}) \begin{pmatrix} 0 & \sqrt{2} & 0 & \cdots & 0 \\ \sqrt{2} & 0 & 1 & 0 & \vdots \\ 0 & 1 & 0 & 1 & \vdots \\ \vdots & 0 & 1 & \ddots & 1 \\ 0 & \cdots & \cdots & 1 & 0 \end{pmatrix}. \tag{5.77}$$

The couplings g^γ_{lm} correspond to the local Kondo interaction $J_\gamma \delta_{l0}\delta_{m0}$.

Substituting the above expressions into Eq. (5.70), one can obtain the functional derivative used to integrate the imaginary-time evolution (5.56) and the real-time evolution (5.59). From the resulting time-dependent covariance matrix Γ, one can study the ground state and the real-time dynamics of the anisotropic Kondo model. We remark that physical quantities can be efficiently obtained from Γ. For instance, the impurity-environment spin correlations can be given as

$$\chi_l^x = \frac{1}{4}\langle \hat{\sigma}_{\text{imp}}^x \hat{\sigma}_l^x \rangle = \frac{1}{4}\sigma_{\text{imp}}^x \sigma_{\alpha\beta}^x (\Gamma_f)_{l\alpha,l\beta}, \tag{5.78}$$

$$\chi_l^y = \frac{1}{4}\langle \hat{\sigma}_{\text{imp}}^y \hat{\sigma}_l^y \rangle = \frac{1}{4}(-i\sigma_{\alpha\beta}^y)(\Gamma_f^{\text{P}})_{l\alpha,l\beta}, \tag{5.79}$$

$$\chi_l^z = \frac{1}{4}\langle \hat{\sigma}_{\text{imp}}^z \hat{\sigma}_l^z \rangle = \frac{1}{4}\sigma_{\text{imp}}^x \sigma_{\alpha\beta}^z (\Gamma_f^{\text{P}})_{l\alpha,l\beta}. \tag{5.80}$$

We note that $\langle \cdots \rangle$ represents an expectation value in the *original* frame. Using Eq. (5.62), the magnetization is given by

$$\langle \hat{\sigma}_{\text{imp}}^z \rangle = \sigma_{\text{imp}}^x \langle \hat{\mathbb{P}}_{\text{bath}} \rangle_G = \sigma_{\text{imp}}^x (-1)^{N_f} \text{Pf}\left[\frac{\Gamma_F}{2}\right]. \tag{5.81}$$

5.4.2 Entanglement Structure of the Variational Ground State

Before comparing our variational approach with other theoretical methods, it is useful to clarify the fact that our variational ground state can naturally encode the entanglement structure in the Kondo-singlet state. To see this, we consider a sector corresponding to zero total spin-z component $\hat{\sigma}_{\text{tot}}^z = \hat{\sigma}_{\text{imp}}^z + \hat{\sigma}_{\text{bath}}^z = 0$ with $\hat{\sigma}_{\text{bath}}^z = \sum_{l=0}^{L} \hat{\sigma}_l^z$.

From Eqs. (5.11) and (5.15) with $\sigma_{\text{imp}}^x = 1$, the many-body state in the *original* frame contains the spin-up impurity $|\uparrow\rangle$ coupled to a bath wavefunction with an even number of spin-up fermions $N_\uparrow$ while the spin-down impurity $|\downarrow\rangle$ is coupled to that with odd $N_\uparrow$. Thus, the projected states $|\Psi_\pm\rangle = \hat{\mathbb{P}}_\pm |\Psi_{\text{bath}}\rangle$ appearing in Eq. (5.15) are eigenstates of $\hat{\sigma}_{\text{bath}}^z$:

$$\hat{\sigma}_{\text{bath}}^z |\Psi_\pm\rangle = \mp |\Psi_\pm\rangle. \tag{5.82}$$

In the deep AFM phase, we numerically check that the projected states $|\Psi_\pm\rangle$ satisfy

$$|||\Psi_\pm\rangle|| = 1/2, \quad \langle \Psi_- | \hat{\sigma}_{\text{bath}}^+ | \Psi_+ \rangle = -1/2. \tag{5.83}$$

These relations together with Eq. (5.15) automatically ensure the fact that the variational ground state is the singlet state:

$$|\Psi_{\text{AFM}}\rangle = \frac{1}{\sqrt{2}}\left(|\uparrow\rangle_{\text{imp}}|\Psi_\downarrow\rangle - |\downarrow\rangle_{\text{imp}}|\Psi_\uparrow\rangle\right), \tag{5.84}$$

where we introduce $|\Psi_\downarrow\rangle = \sqrt{2}|\Psi_+\rangle$ and $|\Psi_\uparrow\rangle = -\sqrt{2}|\Psi_-\rangle$.

We note that, if we specify $|\Psi_{\text{bath}}\rangle$ in Eq. (5.15) to be a single-particle excitation on top of the Fermi sea, Eq. (5.84) reproduces the Yosida ansatz [125–127] (see Eq. (5.88) below). Recently, Ref. [128] has revisited the Yosida ansatz and pointed out that it can contain the most part of the entanglement in the Kondo singlet state,

5.4 Application to the Anisotropic Kondo Model

indicating the ability of our variational states to efficiently capture the impurity-environment entanglement. That said, it should be emphasized that our variational states take into account all the possible two-particle excitations and thus go beyond the Yosida ansatz. This flexibility becomes particularly important when we aim to analyze in- and out-of-equilibrium properties of SIM in a quantitative manner.

Another criterion to verify the Kondo-singlet formation is to test the sum rule [26] of the spin correlations $\chi_l = \langle \hat{\boldsymbol{\sigma}}_{\text{imp}} \cdot \hat{\boldsymbol{\sigma}}_l \rangle / 4$:

$$\sum_{l=0}^{L} \chi_l = \frac{1}{8} \langle \hat{\boldsymbol{\sigma}}_{\text{tot}}^2 - \hat{\boldsymbol{\sigma}}_{\text{imp}}^2 - \hat{\boldsymbol{\sigma}}_{\text{bath}}^2 \rangle = -\frac{3}{4}. \tag{5.85}$$

In the numerical results presented below, the sum rule is satisfied with an error below 0.5% in the AFM regime with $J_\parallel > 0$.

5.4.3 Benchmark Tests with the Matrix-Product States

Matrix-product states (MPS)

We use the MPS to benchmark our variational approach. Let us briefly review the idea of the MPS. The MPS is the ansatz for quantum many-body states that can be written as

$$|\Psi_D\rangle = \sum_{\{i_k\}} \text{Tr}\left(A[0]^{i_0} A[1]^{i_1} \cdots A[N-1]^{i_{N-1}}\right) |i_0 i_1 \cdots i_{N-1}\rangle, \tag{5.86}$$

where $\{|i_k\rangle\}$ is a basis of the Hilbert space at site k and $A[k]^{i_k}$ represents a $D \times D$ matrix labelled by i_k, where D is the bond dimension. The MPS is applicable to analyzing one-dimensional quantum many-body states with a small entanglement such as low-temperature equilibrium states or time-evolving states in short-time regimes [81, 84, 129–133].

In practice, to find the best MPS approximation to the ground state of the anisotropic Kondo model (5.73), we represent the Hamiltonian by a matrix-product operator (MPO) [134] and employ the variational minimization of the energy

$$|\Psi_{\text{MPS}}\rangle = \text{argmin} \frac{\langle \Psi_D | \hat{H} | \Psi_D \rangle}{\langle \Psi_D | \Psi_D \rangle}. \tag{5.87}$$

The minimization can be done by performing the alternating least squares method until the ground-state energy is converged to a desired precision (see Refs. [81, 84] for details). To make comparisons with our variational results, we use a convergence precision 10^{-8}–10^{-6} in units of $t_h = 1$.

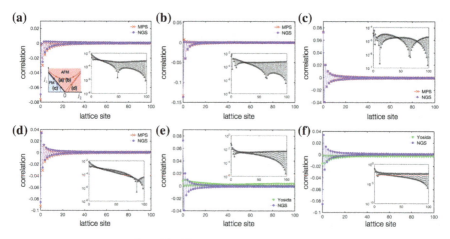

Fig. 5.3 Spin correlations $\chi_l^z = \langle \hat{\sigma}_{\text{imp}}^z \hat{\sigma}_l^z \rangle / 4$ at zero temperature obtained from our non-Gaussian variational state (NGS, blue circles), the matrix-product state (MPS, red crosses), and the Yosida ansatz (green triangulars). The insets plot the deviations between the results obtained from two methods. As shown in the inset of (**a**), we set $(j_\parallel, j_\perp)$ to be (**a**) $(-0.1, 0.4)$, (**b**) $(0.1, 0.4)$, (**c, e**) $(-0.4, 0.1)$ and (**d, f**) $(0.4, 0.1)$. We set $L = 100$ and use the bond dimension $D = 260$ for MPS. Reproduced from Fig. 2 of Ref. [85]. Copyright © 2018 by the American Physical Society

The MPS can also be used for analyzing dynamics by approximating the time-evolution operator by the Suzuki–Trotter decomposition [135, 136] to express it as a product of MPO [132, 134, 137]. The calculation is again done by the alternating least squares method and minimizing the distance between the MPS and the time-evolved state at each time step (see Refs. [81, 84] for details). In general, a real-time evolution will lead to an increase of the entanglement in the quantum state especially in a long-time regime. To maintain the numerical accuracy, we check convergence of the results against an increasing bond dimension at different times.

Yosida ansatz

To illustrate the importance of contributions beyond single-particle excitations, we also make comparisons with the Yosida ansatz [125], which is defined as follows:

$$|\Psi_{\text{Yosida}}\rangle = \begin{cases} \frac{1}{\sqrt{2}} \sum_{n > n_F} d_n \left(|\uparrow\rangle_{\text{imp}} \hat{c}_{n\downarrow}^\dagger - |\downarrow\rangle_{\text{imp}} \hat{c}_{n\uparrow}^\dagger \right) |\text{FS}\rangle & (-J_\parallel \leq J_\perp); \\ \sum_{n > n_F} d_n |\uparrow\rangle_{\text{imp}} \hat{c}_{n\uparrow}^\dagger |\text{FS}\rangle & (-J_\parallel > J_\perp), \end{cases} \quad (5.88)$$

where $|\text{FS}\rangle$ is the half-filled Fermi sea, n labels a bath energy mode and $n > n_F$ represents a mode above the Fermi surface. The first (second) line in Eq. (5.88) represents the AFM-singlet (FM-triplet) state. We minimize the mean energy $\langle \hat{H} \rangle_{\text{Yosida}}$ by optimizing the amplitudes $\{d_n\}$, leading to

5.4 Application to the Anisotropic Kondo Model

$$\epsilon_n d_n - \frac{1}{4} J_{\text{eff}} \psi_{0n}^* \sum_{m > n_F} \psi_{0m} d_m = E_{\text{GS}} d_n, \tag{5.89}$$

where ψ_{ln}'s are expansion coefficients of a fermion at site l in the energy basis, i.e., $\hat{c}_l = \sum_n \psi_{ln} \hat{c}_n$, ϵ_n is a bath energy, and E_{GS} is the variational ground-state energy. The effective Kondo coupling J_{eff} is defined by

$$J_{\text{eff}} = \begin{cases} 2J_\perp + J_\parallel & (-J_\parallel \leq J_\perp), \\ -J_\parallel & (-J_\parallel > J_\perp). \end{cases} \tag{5.90}$$

The variational ground-state energy E_{GS} and the amplitudes $\{d_n\}$ can be obtained from solving the eigenvalue problem (5.89). It is useful to explicitly see how the Yosida ansatz relates to our variational states. Using the disentangling transformation $\hat{U}$ in Eq. (5.13), the singlet state in the AFM regime can be expressed as

$$\hat{U}^{-1} |\Psi_{\text{Yosida}}\rangle = |+_x\rangle \sum_{n > n_F} \frac{d_n}{\sqrt{2}} \left(\hat{c}_{n\downarrow}^\dagger - \hat{c}_{n\uparrow}^\dagger \right) |\text{FS}\rangle. \tag{5.91}$$

The bath state in this transformed frame is just a single-particle excitation on top of the Fermi sea, which belongs to the fermionic Gaussian states. Thus, the Yosida ansatz is a very special subclass of our variational states.

Benchmark tests in equilibrium regimes

Figure 5.3 shows the benchmarking results for the ground-state impurity-bath spin correlations $\chi_l^z = \langle \hat{\sigma}_{\text{imp}}^z \hat{\sigma}_l^z \rangle / 4$. Our variational results quantitatively agree with the MPS results; the observed deviation is at most a few percent of the maximum value at the impurity site. In the deep FM and AFM regimes, we find a particularly good agreement (see Fig. 5.3c, d), in which the deviation is below 1%.

Figure (5.4a, b) plot the ground-state energy E_{var} and the magnetization $\langle \hat{\sigma}_{\text{imp}}^z \rangle$. Our results show quantitative agreement with the MPS results with a deviation below 0.5% in the FM regime, and even achieve lower energies in $j_\parallel \lesssim -0.7$ (see the left inset of Fig. 5.4a). The largest deviation in the ground-state energy (see the right inset of Fig. 5.4a) and the magnetization (see Fig. 5.4b) can be found near the phase boundary. We attribute the observed deviations to a large amount of entanglement that has to be encoded in the many-body wavefunction in this regime. Due to the nonmonotonic RG flows, the Kondo couplings can be very small in their midflows, indicating the formation of a large Kondo cloud. Since the analysis has been done in real space, our variational states should encode a large amount of the entanglement, which may go above the amount that can be generated by the canonical transformation $\hat{U}$. In the AFM phase, the ground-state energies calculated from our variational states and the MPS quantitatively agree with a great accuracy; the deviation is typically below 0.5% in $j_\parallel > 0$. The residual magnetization in the AFM phase is $O(10^{-5})$ in our variational method and $\langle \hat{\sigma}_{\text{imp}}^z \rangle \simeq O(10^{-4})$ in the MPS.

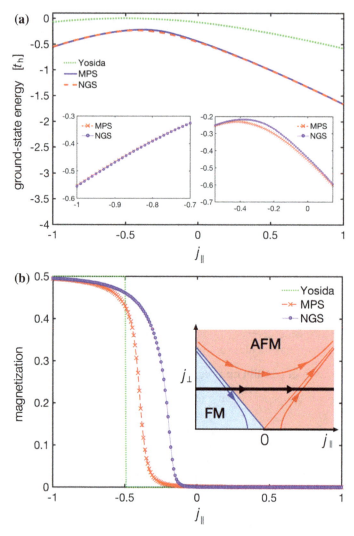

Fig. 5.4 **a** The ground-state energy and **b** the magnetization obtained from our variational state (NGS, blue circles), the matrix-product state (MPS, red crosses), and the Yosida ansatz (green, dashed curves). As schematically illustrated in the inset of (**b**), we use $j_\perp = 0.5$ and alter $j_\parallel$ from -1 to 1. We set $L = 200$ and use the bond dimension $D = 280$ for MPS. Reproduced from Fig. 3 of Ref. [85]. Copyright © 2018 by the American Physical Society

5.4 Application to the Anisotropic Kondo Model

We remark the great efficiency of our variational approach. In the MPS, the number of variational parameters is roughly equal to $4LD^2$ with D and L being the bond dimension and the system size, respectively. Meanwhile, our variational ansatz uses $4L^2$ variational parameters. As the bond dimension D is taken to be about 200-300 and the system size L is typically 100-200, our variational states have achieved the accuracy comparable to the MPS with two or three orders of magnitude fewer variational parameters (and shorter CPU time accordingly) than the corresponding MPS ansatz.

Benchmark tests in out-of-equilibrium regimes

We next show the comparisons for the dynamics of the magnetization. We choose the initial state to be

$$|\Psi(0)\rangle = |\uparrow\rangle_{\text{imp}}|\text{FS}\rangle. \tag{5.92}$$

We then suddenly switch on the impurity-environment coupling at time $t = 0$. Figure 5.5a shows the comparison of the magnetization dynamics at the isotropic points $j_\parallel = j_\perp = \pm 0.35$. In the AFM phase, the Kondo-singlet formation leads to the relaxation of the magnetization to zero. In the FM phase, the oscillation is caused by a particle excitation from the bottom to the top of the band, in which the bandwidth $\mathcal{D} = 4t_h$ determines its period $2\pi\hbar/\mathcal{D}$. The long-time persistence of the oscillation can be understood from the decoupling of the impurity spin in the low-energy limit. Our variational results agree with the MPS results with a deviation below

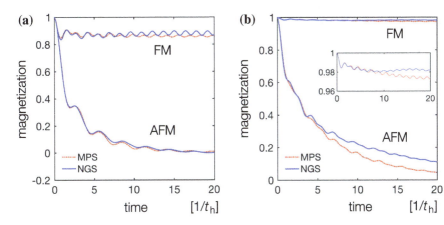

Fig. 5.5 Time evolutions of the magnetization calculated from our variational state (NGS, blue solid curves) and the matrix-product state (MPS, red dashed curves). We set $j_\parallel = j_\perp = 0.35$ for the ferromagnetic (FM) case and $j_\parallel = j_\perp = 0.35$ for the antiferromagnetic case (AFM) in (**a**). In (**b**), we use $(j_\parallel, j_\perp) = (0.1, 0.4)$ for the FM case and $(-0.4, 0.1)$ for the AFM case. We set $L = 100$ and vary the bond dimension $D \in [220, 260]$ of the MPS for which the results converge within the time scale in the plots. Reproduced from Fig. 4 of Ref. [85]. Copyright © 2018 by the American Physical Society

$O(10^{-2})$. Figure 5.5b shows the magnetization dynamics in the anisotropic AFM and FM points. Our variational results agree well with the MPS results in the short and intermediate time regimes. A small discrepancy can be found in the long-time regime ($t \gtrsim 10$). Its typical order is $O(1 \times 10^{-2})$ ($O(6 \times 10^{-2})$) in the FM (AFM) phase. Yet, the results obtained from these two methods still exhibit qualitatively the same features in both phases.

5.4.4 Benchmark Test with the Bethe Ansatz Solution

We next compare our variational approach to the exact solution obtained from the Bethe ansatz (BA) [31–33, 122–124]. While the BA solution is valid only under the infinite-bandwidth assumption, our approach should still reproduce the BA results if the relevant energy scale is much less than the bandwidth $\mathcal{D} = 4t_h$. The BA solution for the zero-temperature magnetization $m = \langle \hat{\sigma}_{\mathrm{imp}}^z \rangle / 2$ can be given as [124]:

$$m_{\mathrm{BA}} = \begin{cases} \frac{1}{4\sqrt{\pi^3}} \sum_{k=0}^{\infty} \frac{(-1)^k}{k!} \left(\frac{\pi}{2}\right)^{2k+1} \left(\frac{k+1/2}{2\pi}\right)^{k-\frac{1}{2}} \left(\frac{h_z}{T_K}\right)^{2k+1} & (h_z/T_K \leq \sqrt{8/(\pi e)}); \\ \frac{1}{2}\left\{1 - \frac{1}{\pi^{3/2}} \int_0^\infty dt \, \frac{\sin(\pi t)}{t} \left[\frac{8}{\pi e}\left(\frac{T_K}{h_z}\right)^2\right] e^{-t(\ln t - 1)} \Gamma\left(t + \frac{1}{2}\right)\right\} & (h_z/T_K > \sqrt{8/(\pi e)}), \end{cases} \quad (5.93)$$

where $\Gamma(\cdot)$ is the Gamma function and the magnetic susceptibility $\chi \equiv \partial m/\partial h_z = 1/(4T_K)$ at $h_z = 0$ determines the Kondo temperature T_K. In Fig. 5.6, we compare the BA solution m_{BA} (dashed black line) to our variational results. The deviation is at most a few percent in the regime $h_z/T_K \lesssim 1$, while it becomes more significant as we increase h_z/T_K due to the finite bandwidth of the lattice model. At a larger Kondo coupling j, the effect of the finite bandwidth begins to take place at a smaller threshold value of h_z/T_K because of a larger T_K.

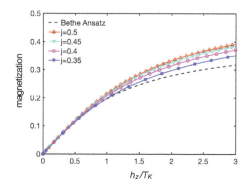

Fig. 5.6 The ground-state magnetization $m = \langle \hat{\sigma}_{\mathrm{imp}}^z \rangle / 2$ obtained from our approach and the Bethe ansatz (dashed black line). We use T_K that is determined from the magnetic susceptibility $\chi = 1/(4T_K)$. Reproduced from Fig. 5 of Ref. [85]. Copyright © 2018 by the American Physical Society

5.4 Application to the Anisotropic Kondo Model

5.4.5 Tests of Nonperturbative Scaling and Universal Behavior

Equilibrium properties

We here test the universal behavior of the Kondo screening length ξ_K in the variational ground state. In practice, we obtain ξ_K as a size of the Kondo screening cloud [71, 139] as follows. First, we calculate the integrated antiferromagnetic correlations $\Sigma_{AF}(l) = \sum_{|m|=0,2,4...}^{l} \chi_m$ with $\chi_m = \langle \hat{\sigma}_{imp} \cdot \hat{\sigma}_m \rangle / 4$ (see the inset of Fig. 5.7a). Second, we determine ξ_K from $f = 1 - \Sigma_{AF}(\xi_K(f))/\Sigma_{AF}(L)$ with $f > 0$ being a

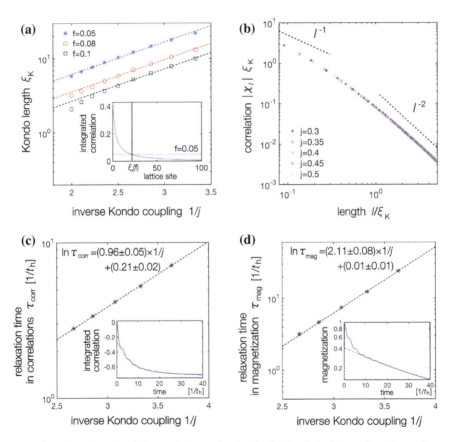

Fig. 5.7 a Screening length ξ_K at different thresholds f. The dashed lines show the scalings $\xi_K \propto e^{1/j}$. (inset) We determine ξ_K as a length in which a fraction $1 - f$ of total correlations exists. **b** The universal scaling of spin correlations. **c, d** Scalings of relaxation times τ in (**c**) the spin correlations and (**d**) the impurity magnetization. (insets) We determine the relaxation times via fitting the tales of $\Sigma_{AF}(L)(t)$ and $\langle \hat{\sigma}_{imp}^z(t) \rangle$ with $a + be^{-t/c}$. The dashed lines show the scalings $\ln \tau \propto 1/j$. We set $L = 400$. Reproduced from Figs. 2 and 3 of Ref. [138]. Copyright © 2018 by the American Physical Society

small threshold value. In Fig. 5.7a, the extracted Kondo length $\xi_K(f)$ is plotted against $1/j$, where its correct nonperturbative scaling $\xi_K \propto T_K^{-1} \propto e^{1/j}$ has been confirmed [5]. Figure 5.7b confirms the universal behavior of χ_l in units of ξ_K at different Kondo couplings j, exhibiting the crossover from l^{-1} to l^{-2} decay at $l/\xi_K \sim 1$ [140–142]. In particular, the correlations show the correct l^{-2} behavior even in a long-distance regime.

Out-of-equilibrium properties

We next study the scalings of the relaxation times τ_{corr} and τ_{mag} for the integrated correlations $\Sigma_{\text{AF}}(L, t)$ and the impurity magnetization $\langle \hat{\sigma}^z_{\text{imp}}(t) \rangle$. The relaxation times are determined from fitting the time evolutions of the corresponding observables with exponential functions (see the caption of Fig. 5.7 and the insets of Fig. 5.7c, d). The extracted relaxation times show the nonperturbative dependence on the Kondo coupling j for both observables but with different scalings $\tau_{\text{corr}} \propto e^{1/j}$ and $\tau_{\text{mag}} \propto e^{2/j}$ (see the main panels of Fig. 5.7c, d). Our findings seem to be consistent with the TEBD results [71], which predict the scalings $\tau_{\text{corr}} \propto e^{(1.5\pm 0.2)/j}$ and $\tau_{\text{mag}} \propto e^{(1.9\pm 0.2)/j}$.[4]

5.4.6 Spatiotemporal Dynamics after the Quench

Finally, we demonstrate that our approach allows one to study previously unexplored nonequilibrium dynamics. We consider the sudden quench of the impurity-bath interactions starting from the initial state (5.92). We plot the spatiotemporal dynamics of the impurity-bath correlations $\chi_l^z(t)$ in three different regimes of the phase diagram (see the left most panel in Fig. 5.8). The quench generates AFM (FM) ballistic spin waves in the FM (AFM) phase (see panels I and II) after which the equilibrium FM (AFM) correlations are formed. Here, the AFM (FM) ballistic spin waves originate from the excess spin associated with the formation of the triplet (singlet) correlations.

Most interestingly, the correlation shows the long-time crossover dynamics from the triplet to singlet behavior in the AFM regime III close to the phase boundary (see panel III and its closeup IV in Fig. 5.8). This peculiar dynamics can be understood from the nonmonotonic RG flows in this regime (c.f. the left most panel in Fig. 5.8); the high (low) energy physics associated with the FM (AFM) renormalized coupling $J_\parallel < 0$ ($J_\parallel > 0$) characterizes the short (long) time dynamics, leading to the crossover from the triplet-to-singlet behavior. In this respect, here one can interpret the real time as an effective inverse RG scale [45]. The predicted spatiotemporal dynamics can be tested with, e.g., quantum gas microscopy [143–147] (see also discussions in Sect. 5.6).

We note that, by employing the infinite-bandwidth approximation, the Kondo problem can be mapped to the spin-boson model [1] that is continuous in space as a result of bosonization. With this mapping, the impurity relaxation and the

---

[4] We remark that a relatively large deviation in the scaling of τ_{corr} has been attributed to the difficulty of taking the adiabatic limit in the Anderson model analyzed in the TEBD study [71].

5.4 Application to the Anisotropic Kondo Model

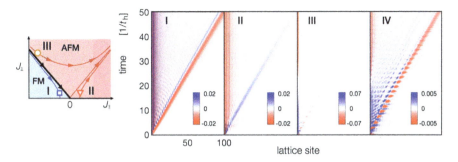

Fig. 5.8 Spatiotemporal dynamics of spin correlations $\chi_l^z(t)$ after the quench in I FM phase, II AFM phase, III easy-plane FM regime. Panel IV is a closeup of panel III. We use $L = 400$ and set $(j_\parallel, j_\perp)$ to be $(-0.5, 0.2)$ in I, $(0.5, 0.2)$ in II, and $(-1.85, 2)$ in III. Reproduced from Fig. 1 of Ref. [138]. Copyright © 2018 by the American Physical Society

spatiotemporal dynamics in a similar setup have been analyzed by the TD-NRG method [65, 70]. However, a strictly linear dispersion and an artificial cut-off energy are necessary in that method. In this respect, our approach is advantageous since it allows one to make a direct quantitative comparison with experimental systems, as it relies neither on the infinite-bandwidth limit nor on bosonization. This aspect is particularly important especially in view of rapid experimental developments to simulate nonequilibrium quantum dynamics using artificial quantum systems [34–39, 148, 149].

5.5 Application to the Two-Lead Kondo Model

5.5.1 Model

We next apply our general variational approach to study the two-lead Kondo model [46], in which the spin impurity interacts with fermions at the centers of the left and right leads (see Fig. 5.9). The Hamiltonian is

$$\hat{H}_{\text{two}} = \sum_{l\eta} \left[-t_{\text{h}} \left(\hat{c}^\dagger_{l\eta\alpha} \hat{c}_{l+1\eta\alpha} + \text{h.c.} \right) + eV_\eta \, \hat{c}^\dagger_{l\eta\alpha} \hat{c}_{l\eta\alpha} \right]$$
$$+ \frac{J}{4} \sum_{\eta\eta'} \hat{\boldsymbol{\sigma}}_{\text{imp}} \cdot \hat{c}^\dagger_{0\eta\alpha} \boldsymbol{\sigma}_{\alpha\beta} \hat{c}_{0\eta'\beta} - \frac{h_z}{2} \hat{\sigma}^z_{\text{imp}}, \tag{5.94}$$

where $\hat{c}^\dagger_{l\eta\alpha}$ ($\hat{c}_{l\eta\alpha}$) is a creation (annihilation) operator of a fermion with spin α and position l on the left ($\eta = \text{L}$) or right ($\eta = \text{R}$) lead, J is the Kondo coupling strength, the hopping $t_{\text{h}} = 1$ sets the energy unit, and eV_η is a chemical potential acting on each lead.

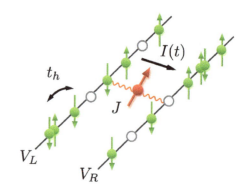

Fig. 5.9 A schematic illustration of the two-lead Kondo model. Our variational approach allows one to calculate the dynamics of the current $I(t)$, the magnetization $\langle \hat{\sigma}^z_{\text{imp}}(t) \rangle$, and the spatiotemporal behavior of fermions in two leads. Reproduced from Fig. 6 of Ref. [85]. Copyright © 2018 by the American Physical Society

To apply our approach, we note that only the following symmetric bath modes are coupled to the impurity:

$$\hat{\Psi}_{0\eta,\alpha} = \hat{c}_{0\eta\alpha}, \quad \hat{\Psi}_{l\eta,\alpha} = \frac{1}{\sqrt{2}} \left(\hat{c}_{l\alpha\eta} + \hat{c}_{-l\eta\alpha} \right) \quad l = 1, 2, \ldots, L. \quad (5.95)$$

The bath Hamiltonian h_2 in this basis can be taken as

$$h_2 = \begin{pmatrix} h_1 + eV_L I_{L+1} & 0 \\ 0 & h_1 + eV_R I_{L+1} \end{pmatrix}, \quad (5.96)$$

where h_1 is the hopping matrix for a single lead (see Eq. (5.77)). The impurity-bath coupling matrix g^γ_{lm} is

$$g = J \begin{pmatrix} 1 & 1 \\ 1 & 1 \end{pmatrix} \otimes \text{diag}_{L+1}(1, 0, \ldots, 0), \quad (5.97)$$

where $\text{diag}_d(v)$ denotes a $d \times d$ diagonal matrix with elements v, and we consider the isotropic interaction for the sake of simplicity. Substituting Eqs. (5.96) and (5.97) into the functional derivative (5.70), we integrate the real-time evolution (5.59) to analyze transport properties and out-of-equilibrium dynamics of the two-lead Kondo model.

5.5.2 Spatiotemporal Dynamics of the Environment after the Quench

We consider the quench protocol starting from the initial state

$$|\Psi(0)\rangle = |\uparrow\rangle_{\text{imp}} |\text{FS}\rangle_L |\text{FS}\rangle_R, \quad (5.98)$$

5.5 Application to the Two-Lead Kondo Model

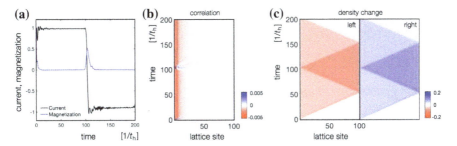

Fig. 5.10 **a** Time evolutions of the magnetization $\langle \hat{\sigma}^z_{\text{imp}}(t)\rangle$ (blue dashed curve) and the current $I(t)$ (black solid curve). **b** Spatiotemporal dynamics of the spin correlations $\chi_l^z(t)$. **c** Spatiotemporal dynamics of density changes relative to the initial values in each lead. We use $L = 100$ for each lead, and set $V_L = -V_R = 0.25$ and $j = 0.4$. Reproduced from Fig. 7 of Ref. [85]. Copyright © 2018 by the American Physical Society

where $|\text{FS}\rangle_{\text{L,R}}$ are the half-filled Fermi sea of each lead. We suddenly switch on the Kondo coupling J and bias potentials $V_L = V/2$ and $V_R = -V/2$ at time $t = 0$. A bias is taken to be $V > 0$ without loss of generality. We calculate the dynamics of the current $I(t)$ between two leads[5]:

$$I(t) = i\frac{eJ}{4\hbar}\left[\langle \hat{\sigma}_{\text{imp}} \cdot \hat{c}^\dagger_{0L\alpha}\boldsymbol{\sigma}_{\alpha\beta}\hat{c}_{0R\beta}\rangle - \text{h.c.}\right]$$
$$= -\frac{eJ}{2\hbar}\text{Im}\left[\sigma^x_{\text{imp}}\sigma^x_{\alpha\beta}(\Gamma_f)_{0L\alpha,0R\beta} + (-i\sigma^y + \sigma^x_{\text{imp}}\sigma^z)_{\alpha\beta}(\Gamma^P_f)_{0L\alpha,0R\beta}\right]. \quad (5.99)$$

The result is shown in Fig. 5.10a as the black solid curve; the current relaxes to a steady-state value after a short transient regime. This relaxation time is comparable to the decay time of the magnetization and thus can be interpreted as the time scale for the formation of the Kondo-singlet state (see the blue dashed curve in Fig. 5.10a). The spatial dynamics of the spin correlation $\chi_l^z(t)$ and the density change at each lead are shown in Fig. 5.10b and c, respectively. After the ballistic density waves reflect at the ends of leads, they propagate back to the impurity site. At the moment they pass through the impurity site, the magnetization shows the recurrence while the current flips its sign as shown in Fig. 5.10a. The density will reproduce the initial homogeneous profile when the density waves again propagate back to the impurity site after the second reflection.

The results presented here clearly demonstrate the ability of our variational approach to accurately predict spatiotemporal dynamics in a long-time regime, which has been difficult to achieve in the previous approaches. In a global quench procedure considered here, a substantial energy (which linearly scales with the system size) will be acquired, leading to a linear increase of the entanglement entropy in

---

[5] We note that $\langle \cdots \rangle$ represents an expectation value with respect to a quantum state in the original frame and that $\hbar$ is placed back when the current and conductance are studied.

time. This imposes the fundamental difficulty on, e.g., tDMRG method especially in a long-time regime [61].

5.5.3 Transport Properties

Differential conductance

We next study transport properties by analyzing the differential conductance. For a given bias V_0, we calculate a steady-state value of the current $\bar{I}(V_0)$ by taking the time average over the plateau regime. The differential conductance G is then obtained from [58, 59, 61, 69]

$$G(V_0) \simeq \frac{\bar{I}(V_0 + \Delta V) - \bar{I}(V_0 - \Delta V)}{2\Delta V}, \quad (5.100)$$

where ΔV is a small modulation of the bias potential. To benchmark our results, we first test the quadratic behavior $G_0(1 - c_B(h_z/T_K)^2)$ with $c_B = \pi^2/16$ at a low magnetic field h_z and the logarithmic behavior $\pi^2 G_0/(16\ln^2(h_z/T_K))$ at a high magnetic field [19, 150–157]. As shown in Fig. 5.11a and in its inset, our results reproduce the correct quadratic behavior at a low field with great accuracy. They are also consistent with the logarithmic scaling at a high field. Figure 5.11b plots the differential conductance G against the Kondo coupling j. Its nonmonotonic behavior with respect to j is a nontrivial feature originating from the finite bandwidth of the lattice model; in the limit of a large j, the formation of the bound state localized at the impurity site will prevent other electrons from approaching the junction, leading to a decrease of G in the limit of $j \to \infty$. We remark that the confirmation of this nonmonotonic feature of G is a nontrivial test which the conventional approach has failed to pass [80, 150, 158].

At a finite bias V, the differential conductance exhibits the nonlinear behavior (see Fig. 5.11c). We make two remarks on the numerical results presented here. First, the current fluctuation in a steady-state regime causes a numerical error which can easily mask small changes of G in a perturbative regime $V \ll T_K$. As a result, a precise test of the quadratic behavior of G at small bias [58, 59, 61, 69] is rather difficult in our real-time calculations. This difficulty should be circumvented by using a different quench operation. Second, numerical results on the current and conductance will no longer be faithful in a high-bias regime $V \gg T_K$ at which the bias can be of the order of the Fermi energy and the finite bandwidth. Such a difficulty has been commonly found in numerical methods performed in real-space basis [61, 69] and can be circumvented by relying on the infinite-bandwidth approximation by using the momentum basis. We still emphasize that, in the intermediate regime $V \sim T_K$, the present implementation of our variational approach already provides faithful results.

At a finite bias and magnetic field, the conductance exhibits a peak around $h_z = V$ (see Fig. 5.11d), which is caused by the level matching of the Fermi surfaces in each

5.5 Application to the Two-Lead Kondo Model

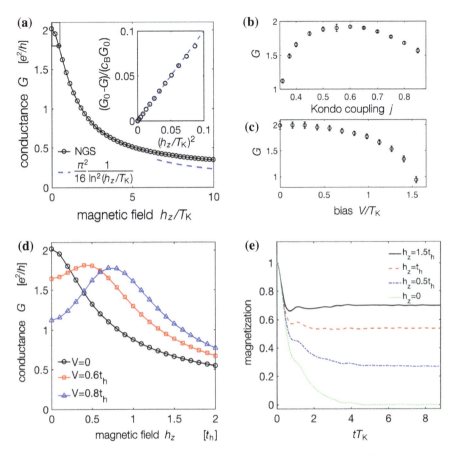

Fig. 5.11 Differential conductance G plotted against (a,d) h_z/T_K, (b) j and (c) V/T_K. (inset, **a**) Our results (black open circles) are compared with the low-field asymptotic scaling (blue dashed line). In (**a–d**), we set $L = 200$ for each lead and use **a** $j = 0.35$ and $V = 0$, **b** $h_z = 0$ and $V = 0.8t_h$, **c** $h_z = 0$ and $j = 0.4$, and **d** $j = 0.35$. **e** Dynamics of the magnetization $\langle \hat{\sigma}_{\text{imp}}^z(t) \rangle$ at finite h_z. In (**e**), we set $L = 100$ for each lead and use $j = 0.35$ and $V = 0.8$. Reproduced from Fig. 4 of Ref. [138] and Fig. 9 of [85]. Copyright © 2018 by the American Physical Society

lead. Due to a partial destroy of the Kondo singlet (which can be inferred from the nonzero magnetization in Fig. 5.11e), the peak values of G are smaller than the unitarity limit $2e^2/h$. These characteristic features are consistent with the analytical results at the Toulouse point [80] and with the previous findings in the Anderson model [58, 69, 108].

Finally, we make a remark on the asymmetric behavior of the conductance against the bias and magnetic field, whose origin has not been fully understood here. As we mentioned above, several features found in our results are consistent with the results at the Toulouse point. These include the peak of G at $h_z \sim V$ and the asymmetry in the conductance behavior at $h_z = 0$ with changing V and the one at $V = 0$ with changing h_z. In fact, the latter has been argued to be a nontrivial feature that is absent in the conventional bosonization approach [80]. Yet, we have found discrepancies between the asymmetric behavior at the Toulouse point [80] and that found in our results (which are obtained with the finite bandwidth and not in the Toulouse limit). It merits further study to elucidate the origin of the discrepancies observed here.

5.6 Experimental Implementation in Ultracold Gases

We here briefly discuss how one can use ultracold alkaline-earth atoms to experimentally study the predicted nonequilibrium dynamics in the anisotropic Kondo model. We consider fermionic atoms prepared in the ground state $|g\rangle$ and trap them in an optical lattice. The ground-state atom has two internal degrees of freedom $|\uparrow\rangle$ and $|\downarrow\rangle$ corresponding to two spin states. At the initial time, two internal states are equally populated and thus the Fermi sea is formed. A weak laser pulse is then used to excite a single atom into the excited state $|e\rangle$ (see Fig. 5.12a). Because $|e\rangle$ has a different polarizability from $|g\rangle$, we can create a deep optical lattice acting only on an excited atom and freeze its orbital motion, mimicking the localized impurity. We denote two internal states of the excited atom by $|\Uparrow\rangle$ and $|\Downarrow\rangle$. In contrast, surrounding ground-state atoms can move and play a role as itinerant fermions. The total system is governed by the following Hamiltonian:

$$\hat{H}_{\text{tot}} = \sum_{k\sigma} \epsilon_k \hat{g}_{k\sigma}^\dagger \hat{g}_{k\sigma} + \frac{V}{\sqrt{L}} \sum_{k\sigma} \hat{g}_{0\sigma}^\dagger \hat{g}_{k\sigma} + U(\hat{n}_{g0\uparrow} + \hat{n}_{g0\downarrow})(\hat{n}_{e0\Uparrow} + \hat{n}_{e0\Downarrow})$$

$$+ U_{\text{ex}} \sum_{\sigma \neq \sigma'} \hat{g}_{0\sigma'}^\dagger \hat{e}_{0\sigma}^\dagger \hat{e}_{0\sigma'} \hat{g}_{0\sigma} - \frac{\delta_g}{2} \sum_{k\sigma} \sigma \hat{n}_{gk\sigma} - \frac{\delta_e}{2} \sum_{\sigma} \sigma \hat{n}_{e0\sigma}, \quad (5.101)$$

where ϵ_k is the energy dispersion with wavevector k, $\hat{g}_{k\sigma}$ annihilates a ground-state atom with wavevector k and spin $\sigma = \uparrow, \downarrow$, $\hat{g}_{0\sigma}$ ($\hat{e}_{0\sigma}$) annihilates the ground-state (excited-state) atom at the impurity site $l = 0$, $\hat{n}_{gk\sigma} = \hat{g}_{k\sigma}^\dagger \hat{g}_{k\sigma}$ and $\hat{n}_{e0\sigma} = \hat{e}_{0\sigma}^\dagger \hat{e}_{0\sigma}$ are the occupation numbers of ground- and excited-state atoms at each mode, $V = \sqrt{z} t_h$ is the mixing term with z being the nearest-neighbor coordination number and t_h being a hopping parameter, $U = (U_{eg}^- + U_{eg}^+)/2$ and $U_{\text{ex}} = (U_{eg}^- - U_{eg}^+)/2$ are the on-site energy and spin-exchange interactions with U_{eg}^- (U_{eg}^+) being proportional to the triplet (singlet) scattering length a_{eg}^- (a_{eg}^+), and δ_g (δ_e) is the Zeeman energy in the ground (excited) state.

In experimentally realizable situations (see e.g., Ref. [34]), due to the large scattering lengths $a_{eg}^\pm$, the system can satisfy the following condition:

$$\epsilon_k, \delta_{e,g}, V \ll U, U_{\text{ex}}. \quad (5.102)$$

In this limit, following the standard perturbation theory, we can show that the original Hamiltonian (5.101) reduces to the effective Hamiltonian

5.6 Experimental Implementation in Ultracold Gases

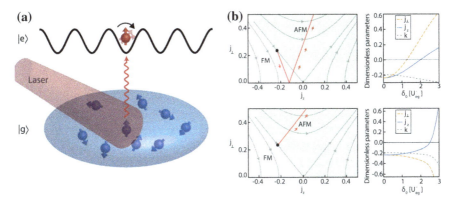

Fig. 5.12 A possible experimental implementation of the anisotropic Kondo model in ultracold fermionic alkaline-earth atoms. **a** At the initial time, two spin states of atoms in the ground state $|g\rangle$ (blue) are equally populated, forming the Fermi sea. A laser pulse is then used to excite a single atom into the excited state $|e\rangle$ (red), which is tightly localized by a deep optical lattice and acts as a localized impurity. **b** Parameter regimes explored by periodically driving the Zeeman field $\delta(\tau) = \delta_0 \cos(\omega\tau)$. The dimensionless Kondo couplings $j_\parallel = \rho_F J_\parallel$ and $|j_\perp| = \rho_F |J_\perp|$ with ρ_F being the density of states at the Fermi sea are determined for (top panels) $\omega = 0.8 U_{eg}^-$ and (bottom) $\omega = 0.95 U_{eg}^-$ with increasing δ_0. Parameters are chosen to be $t_h = 0.35 U_{eg}^-$ and $U_{eg}^+ = 15 U_{eg}^-$ as realized in Ref. [34]. Reproduced from Figs. 1 and 5 of Ref. [147]. Copyright © 2018 by the American Physical Society

$$\hat{H}_K = \sum_{k\sigma}(\epsilon_k - \sigma\delta_g/2)\hat{g}^\dagger_{k\sigma}\hat{g}_{k\sigma} + K\sum_{kk'\sigma}\hat{g}^\dagger_{k\sigma}\hat{g}_{k'\sigma} + \frac{J_\perp}{4}\sum_{\gamma=x,y}\hat{\sigma}^\gamma_e\sum_{\sigma\sigma'}\hat{g}^\dagger_{0\sigma}\sigma^\gamma_{\sigma\sigma'}\hat{g}_{0\sigma'}$$
$$+ \frac{J_\parallel}{4}\hat{\sigma}^z_e\sum_{\sigma\sigma'}\hat{g}^\dagger_{0\sigma}\sigma^z_{\sigma\sigma'}\hat{g}_{0\sigma'} - \frac{h_e}{2}\hat{\sigma}^z_e - \frac{h_g}{2}\sum_{\sigma\sigma'}\hat{g}^\dagger_{0\sigma}\sigma^z_{\sigma\sigma'}\hat{g}_{0\sigma'}, \tag{5.103}$$

where $\hat{\sigma}_e$ is the localized spin-1/2 operator and parameters $J_{\perp,\parallel}$ and K can be obtained from the microscopic parameters as functions of $\delta = \delta_e - \delta_g$:

$$J_\perp(\delta) = \frac{2V^2 U_{ex}(U^2 - U_{ex}^2)}{(U^2 - U_{ex}^2)^2 - U\delta^2}, \quad J_\parallel(\delta) = \frac{2V^2 U_{ex}(U^2 - U_{ex}^2)}{(U^2 - U_{ex}^2)^2 - U^2\delta^2},$$
$$K(\delta) = -\frac{V^2}{2}\frac{2U - U_{ex}}{U^2 - U_{ex}^2} - \frac{\delta^2}{4}\frac{UU_{ex}}{(U^2 - U_{ex}^2)^2}J_\perp(\delta),$$
$$h_e(\delta) = -h_g(\delta) = \delta\left(\frac{1}{2} - \frac{V^2 U_{ex}^2}{(U^2 - U_{ex}^2)^2 - U^2\delta^2}\right). \tag{5.104}$$

The second term in the first line of the right-hand side of Eq. (5.103) can be eliminated using a basis transformation of the bath modes and has no significant effect on the low-energy Kondo physics. The terms including $J_\perp$ and $J_\parallel$ describe the anisotropic Kondo couplings while those including h_e and h_g represent the magnetic fields acting on the impurity and the bath. In particular, at the SU(2)-symmetric point ($\delta = 0$), we can simplify the Kondo coupling as

$$J = V^2 \left(\frac{1}{U_{eg}^+} - \frac{1}{U_{eg}^-} \right). \tag{5.105}$$

To control the Kondo couplings and the magnetic fields separately, one can use a periodically modulated Zeeman field $\delta(\tau) = \delta_0 \cos(\omega\tau) + \delta_1$. Since the driving frequency is much faster than the low-energy dynamics, we can employ the standard Floquet expansion and obtain the parameters in the effective Hamiltonian. For instance, if we set the static Zeeman field to be zero $\delta_1 = 0$, the magnetic fields h_g and h_e vanish because they are odd functions of δ and thus cancel out after taking the time average. Typical parameter regimes that can be explored in the experimental setup of Ref. [34] are plotted in Fig. 5.12b. In particular, one can go across the phase transition line via the periodic-drive control. Thus, the spatiotemporal dynamics (such as the long-time crossover dynamics near the phase boundary) predicted in the previous section can be experimentally tested by, for example, using the site-resolved measurements with quantum gas microscopy.

5.7 Generalization to a Bosonic Environment

5.7.1 General Formalism

We here generalize our variational approach to a bosonic environment. This can be done by combining the canonical transformation with the bosonic Gaussian states, as the construction of the disentangling transformation presented in Sect. 5.2 does not rely on particle species. To be concrete, we consider the following bosonic quantum impurity Hamiltonian

$$\hat{H}_b = \sum_{lm\alpha} \hat{b}_{l\alpha}^\dagger h_{lm} \hat{b}_{l\alpha} + \hat{\mathbf{s}}_{\text{imp}} \cdot \hat{\mathbf{\Sigma}}_b, \tag{5.106}$$

where the first term denotes a noninteracting bosonic environment with $\hat{b}_{l\alpha}^\dagger$ ($\hat{b}_{l\alpha}$) being a bosonic creation (annihilation) operator corresponding to a bath mode $l = 1, 2, \ldots, N_b$ and the spin-z component $\alpha = \uparrow, \downarrow$ and h_{lm} being an arbitrary $N_b \times N_b$ Hermitian matrix describing a single-particle Hamiltonian of a bath. The second term describes a generic impurity-bath interaction, in which we introduce the bosonic bath-spin density operator including the $N_b \times N_b$ interaction couplings g_{lm}^γ labeled by $\gamma = x, y, z$ as

$$\hat{\Sigma}_b^\gamma = \frac{1}{2} \sum_{lm\alpha\beta} g_{lm}^\gamma \hat{b}_{l\alpha}^\dagger \sigma_{\alpha\beta}^\gamma \hat{b}_{m\beta}. \tag{5.107}$$

5.7 Generalization to a Bosonic Environment

Disentangling transformation

To disentangle the impurity spin, we can use the unitary transformation in Sect. 5.2 by just replacing the fermionic operators with the bosonic ones:

$$\hat{U}_b = \exp\left[\frac{i\pi}{4}\hat{\sigma}_{\text{imp}}^y \hat{\mathbb{P}}_b\right] = \frac{1}{\sqrt{2}}\left(1 + i\hat{\sigma}_{\text{imp}}^y \hat{\mathbb{P}}_b\right), \quad (5.108)$$

where we denote the bosonic bath parity operator as

$$\hat{\mathbb{P}}_b = \exp\left[\frac{i\pi}{2}\left(\sum_l \hat{\sigma}_{b,l}^z + \hat{N}_B\right)\right] \equiv e^{i\pi \hat{N}_{B,\uparrow}}. \quad (5.109)$$

We here represent the bath-spin density of mode l and the total particle number of bosons by

$$\hat{\sigma}_{b,l}^\gamma \equiv \sum_{\alpha\beta} \hat{b}_{l\alpha}^\dagger \sigma_{\alpha\beta}^\gamma \hat{b}_{l\beta}, \quad \hat{N}_B = \sum_{l\alpha} \hat{b}_{l\alpha}^\dagger \hat{b}_{l\alpha}, \quad (5.110)$$

and $\hat{N}_{B,\uparrow}$ is the number of spin-up bosons. We can then transform the Hamiltonian as

$$\hat{\tilde{H}}_b = \hat{U}_b^\dagger \hat{H}_b \hat{U}_b = \hat{\tilde{H}}_{b,0} + \hat{\tilde{H}}_{b,1}, \quad (5.111)$$

where $\hat{\tilde{H}}_{b,0}$ is the quadratic part of the transformed Hamiltonian

$$\hat{\tilde{H}}_{b,0} = \sum_{lm\alpha} \hat{b}_{l\alpha}^\dagger h_{lm} \hat{b}_{m\alpha} + s_{\text{imp}}^x \hat{\Sigma}_b^x, \quad (5.112)$$

while $\hat{\tilde{H}}_{b,1}$ is its interacting part

$$\hat{\tilde{H}}_{b,1} = \hat{\mathbb{P}}_b \left(-\frac{i\hat{\Sigma}_b^y}{2} + \hat{s}_{\text{imp}}^x \hat{\Sigma}_b^z\right). \quad (5.113)$$

Bosonic Gaussian states

In the transformed frame, we approximate the bath wavefunction by the bosonic Gaussian states. The bosonic Gaussian states are represented by exponentials of a function of bosonic operators up to quadratic terms. In analogy with the Majorana representation for the fermionic case, it is useful to introduce the position and momentum representations of the bosonic operators:

$$\hat{\phi} = (\hat{x}_{1\uparrow}, \cdots, \hat{x}_{N_b\uparrow}, \hat{x}_{1\downarrow}, \cdots, \hat{x}_{N_b\downarrow}, \hat{p}_{1\uparrow}, \cdots, \hat{p}_{N_b\uparrow}, \hat{p}_{1\downarrow}, \cdots, \hat{p}_{N_b\downarrow})^T, \quad (5.114)$$

$$\hat{x}_{l\alpha} = \hat{b}_{l\alpha} + \hat{b}_{l\alpha}^\dagger, \quad \hat{p}_{l\alpha} = i(\hat{b}_{l\alpha}^\dagger - \hat{b}_{l\alpha}). \quad (5.115)$$

The bosonic Gaussian state $|\Psi_{b,G}\rangle$ is then fully characterized by its mean field $\boldsymbol{\phi}$ and covariance matrix Γ_b:

$$\boldsymbol{\phi} = \langle \hat{\boldsymbol{\phi}} \rangle_{b,G}, \quad \Gamma_b = \frac{1}{2} \langle \{ \delta\hat{\boldsymbol{\phi}}, \delta\hat{\boldsymbol{\phi}}^T \} \rangle_{b,G}, \tag{5.116}$$

where $\delta\hat{\boldsymbol{\phi}} = \hat{\boldsymbol{\phi}} - \boldsymbol{\phi}$, and $\langle \cdots \rangle_{b,G}$ denotes the expectation value with respect to $|\Psi_{b,G}\rangle$. We can explicitly express $|\Psi_{b,G}\rangle$ by

$$|\Psi_{b,G}\rangle = e^{i\theta_0} e^{\frac{i}{2}\hat{\boldsymbol{\phi}}^T \sigma \boldsymbol{\phi}} e^{-\frac{i}{4}\hat{\boldsymbol{\phi}}^T X_b \hat{\boldsymbol{\phi}}} |0\rangle, \tag{5.117}$$

where $\sigma = i\sigma^y \otimes I_{2N_b}$ and the matrix X_b is related to the covariance matrix Γ_b via

$$\Gamma_b = e^{\sigma X_b} \left(e^{\sigma X_b} \right)^T. \tag{5.118}$$

We here include the phase factor θ_0 that will be necessary when we calculate the spectrum function later.

The variational energy is given by

$$E_{b,\text{var}} = \langle \hat{\tilde{H}}_b \rangle_{b,G} = \frac{1}{4}\text{Tr}\left[\mathcal{H}_{b,0}^T \Gamma_b\right] + \frac{1}{4}\boldsymbol{\phi}^T \mathcal{H}_{b,0} \boldsymbol{\phi} + \frac{1}{4}\text{Tr}\left[\mathcal{G}^T \Gamma_b^P\right] - \frac{1}{4}\text{Tr}\left[\mathcal{H}_{b,0}\right], \tag{5.119}$$

where $\mathcal{G} = -i\sigma^y \otimes g_{lm}^y + \sigma_{\text{imp}}^x \sigma^z \otimes g_{lm}^z$ is a matrix including the impurity-bath couplings and we denote the quadratic part $\mathcal{H}_{b,0}$ as

$$\mathcal{H}_{b,0} = \mathcal{S}[I_2 \otimes h_0], \quad h_0 = I_2 \otimes \text{diag}(h_{lm}) + (\sigma_{\text{imp}}^x/4)\sigma^x \otimes g_{lm}^x. \tag{5.120}$$

We here define the symmetrization $\mathcal{S}[\cdot]$ of a matrix by

$$\mathcal{S}[A] \equiv \text{Re}\left[\frac{A + A^T}{2}\right]. \tag{5.121}$$

The matrix Γ_b^P is defined in the similar manner as in the fermionic case:

$$\Gamma_b^P = \langle \hat{\mathbb{P}}_b \hat{\mathbf{b}}^\dagger \hat{\mathbf{b}} \rangle_{GS}, \quad \hat{\mathbf{b}} = (\hat{b}_{1\uparrow}, \cdots, \hat{b}_{N_B\uparrow}, \hat{b}_{1\downarrow}, \cdots, \hat{b}_{N_B\downarrow}), \tag{5.122}$$

which can be represented as

$$\Gamma_b^P = -\Sigma_z \langle \hat{\mathbb{P}}_b \rangle_{GS} (I_{2N_b}, -iI_{2N_b}) \left(\Gamma_B^{-1}\right)^T \left[\frac{1}{2}(\Gamma_b - I_{4N_b}) + \boldsymbol{\phi}\boldsymbol{\phi}^T \Gamma_B^{-1}\right] \begin{pmatrix} I_{2N_b} \\ iI_{2N_b} \end{pmatrix}, \tag{5.123}$$

5.7 Generalization to a Bosonic Environment

where the mean value of the bath parity can be expressed as

$$\langle \hat{\mathbb{P}}_b \rangle_{b,G} = \frac{1}{\sqrt{\det(\Gamma_B/2)}} e^{-\frac{1}{2}\phi^T \Gamma_B^{-1}(1+\Lambda)\phi}, \quad (5.124)$$

and we introduce the matrix Γ_B as

$$\Gamma_B = (I_{4N_b} + \Lambda)\Gamma_b + I_{4N_b} - \Lambda \quad (5.125)$$

with $\Lambda = I_2 \otimes \Sigma_z$ and $\Sigma_z = \sigma^z \otimes I_{N_b}$.

Variational time-evolution equations

Employing the time-dependent variational principle, we can derive the equations of motions for the mean field ϕ and the covariance matrix Γ_b in the same manner as performed in Sect. 5.3.2. The results are

$$\frac{d\phi}{dt} = -\Gamma_b h_\phi, \quad \frac{d\Gamma_b}{dt} = \sigma h_\Gamma \sigma - \Gamma_b h_\Gamma \Gamma_b \quad (5.126)$$

for the imaginary-time evolution and

$$\frac{d\phi}{dt} = \sigma h_\phi, \quad \frac{d\Gamma_b}{dt} = \sigma h_\Gamma \Gamma_b - \Gamma_b h_\Gamma \sigma \quad (5.127)$$

for the real-time evolution. Here, we introduce the functional derivatives

$$h_\phi = 2\frac{\delta E_{b,\text{var}}}{\delta \phi}, \quad h_\Gamma = 4\frac{\delta E_{b,\text{var}}}{\delta \Gamma_b}. \quad (5.128)$$

From Eqs. (5.119), (5.123) and (5.124), after doing some calculations we can obtain their analytical expressions as follows:

$$h_\phi = \mathcal{H}_{b,0}\phi - \frac{1}{2}\text{Tr}\left[\mathcal{G}^T \Gamma_b^P\right] \Gamma_B^{-1}(1+\Lambda)\phi$$
$$- \langle \hat{\mathbb{P}}_b \rangle \Gamma_B^{-1} S\left[\begin{pmatrix} I_{2N_b} \\ -iI_{2N_b} \end{pmatrix} \Sigma_z \mathcal{G}(I_{2N_b}, iI_{2N_b})\right](\Gamma_B^{-1})^T \phi, \quad (5.129)$$

$$h_\Gamma = \mathcal{H}_{b,0} + S\left[\frac{1}{2}\text{Tr}\left[\mathcal{G}^T \Gamma_b^P\right] \Gamma_B^{-1}(1+\Lambda)\left(\Upsilon_b - I_{4N_b}\right)\right.$$
$$+ \langle \hat{\mathbb{P}}_b \rangle_{GS} \Gamma_B^{-1} \begin{pmatrix} I_{2N_b} \\ -iI_{2N_b} \end{pmatrix} \Sigma_z \mathcal{G}(I_{2N_b}, iI_{2N_b})(\Gamma_B^{-1})^T \left(2\Upsilon_b - I_{4N_b}\right)\bigg], \quad (5.130)$$

where we introduce

$$\Upsilon_b = \phi\phi^T \Gamma_B^{-1}(1+\Lambda). \quad (5.131)$$

Calculation of the phase factor

If one wants to calculate the spectral function or the overlap between the two time-dependent wavefunctions, one has to take into account the phase factor $\theta_0(t)$ of the time-evolving wavefunction (see Eq. (5.117)). To be specific, let us consider the overlap $S(t)$:

$$S(t) = \langle \Psi_0 | \Psi_t \rangle, \quad |\Psi_t\rangle = e^{-i\hat{H}_b t}|\Psi_0\rangle, \tag{5.132}$$

which can be directly related to the absorption spectrum as discussed in the next subsection (see Eq. (5.145)). In the present variational approach, $S(t)$ is expressed as

$$S(t) = e^{i\theta_0(t)} e^{-\frac{1}{2}|\delta\beta(t)|^2} e^{\delta\beta^*(t) \Xi_b(t) \delta\beta^*(t)}, \tag{5.133}$$

where we denote

$$\delta\beta(t) = \beta(t) - \beta(0), \quad \beta = \frac{1}{2} Y^\dagger \phi \tag{5.134}$$

with

$$Y = \begin{pmatrix} I_{2N_b} & I_{2N_b} \\ -iI_{2N_b} & iI_{2N_b} \end{pmatrix}. \tag{5.135}$$

The real-time evolution equations of the variational parameters $\theta_0(t)$ and $\Xi_b(t)$ can be obtained as

$$\frac{d\theta_0}{dt} = -\langle \hat{\tilde{H}}_b \rangle_{b,G} + \frac{1}{4}\delta\phi^T h_\phi + \frac{1}{4}\mathrm{Tr}\,[h_\Gamma \Gamma_b] - \frac{1}{2}\mathrm{Tr}[h_1] - \mathrm{Tr}\left[h_2^\dagger \Xi_b\right], \tag{5.136}$$

$$i\frac{d\Xi_b}{dt} = \frac{1}{2}h_2 + h_1 \Xi_b + \Xi_b h_1^T + 2\Xi_b h_2^\dagger \Xi_b, \tag{5.137}$$

where we introduce

$$\delta\phi = \phi(t) - \phi(0), \quad \begin{pmatrix} h_1 & h_2 \\ h_2^\dagger & h_1^T \end{pmatrix} = \frac{1}{2} Y^\dagger h_\Gamma Y. \tag{5.138}$$

5.7.2 Application to the Rydberg Central Spin Problem

We here apply the variational approach developed in the previous section to a novel type of a strongly correlated system, namely, the Rydberg central spin problem (Fig. 5.13). We consider a Rydberg impurity immersed in a three-dimensional Bose gas [109–111, 159], which can be prepared by exciting a small fraction of atoms in the original BEC to a Rydberg state. Due to the scattering between the valence electron of the Rydberg impurity and the surrounding bosons, the impurity effectively creates a long-range oscillating potential for the bosons, which is known as the Fermi pseudo-potential [160].

5.7 Generalization to a Bosonic Environment

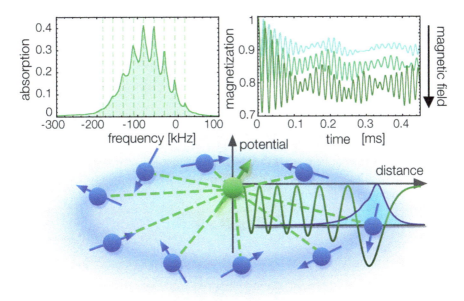

Fig. 5.13 A Rydberg impurity interacts with high-density bosons via triplet and singlet scattering channels. The Rydberg electron spin plays a role as the central spin that interacts with surrounding mobile spins via long-range interactions created by molecular potentials. The formation of the many-body-dressed molecular state manifests itself as the multiple sharp peaks in the absorption spectrum (top left panel). The Rydberg central spin exhibits a long-lasting precession manipulatable by an external magnetic field (top right panel). Reproduced from Fig. 1 of Ref. [161]. Copyright © 2019 by the American Physical Society

Choosing a certain Rydberg state, the long-range potential created by the Rydberg impurity can depend on the internal degrees of freedom of the surrounding bosons [111]. The resulting Hamiltonian can be written as

$$\hat{H} = \sum_{\alpha=\Uparrow,\Downarrow} \int d\mathbf{r} \hat{b}_\alpha^\dagger(\mathbf{r}) \hat{h}_0 \hat{b}_\alpha(\mathbf{r}) + \hat{\mathbf{S}}_\mathrm{i} \cdot \int d\mathbf{r} g(\mathbf{r}) \hat{\mathbf{S}}_\mathbf{r}, \tag{5.139}$$

$$\hat{h}_0 \equiv -\frac{\nabla^2}{2m} + V_0(\mathbf{r}), \quad \hat{\mathbf{S}}_\mathbf{r} \equiv \sum_{\alpha,\beta=\Uparrow,\Downarrow} \hat{b}_\alpha^\dagger(\mathbf{r}) \left(\frac{\boldsymbol{\sigma}}{2}\right)_{\alpha\beta} \hat{b}_\beta(\mathbf{r}), \tag{5.140}$$

where $\hat{b}_\alpha^\dagger(\mathbf{r})$ ($\hat{b}_\alpha(\mathbf{r})$) creates (annihilates) a boson with an internal state $\alpha = \Uparrow, \Downarrow$ at position $\mathbf{r}$ and $\hat{\mathbf{S}}_\mathrm{i} = \hat{\boldsymbol{\sigma}}_\mathrm{i}/2$ is the electron spin of the Rydberg impurity. We omit the hyperfine coupling and neglect the p-wave scattering contribution for the sake of simplicity. We denote

$$V_0(\mathbf{r}) = \frac{3V_T(\mathbf{r}) + V_S(\mathbf{r})}{4}, \quad g(\mathbf{r}) = V_T(\mathbf{r}) - V_S(\mathbf{r}), \tag{5.141}$$

as linear superpositions of the triplet and singlet scattering potentials V_T and V_S that are isotropic. It is convenient to define the following single-particle computational basis:

$$\hat{h}_0 \psi_l(\mathbf{r}) = \epsilon_l \psi_l(\mathbf{r}), \quad \hat{b}_{l\alpha}^\dagger = \int d\mathbf{r} \psi_l(\mathbf{r}) \hat{b}_\alpha^\dagger(\mathbf{r}), \tag{5.142}$$

where $l = (n, L, M)$ denotes a set of quantum numbers including a principal quantum number n, an angular momentum L, and a magnetic quantum number m. Hereafter, we can focus on the $L = M = 0$ sector and identify l as n because the initial state resides in this sector and the interaction parameter $g(\mathbf{r})$ is isotropic. We can then rewrite the Hamiltonian as

$$\hat{H} = \sum_{l\alpha} \epsilon_l \hat{b}_{l\alpha}^\dagger \hat{b}_{l\alpha} + \hat{\mathbf{S}}_i \cdot \hat{\boldsymbol{\Sigma}}, \tag{5.143}$$

$$\hat{\boldsymbol{\Sigma}} = \frac{1}{2} \sum_{lm\alpha\beta} g_{lm} \hat{b}_{l\alpha}^\dagger (\boldsymbol{\sigma})_{\alpha\beta} \hat{b}_{m\beta}, \quad g_{lm} \equiv \int d\mathbf{r} g(\mathbf{r}) \psi_l^*(\mathbf{r}) \psi_m(\mathbf{r}). \tag{5.144}$$

To analyze the out-of-equilibrium dynamics of this system, we use our general variational approach developed in the previous subsection. It should be challenging (if not impossible) to analyze this system with the other theoretical approaches due to the long-range nature of the impurity-bath couplings. An experimentally observable quantity is the absorption spectrum $A(\omega)$, which is the Fourier spectrum of the overlap:

$$A(\omega) = \text{Re} \int_0^\infty e^{i\omega t} S(t), \quad S(t) = \langle \Psi_0 | \Psi_t \rangle, \quad |\Psi_t\rangle = e^{-i\hat{H}t} |\Psi_0\rangle. \tag{5.145}$$

The overlap $S(t)$ can be calculated from Eq. (5.133). To obtain the time evolutions of the mean field $\boldsymbol{\phi}$ and the covariance matrix Γ_b, we integrate Eq. (5.127) by substituting the parameters ϵ_l and g_{lm} into general expressions of the functional derivatives in Eqs. (5.129) and (5.130). The phase factor of the variational state can be calculated by integrating Eqs. (5.136) and (5.137).

The initial state is chosen as

$$|\Psi_0\rangle = |\uparrow\rangle_{\text{imp}} |\text{BEC}_\downarrow\rangle, \tag{5.146}$$

which can be prepared in an experiment of Tilman Pfau's group [109–111]. Here, only the Rydberg impurity spin is prepared in the spin-up state while all the surrounding bosons are prepared in the spin-down state and occupy the lowest-energy single-particle state in free space. This condition results in the conservation law

5.7 Generalization to a Bosonic Environment

$\hat{\sigma}_i^z + 2\hat{N}_{B,\Uparrow} = 1$.[6] For this initial condition, the problem will be exactly solvable in the limit of an infinite mass $m \to \infty$, in which the system reduces to the central spin model [162] (see Appendix B).

The corresponding initial conditions for the variational parameters are

$$\boldsymbol{\phi}(t=0) = \sqrt{N_B} Y \begin{pmatrix} \int d\mathbf{r} \psi_l^*(\mathbf{r}) \psi_0(\mathbf{r}) \delta_{\alpha \Downarrow} \\ \int d\mathbf{r} \psi_l(\mathbf{r}) \psi_0^*(\mathbf{r}) \delta_{\alpha \Downarrow} \end{pmatrix}, \quad \Gamma_b(t=0) = I_{4N_b}, \quad (5.147)$$

where N_B is the total number of surrounding bosons in the system and ψ_0 denotes the wavefunction of the lowest-energy single particle state without the Rydberg potential. In practice, we impose a large distance cutoff R to obtain the initial coefficients as

$$\int d\mathbf{r} \psi_l^*(\mathbf{r}) \psi_0(\mathbf{r}) = \int_0^R dr \chi_l^*(r) \sqrt{\frac{2}{R}} \sin\left(\frac{\pi}{R} r\right), \quad (5.148)$$

where $\chi_l(r)$ denotes the radius wavefunction of an eigenstate l. The particle number N_B can be related to the density ρ via

$$\rho = N_B |\psi_0(\mathbf{r}=0)|^2 \iff N_B = \left(\frac{4}{3}\pi R^3\right) \frac{\rho}{(2\pi^2/3)}. \quad (5.149)$$

We plot numerical results on the absorption spectra $A(\omega)$, the magnetization dynamics $m_z(t) = \langle \hat{\sigma}^z(t) \rangle$ and their spectra $\tilde{m}_z(\omega)$ at different densities of bosons in Fig. 5.14. Here we assume using ^{87}Rb atoms and exciting an atom into Rb(87s). For the sake of comparisons, we also plot the results for the solvable case with an infinite mass in the top panels. As the density increases, the spectrum approaches a Gaussian profile due to stochastic occupations of bound-state energies by a large number of bosons [159]. The mean frequency can be estimated from

$$\omega_{\text{mean}} = N_B \left\langle -\frac{\nabla^2}{2m} + \frac{V_T + V_S}{2} \right\rangle_0, \quad (5.150)$$

where $\langle \cdots \rangle_0$ denotes an expectation value with respect to the initial wavefunction $\psi_0(\mathbf{r})$.

We find two features that are unique to the finite-mass case. Firstly, one can find sharp peaks present in the spectra, which are positioned with a roughly equal-spacing frequency. The spacing frequency of peaks in the spectra is roughly equal to the dominant bound-state energy of the potential $(V_T + V_S)/2$. The appearance of the potential $(V_T + V_S)/2$ can be understood from the mean-field Hamiltonian experienced by the spin-$\Downarrow$ bath bosons obtained by neglecting the spin-exchange coupling:

---

[6]In practical calculations, this spin conservation can be violated due to numerical errors associated with integrating nonlinear variational equations. To remedy this, we implement the penalty term that can ensure the spin conservation with an error below 1% (see Appendix A).

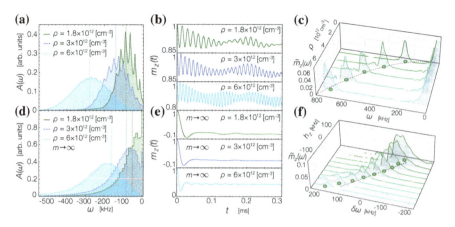

Fig. 5.14 a,d Absorption spectra $A(\omega)$ at different particle densities ρ of (**a**) mobile and (**d**) immobile (i.e., infinite mass $m \to \infty$) environmental atoms. Dashed lines indicate the mean-field shifts Δ_{mean} (c.f. Eq. (5.150)) of the spectra. **b, e** Central-spin dynamics $m_z(t) = \langle \hat{\sigma}_e(t) \rangle$ after the quench with (**b**) mobile and (**e**) immobile environmental spins. The results for the infinite-mass case are obtained by taking the ensemble average over 10^5 different realizations of atomic configurations (see Appendix B). (**c, f**) The Fourier spectra $\tilde{m}_z(\omega)$ of the central-spin dynamics $m_z(t)$ (**c**) at different particle densities ρ with a zero magnetic field and (**f**) at different magnetic fields h_z with $\rho = 1.8 \times 10^{12}$ cm^{-3}. The black dashed curve and line at the bottom planes indicate the square root scaling $\omega \propto \sqrt{\rho}$ in (**c**) and the linear relation $\delta\omega = -h_z$ in (**f**), respectively. The circles at the bottom planes indicate the mean frequencies of the spectra around the peak values. Reproduced from Figs. 2 and 4 of Ref. [161]. Copyright © 2019 by the American Physical Society

$$\hat{H}'_0 = \int d\mathbf{r}\, \hat{b}^\dagger_\downarrow(\mathbf{r}) \hat{h}_0 \hat{b}_\downarrow(\mathbf{r}) + \hat{S}^z_i \cdot \int d\mathbf{r}\, g(\mathbf{r}) \hat{S}^z_{\mathbf{r}} \tag{5.151}$$

$$= \int d\mathbf{r}\, \hat{b}^\dagger_\downarrow(\mathbf{r}) \left[-\frac{\nabla^2}{2m} + \frac{V_T(\mathbf{r}) + V_S(\mathbf{r})}{2} \right] \hat{b}_\downarrow(\mathbf{r}), \tag{5.152}$$

where we use the initial condition $\hat{S}^z_i = 1/2$ and $\hat{S}^z_{\mathbf{r}} = -\hat{b}^\dagger_\downarrow(\mathbf{r}) \hat{b}_\downarrow(\mathbf{r})/2$. To elucidate this, we calculate the spectrum by just quenching the noninteracting Hamiltonian $\hat{H}'_0$. The same feature such as the equal spacing corresponding to the most dominant bound-state energy for $(V_T + V_S)/2$ also emerges in this simple model. Yet, we still find the nonzero difference between this single-particle bound-state energy and the equal-spacing frequency found in the interacting model, which can be interpreted as a many-body shift of the bound-state energy. This many-body feature can be explicitly demonstrated by plotting the correlation $C(\nu) = \int d\omega\, \delta A(\omega) \delta A(\omega + \nu)$ of the spectrum with detuning ν, where $\delta A(\omega)$ denotes the absorption spectrum subtracted from the fitted-Gaussian profile (see Fig. 5.15).

Secondly, the magnetization $m_z(t)$ exhibits long-lasting fast oscillations that are absent in the infinite-mass case due to an incoherent summation over atomic positions (see Appendix B). The nondecaying magnetization is one of the key features in the central spin problem with an initially fully polarized environment; only a small portion of a many-body state with the opposite central spin can be admixed due to a

5.7 Generalization to a Bosonic Environment

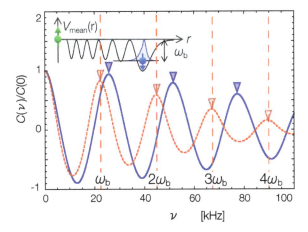

Fig. 5.15 Correlation $C(\nu)$ of the absorption spectrum with detuning ν at $\rho = 3 \times 10^{12}$ cm^{-3} (main panel). The blue solid curve (red dashed curve) shows the result obtained by quenching the full interacting Hamiltonian $\hat{H}$ (the quadratic Hamiltonian $\hat{H}'_0$). The red dashed vertical lines indicate multiple values of the deepest bound-state energy ω_b of the mean Rydberg potential $V_{\text{mean}} = (V_T + V_S)/2$ (c.f. inset), which match with the peak positions of the quadratic result but not with the interacting one, indicating the many-body dressing of the Rydberg molecular state. Reproduced from Fig. 3 of Ref. [161]. Copyright © 2019 by the American Physical Society

large energy cost to flip the central spin immersed in a polarized environment. The Rydberg spin exhibits a long-lasting precession whose frequency ω_{mag} becomes high as density is increased. To further investigate the dependence of ω_{mag} on density ρ, in Fig. 5.15c we plot the Fourier spectra $\tilde{m}_z(\omega)$ of the dynamics $m_z(t)$. As inferred from the dashed curve at the bottom of the plot, we find the square root scaling $\omega_{\text{mag}} \propto \sqrt{\rho}$ that is distinguished from the conventional linear scaling found in studies of the central spin problem. The nonanalytic behavior implies that a nonperturbative treatment (as performed here) is essential to understanding of the Rydberg central spin problem. The precession frequency and amplitude of the central spin can be manipulated by magnetic field h_z. Figure 5.15f shows Fourier spectra $\tilde{m}_z(\omega)$ at different h_z. As magnetic field is applied, the precession frequency shifts linearly with h_z from the zero-field value (see black dashed line at the bottom in Fig. 5.15f) while it eventually exhibits the nonlinear behavior at a large field. The precession amplitude is enhanced (suppressed) when a magnetic field is applied in such a way that the resonance is approached (departed) (c.f. Figs. 5.13 and 5.15f). These magnetic-field dependences are consistent with those found in the conventional central spin problem [163], suggesting a possibility to control the electron spin of dense Rydberg gases in an analogous way as in solid-state qubits [112, 162, 164–167].

5.8 Conclusions and Outlook

In this chapter, we have studied in- and out-of-equilibrium physics of a strongly correlated open quantum system by focusing on its most fundamental paradigm, a quantum impurity. There, the presence of the strong system-environment entanglement plays a central role in understanding the physics. Such a nontrivial role of the system-environment entanglement has not been addressed in the studies of the first part of this Thesis, as they rely on the assumption that an environment (or a meter) is memoryless and thus the system-environment correlation is (almost) absent.

More specifically, we have developed a versatile and efficient theoretical approach to study the ground-state properties and nonequilibrium dynamics of generic quantum spin-impurity systems [85, 138, 147, 161, 168]. This variational approach has been motivated by the original papers by Tomonaga [86] and Lee, Low and Pines [87]. A key idea is to introduce a canonical transformation that decouples the impurity and the bath degrees of freedom such that the impurity dynamics can be made completely frozen in the transformed frame. We have constructed such a decoupling transformation for spin-impurity models by employing the conserved parity operator in the total Hamiltonian. We have also discussed its generalization to two-impurity systems and the Anderson model. Combining the constructed canonical transformation with the fermionic Gaussian states, we have presented a family of variational states that can efficiently represent nontrivial entanglement between the impurity spin and the environment. Integrating the imaginary- and real-time evolutions projected on this variational manifold, one can study ground-state and dynamical properties of spin-impurity models on demand. We have throughly benchmarked our variational approach with the MPS-based method, the Bethe-ansatz solution, the known nonperturbative scalings and conductance behavior. A remarkable efficiency has been achieved in our approach, as we found accuracy comparable to the MPS-based method with several orders of magnitude fewer variational parameters. The simplicity of our variational approach will allow one to provide physical insights into challenging problems of strongly correlated open quantum systems.

Our approach has already found applications to exploring new types of nonequilibrium phenomena that have not been studied in the other methods. Examples include the long-time crossover dynamics in the FM easy-plane in the anisotropic Kondo model and the spatiotemporal environment dynamics in the two-lead Kondo model. Such long-time spatiotemporal dynamics are difficult (if not impossible) to obtain in the previous approaches. We propose to use quantum gas microscopy to experimentally test the predicted dynamics. We also generalize our approach to a bosonic environment and apply it to the novel type of the impurity system that can be realized by Rydberg atoms. The predicted absorption spectrum can be tested with the state-of-the-art experimental techniques of ultracold atoms. In these respects, one of the remarkable features in our approach is that it enables one to make quantitative comparisons between theory and experiments under nonequilibrium conditions even in long-time regimes. This advantage is particularly important in view of recent developments of simulating quantum dynamics such as in ultracold atoms [34, 143–146, 149, 169–179] and quantum dots [35–39].

5.8 Conclusions and Outlook

The present variational approach can be readily generalized in many ways. For example, our formulation can be applied to systems associated with disorder [180], an interacting bath (e.g., as in the Kondo-Hubbard model [181]), multiple channels [18, 38, 182], and long-range interactions [183, 184]. The canonical transformations introduced in Sects. 5.2.2 and 5.2.3 can be used to analyze two-impurity systems and the Anderson-type Hamiltonian. The present approach can be also generalized to analyze quantum pumping and driven systems, in which previous studies have mainly focused on noninteracting systems [185, 186]. It merits a further study to analyze the full-counting distribution in charge transport based on our approach. Most of previous studies in this direction have been restricted to (bosonized) one-dimensional models [187] and noninteracting systems [188]. Testing the maximally fast information scrambling [189] in the multi-channel systems [72] and quasi-integrable systems [190] will be a particularly interesting direction. On another front, our approach could be used as a new type of impurity solver for DMFT [21]. Another promising direction is to generalize the present approach to Gaussian density matrices such that it can be applied to dissipative systems [191] and finite-temperature systems.

Appendix A: Penalty Term to Ensure the Spin Conservation

We provide details on the penalty term used in Sect. 5.7.2 to ensure the spin conservation of the impurity and the bosonic bath. There, we consider the initial state that is an eigenstate of the quantity $\hat{\sigma}_{\text{imp}}^z + \hat{\sigma}_{\text{bath}}^z + \hat{N}_B \equiv \hat{\sigma}_{\text{imp}}^z + 2\hat{N}_{B,\uparrow}$ with eigenvalue 1, where $\hat{\sigma}_{\text{imp}}^z$ and $\hat{\sigma}_{\text{bath}}^z$ are the spin-z components of the impurity spin and the (pseudo spin-1/2) bosonic particles, respectively, and $\hat{N}_B$ is the total number of bosons. Since both the total spin-z component $\hat{\sigma}_{\text{imp}}^z + \hat{\sigma}_{\text{bath}}^z$ and the particle number $\hat{N}_B$ are conserved, the above quantity should also be conserved during the variational time evolution. However, in practical calculations, its value during the real-time evolution is not necessarily kept exactly at the initial value $\hat{\sigma}_{\text{imp}}^z + 2\hat{N}_{B,\uparrow} = 1$ due to accumulated numerical errors caused by a highly nonlinear nature of the variational equations, especially in a high-density regime of bosons.

As the spin conservation is essential to understand the physics in the Rydberg Kondo problem in Sect. 5.7.2, we have to make sure that the conservation is satisfied during the variational evolution. This can be achieved by adding the following penalty term to the Hamiltonian:

$$\hat{V} = \lambda \left(\hat{\sigma}_{\text{imp}}^z + 2\hat{N}_{B,\uparrow} - 1 \right)^2, \tag{A.1}$$

where we choose λ as an appropriate value to ensure the spin conservation. In the transformed frame, it is written as

$$\hat{\tilde{V}} = \hat{U}_b^\dagger \hat{V} \hat{U}_b = \lambda \left(\hat{\sigma}_{\text{imp}}^x \hat{\mathbb{P}}_b + 2\hat{N}_{B,\uparrow} - 1 \right)^2. \tag{A.2}$$

Its expectation value with respect to the bosonic Gaussian state is

$$\langle \hat{\tilde{V}} \rangle_{b,G} = \lambda \left(2 - 4\langle N_{B,\uparrow} \rangle_{b,G} + 4\langle \hat{N}_{B,\uparrow}^2 \rangle_{b,G} - 2\sigma^x_{\text{imp}} \langle \hat{\mathbb{P}}_b \rangle_{b,G} + 4\sigma^x_{\text{imp}} \text{Tr}\left[P_\uparrow \Gamma_b^P\right] \right)$$

$$= \lambda \bigg(2 - \text{Tr}[P_\uparrow (\Gamma_b - I_{4N_b})] - \phi^T P_\uparrow \phi$$

$$+ \frac{1}{4}\left(\text{Tr}[P_\uparrow (\Gamma_b - I_{4N_b})] + \phi^T P_\uparrow \phi\right)^2 + \frac{1}{2}\left(\text{Tr}[P_\uparrow \Gamma_b P_\uparrow \Gamma_b] - \text{Tr}[P_\uparrow]\right) + \phi^T P_\uparrow \Gamma_b P_\uparrow \phi$$

$$- 2\sigma^x_{\text{imp}} \langle \hat{\mathbb{P}}_b \rangle_{b,G} + 4\sigma^x_{\text{imp}} \text{Tr}\left[P_\uparrow \Gamma_b^P\right] \bigg). \tag{A.3}$$

where $P_\uparrow = I_2 \otimes ((\sigma^z + 1)/2 \otimes I_{N_b})$. After some calculations, we can obtain its functional derivative with respect to the mean field ϕ as

$$h^V_\phi = 2\frac{\delta \langle \hat{\tilde{V}} \rangle_{b,G}}{\delta \phi}$$

$$= \lambda \bigg[2\left(-2 + \text{Tr}[P_\uparrow (\Gamma_b - I_{4N_b})] + \phi^T P_\uparrow \phi\right) P_\uparrow \phi + 4 P_\uparrow \Gamma_b P_\uparrow \phi$$

$$+ \frac{\delta}{\delta \phi}(-4\sigma^x_{\text{imp}} \langle \hat{\mathbb{P}}_b \rangle_{b,G} + 8\sigma^x_{\text{imp}} \text{Tr}\left[P_\uparrow \Gamma_b^P\right]) \bigg], \tag{A.4}$$

where the last term is given by

$$\frac{\delta}{\delta \phi}(-4\sigma^x_{\text{imp}} \langle \hat{\mathbb{P}}_b \rangle_{b,G} + 8\sigma^x_{\text{imp}} \text{Tr}\left[P_\uparrow \Gamma_b^P\right])$$

$$= \sigma^x_{\text{imp}} \bigg[\left(4\langle \hat{\mathbb{P}}_b \rangle_{b,G} - 8\text{Tr}\left[P_\uparrow \Gamma_b^P\right]\right) \Gamma_B^{-1}(I_{4N_b} + \Lambda)\phi$$

$$- 16\langle \hat{\mathbb{P}}_b \rangle_{b,G} \Gamma_B^{-1} \mathcal{S}\left[\begin{pmatrix} I_{2N_b} \\ -iI_{2N_b} \end{pmatrix} \Sigma_z P_\uparrow (I_{2N_b}, iI_{2N_b})\right] (\Gamma_B^{-1})^T \phi \bigg]. \tag{A.5}$$

In the similar manner, the functional derivative with respect to the covariance matrix Γ_b can be obtained as

$$h^V_\Gamma = 4\frac{\delta \langle \hat{\tilde{V}} \rangle_{b,G}}{\delta \Gamma_b}$$

$$= \lambda \bigg[2\left(-2 + \text{Tr}[P_\uparrow (\Gamma_b - I_{4N_b})] + \phi^T P_\uparrow \phi\right) P_\uparrow + 4 P_\uparrow (\Gamma_b + \phi \phi^T) P_\uparrow$$

$$+ \frac{\delta}{\delta \Gamma_b}(-8\sigma^x_{\text{imp}} \langle \hat{\mathbb{P}}_b \rangle_{b,G} + 16\sigma^x_{\text{imp}} \text{Tr}\left[P_\uparrow \Gamma_b^P\right]) \bigg], \tag{A.6}$$

5.8 Conclusions and Outlook

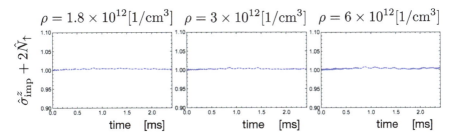

Fig. 5.16 Typical time evolutions of the conserved quantity $\hat{\sigma}_{\text{imp}}^z + \hat{N}_{B,\uparrow}$ at different densities of bosons (see Sect. 5.7.2 for details about the quench protocol)

where the last term is given by

$$\frac{\delta}{\delta \Gamma_b}(-8\sigma_{\text{imp}}^x \langle \hat{\mathbb{P}}_b \rangle_{b,G} + 16\sigma_{\text{imp}}^x \text{Tr}\left[P_\uparrow \Gamma_b^P\right])$$

$$= \sigma_{\text{imp}}^x \mathcal{S}\bigg[(-4\langle \hat{\mathbb{P}}_b \rangle_{b,G} + 8\text{Tr}\left[P_\uparrow \Gamma_b^P\right])\Gamma_B^{-1}(\mathbb{I}_{4N_b} + \Lambda)\left(\Upsilon_b - \mathbb{I}_{4N_b}\right)$$

$$+ 16\langle \hat{\mathbb{P}}_b \rangle_{b,G}\Gamma_B^{-1}\begin{pmatrix}\mathbb{I}_{2N_b}\\-i\mathbb{I}_{2N_b}\end{pmatrix}\Sigma_z P_\uparrow(\mathbb{I}_{2N_b}, i\mathbb{I}_{2N_b})(\Gamma_B^{-1})^\text{T}\left(2\Upsilon_b - \mathbb{I}_{4N_b}\right)\bigg]. \quad (A.7)$$

Figure 5.16 shows typical time evolutions of the quantity $\hat{\sigma}_{\text{imp}}^z + \hat{N}_{B,\uparrow}$, which is conserved with an error below $\sim 1\%$.

Appendix B: Exactly Solvable Dynamics of the Central Spin Problem

Here we explain how the numerical results for the infinite-mass case in Fig. 5.14 in this chapter have been obtained. In the limit of an infinite mass $m \to \infty$ in the Hamiltonian (5.143), the quench dynamics starting from the initial condition (5.146) is exactly solvable via the Laplace transformation [162]. In practice, to obtain the absorption spectrum and the magnetization dynamics, we proceed as follows.

First of all, we randomly generate a set of positions of atoms $\{r_i\}_{i=1}^N$ according to the initial wavefunction $\prod_{i=1}^N \psi_0(\mathbf{r}_i)$. We then obtain the corresponding values of the parameters

$$V_0 \equiv \sum_{i=1}^N V_0(\mathbf{r}_i) = \sum_{i=1}^N \frac{3V_T(\mathbf{r}_i) + V_S(\mathbf{r}_i)}{4}, \quad g_i \equiv g(\mathbf{r}_i) = V_T(\mathbf{r}_i) - V_S(\mathbf{r}_i) \quad \text{(B.1)}$$

in the (infinite-mass) Hamiltonian

$$\hat{H}^{\{r_i\}}_{m\to\infty} = V_0 + \hat{S}_{imp} \cdot \sum_{i=1}^{N} g_i \hat{S}_i. \tag{B.2}$$

Since the total magnetization is conserved, the time-evolution is described by the following wavefunction:

$$|\Psi(t)\rangle = \xi_0(t)|\uparrow\rangle_{imp}|\downarrow\rangle_1 \cdots |\downarrow\rangle_N + \sum_{i=1}^{N} \xi_i(t)|\downarrow\rangle_{imp}|\downarrow\rangle_1 \cdots |\uparrow\rangle_i \cdots |\downarrow\rangle_N. \tag{B.3}$$

Then, the Schrödinger equation becomes

$$i\dot{\xi}_0(t) = \left(V_0 - \frac{G}{4}\right)\xi_0 + \frac{1}{2}\sum_{i=1}^{N} g_i \xi_i, \tag{B.4}$$

$$i\dot{\xi}_i(t) = \left(V_0 + \frac{G-2g_i}{4}\right)\xi_i + \frac{1}{2}g_i\xi_0, \tag{B.5}$$

where $G = \sum_{i=1}^{N} g_i$. This equation can be analytically solved by the Laplace transformation

$$\tilde{\xi}(s) = \int_0^\infty dt\, \xi(t) e^{-st}, \tag{B.6}$$

$$\frac{d\xi}{dt}(t) \to s\tilde{\xi}(s) - \xi(0), \tag{B.7}$$

and by using the initial condition:

$$\xi_0(0) = 1, \quad \xi_i(0) = 0 \ (i = 1, 2, \ldots N). \tag{B.8}$$

The result is

$$\xi_0(t) = \frac{1}{2\pi i}\int_\Gamma d\omega\, e^{-i\omega t - i\Omega t}\left[\omega + \frac{G}{2} - \frac{1}{4}\sum_{i=1}^{N}\frac{g_i^2}{\omega + \frac{g_i}{2}}\right]^{-1}, \tag{B.9}$$

where $\Omega = V_0 + \frac{G}{4}$ and the contour Γ is chosen so that all the poles in the integral lie above it. Using lHopital's rule and performing the integration, we obtain

$$\xi_0(t) = e^{-i\Omega t}\sum_{l=1}^{N+1}\frac{e^{-i\omega_l t}}{1 + \frac{1}{4}\sum_{i=1}^{N}\frac{g_i^2}{(\omega_l + g_i/2)^2}} \equiv e^{-i\Omega t}\sum_{l=1}^{N+1} w_l e^{-i\omega_l t}, \tag{B.10}$$

5.8 Conclusions and Outlook

where $\{\omega_l\}_{l=1}^{N+1}$ is a set of poles in the integration, which can in practice be obtained by diagonalizing $\sum_{i=1}^{N} g_i \hat{\mathbf{S}}_{\text{imp}} \cdot \hat{\mathbf{S}}_i$ within the sector of interest here. In fact, Eq. (B.10) directly gives the time evolution of the overlap:

$$S^{\{r_i\}}(t) = \langle \Psi_0 | e^{-i\hat{H}_{m\to\infty}^{\{r_i\}} t} | \Psi_0 \rangle = \xi_0(t). \tag{B.11}$$

Thus, the absorption spectrum is

$$A^{\{r_i\}}(\omega) = \sum_{l=1}^{N+1} \delta(\omega - \Omega - \omega_l) w_l. \tag{B.12}$$

We then repeat the same calculation of $A^{\{r_i\}}(\omega)$ for different realizations of atomic configurations $\{r_i\}$ and take the ensemble average to obtain

$$A(\omega) = \sum_{\{r_i\}} P(\{r_i\}) A^{\{r_i\}}(\omega), \tag{B.13}$$

where $P(\{r_i\})$ is the distribution function of $\{r_i\}$ determined from the initial state. The magnetization can be obtained from (c.f. Eq. (B.3))

$$m_z(t) \equiv \langle \hat{\sigma}_{\text{imp}}^z(t) \rangle = |\xi_0(t)|^2 - \sum_{i=1}^{N} |\xi_i(t)|^2 = 2|\xi_0(t)|^2 - 1. \tag{B.14}$$

There are several important features which we can infer from the above analysis. First, since the initial state is the equal superposition of the triplet and singlet states, the mean value of $A(\omega)$ in a high-density regime can be estimated from

$$\omega_{\text{mean}}^{m\to\infty} = N \frac{\langle V_T + V_S \rangle_0}{2}, \quad \langle \cdots \rangle_0 \equiv \int d\mathbf{r} |\psi_0(\mathbf{r})|^2 \cdots . \tag{B.15}$$

As density increases, $A(\omega)$ will approach to a Gaussian distribution with this mean value. Second, the Fourier spectrum of the magnetization $m_z(t)$ essentially shares the same information with the absorption spectrum (c.f. Eq. (B.14)):

$$\tilde{m}_z(\omega) = 2 \int d\omega' A^*(\omega' - \omega) A(\omega). \tag{B.16}$$

If $A(\omega)$ is a Gaussian distribution, $\tilde{m}_z(\omega)$ is also a Gaussian distribution but with zero mean, indicating that its Fourier transform $m_z(t)$ will exhibit decoherence to a finite magnetization. This decoherence can be understood as a consequence of the incoherent summation over different atomic positions. These features have been confirmed in the numerical results presented in Fig. 5.14.

References

1. Leggett AJ, Chakravarty S, Dorsey AT, Fisher MPA, Garg A, Zwerger W (1987) Dynamics of the dissipative two-state system. Rev Mod Phys 59:1–85
2. Loss D, DiVincenzo DP (1998) Quantum computation with quantum dots. Phys Rev A 57:120–126
3. Zhang W, Konstantinidis N, Al-Hassanieh KA, Dobrovitski VV (2007) Modelling decoherence in quantum spin systems. J Phys Cond Matt 19:083202
4. Andres K, Graebner JE, Ott HR (1975) 4f-virtual-bound-state formation in $CeAl_3$ at low temperatures. Phys Rev Lett 35:1779–1782
5. Hewson AC (1997) The Kondo problem to heavy fermions. Cambridge University Press, Cambridge New York
6. Löhneysen HV, Rosch A, Vojta M, Wölfle P (2007) Fermi-liquid instabilities at magnetic quantum phase transitions. Rev Mod Phys 79:1015
7. Gegenwart P, Si Q, Steglich F (2008) Quantum criticality in heavy-fermion metals. Nat Phys 4:186–197
8. Si Q, Steglich F (2010) Heavy fermions and quantum phase transitions. Science 329:1161–1166
9. Glazman L, Raikh M (1988) Resonant Kondo transparency of a barrier with quasilocal impurity states. JETP Lett 47:452–455
10. Ng TK, Lee PA (1988) On-site coulomb repulsion and resonant tunneling. Phys Rev Lett 61:1768
11. Meir Y, Wingreen NS, Lee PA (1993) Low-temperature transport through a quantum dot: the Anderson model out of equilibrium. Phys Rev Lett 70:2601–2604
12. Liang W, Shores MP, Bockrath M, Long JR, Park H (2002) Kondo resonance in a single-molecule transistor. Nature 417:725–729
13. Yu LH, Natelson D (2004) The Kondo effect in C60 single-molecule transistors. Nano Lett 4:79–83
14. Goldhaber-Gordon D, Göres J, Kastner MA, Shtrikman H, Mahalu D, Meirav U (1998) From the Kondo regime to the mixed-valence regime in a single-electron transistor. Phys Rev Lett 81:5225–5228
15. Cronenwett SM, Oosterkamp TH, Kouwenhoven LP (1998) A tunable Kondo effect in quantum dots. Science 281:540–544
16. Simmel F, Blick RH, Kotthaus JP, Wegscheider W, Bichler M (1999) Anomalous Kondo effect in a quantum dot at nonzero bias. Phys Rev Lett 83:804–807
17. van der Wiel WG, Franceschi SD, Fujisawa T, Elzerman JM, Tarucha S, Kouwenhoven LP (2000) The Kondo effect in the unitary limit. Science 289:2105–2108
18. Potok RM, Rau IG, Shtrikman H, Oreg Y, Goldhaber-Gordon D (2007) Observation of the two-channel Kondo effect. Nature 446:167–171
19. Kretinin AV, Shtrikman H, Goldhaber-Gordon D, Hanl M, Weichselbaum A, von Delft J, Costi T, Mahalu D (2011) Spin-$\frac{1}{2}$ Kondo effect in an InAs nanowire quantum dot: unitary limit, conductance scaling, and Zeeman splitting. Phys Rev B 84:245316
20. Kretinin AV, Shtrikman H, Mahalu D (2012) Universal line shape of the Kondo zero-bias anomaly in a quantum dot. Phys Rev B 85:201301
21. Georges A, Kotliar G, Krauth W, Rozenberg MJ (1996) Dynamical mean-field theory of strongly correlated fermion systems and the limit of infinite dimensions. Rev Mod Phys 68:13–125
22. Kondo J (1964) Resistance minimum in dilute magnetic alloys. Prog Theor Phys 32:37–49
23. Anderson P (1970) A poor man's derivation of scaling laws for the Kondo problem. J Phys C 3:2436
24. Wilson KG (1975) The renormalization group: critical phenomena and the Kondo problem. Rev Mod Phys 47:773–840
25. Bulla R, Tong N-H, Vojta M (2003) Numerical renormalization group for bosonic systems and application to the sub-ohmic spin-boson model. Phys Rev Lett 91:170601

26. Borda L (2007) Kondo screening cloud in a one-dimensional wire: numerical renormalization group study. Phys Rev B 75:041307
27. Bulla R, Costi TA, Pruschke T (2008) Numerical renormalization group method for quantum impurity systems. Rev. Mod. Phys. 80:395–450
28. Saberi H, Weichselbaum A, von Delft J (2008) Matrix-product-state comparison of the numerical renormalization group and the variational formulation of the density-matrix renormalization group. Phys Rev B 78:035124
29. Borda L, Garst M, Kroha J (2009) Kondo cloud and spin-spin correlations around a partially screened magnetic impurity. Phys Rev B 79:100408
30. Büsser CA, Martins GB, Costa Ribeiro L, Vernek E, Anda EV, Dagotto E (2010) Numerical analysis of the spatial range of the Kondo effect. Phys Rev B 81:045111
31. Kawakami N, Okiji A (1981) Exact expression of the ground-state energy for the symmetric Anderson model. Phys Lett A 86:483–486
32. Andrei N, Furuya K, Lowenstein JH (1983) Solution of the Kondo problem. Rev Mod Phys 55:331–402
33. Schlottmann P (1989) Some exact results for dilute mixed-valent and heavy-fermion systems. Phys Rep 181:1–119
34. Riegger L, Darkwah Oppong N, Höfer M, Fernandes DR, Bloch I, Fölling S (2018) Localized magnetic moments with tunable spin exchange in a gas of ultracold fermions. Phys Rev Lett 120:143601
35. De Franceschi S, Hanson R, van der Wiel WG, Elzerman JM, Wijpkema JJ, Fujisawa T, Tarucha S, Kouwenhoven LP (2002) Out-of-equilibrium Kondo effect in a mesoscopic device. Phys Rev Lett 89:156801
36. Türeci HE, Hanl M, Claassen M, Weichselbaum A, Hecht T, Braunecker B, Govorov A, Glazman L, Imamoglu A, von Delft J (2011) Many-body dynamics of exciton creation in a quantum dot by optical absorption: a quantum quench towards Kondo correlations. Phys Rev Lett 106:107402
37. Latta C, Haupt F, Hanl M, Weichselbaum A, Claassen M, Wuester W, Fallahi P, Faelt S, Glazman L, von Delft J, Türeci HE, Imamoglu A (2011) Quantum quench of Kondo correlations in optical absorption. Nature 474:627–630
38. Iftikhar Z, Jezouin S, Anthore A, Gennser U, Parmentier FD, Cavanna A, Pierre F (2015) Two-channel Kondo effect and renormalization flow with macroscopic quantum charge states. Nature 526:233–236
39. Desjardins MM, Viennot JJ, Dartiailh MC, Bruhat LE, Delbecq MR, Lee M, Choi M-S, Cotter A, Kontos T (2017) Observation of the frozen charge of a Kondo resonance. Nature 545:71–74
40. Schmidt TL, Werner P, Mühlbacher L, Komnik A (2008) Transient dynamics of the Anderson impurity model out of equilibrium. Phys Rev B 78:235110
41. Werner P, Oka T, Millis AJ (2009) Diagrammatic Monte Carlo simulation of nonequilibrium systems. Phys Rev B 79:035320
42. Schiró M, Fabrizio M (2009) Real-time diagrammatic Monte Carlo for nonequilibrium quantum transport. Phys Rev B 79:153302
43. Werner P, Oka T, Eckstein M, Millis AJ (2010) Weak-coupling quantum Monte Carlo calculations on the Keldysh contour: theory and application to the current-voltage characteristics of the Anderson model. Phys Rev B 81:035108
44. Cohen G, Gull E, Reichman DR, Millis AJ, Rabani E (2013) Numerically exact long-time magnetization dynamics at the nonequilibrium Kondo crossover of the Anderson impurity model. Phys Rev B 87:195108
45. Nordlander P, Pustilnik M, Meir Y, Wingreen NS, Langreth DC (1999) How long does it take for the Kondo effect to develop? Phys Rev Lett 83:808–811
46. Kaminski A, Nazarov YV, Glazman LI (2000) Universality of the Kondo effect in a quantum dot out of equilibrium. Phys Rev B 62:8154–8170
47. Hackl A, Kehrein S (2008) Real time evolution in quantum many-body systems with unitary perturbation theory. Phys Rev B 78:092303

48. Keil M, Schoeller H (2001) Real-time renormalization-group analysis of the dynamics of the spin-boson model. Phys Rev B 63:180302
49. Pletyukhov M, Schuricht D, Schoeller H (2010) Relaxation versus decoherence: spin and current dynamics in the anisotropic Kondo model at finite bias and magnetic field. Phys Rev Lett 104:106801
50. Hackl A, Roosen D, Kehrein S, Hofstetter W (2009) Nonequilibrium spin dynamics in the ferromagnetic Kondo model. Phys Rev Lett 102:196601
51. Hackl A, Vojta M, Kehrein S (2009) Nonequilibrium magnetization dynamics of ferromagnetically coupled Kondo spins. Phys Rev B 80:195117
52. Tomaras C, Kehrein S (2011) Scaling approach for the time-dependent Kondo model. Europhys Lett 93:47011
53. Bera S, Nazir A, Chin AW, Baranger HU, Florens S (2014) Generalized multipolaron expansion for the spin-boson model: environmental entanglement and the biased two-state system. Phys Rev B 90:075110
54. Florens S, Snyman I (2015) Universal spatial correlations in the anisotropic Kondo screening cloud: analytical insights and numerically exact results from a coherent state expansion. Phys Rev B 92:195106
55. Blunden-Codd Z, Bera S, Bruognolo B, Linden N-O, Chin AW, von Delft J, Nazir A, Florens S (2017) Anatomy of quantum critical wave functions in dissipative impurity problems. Phys Rev B 95:085104
56. White SR, Feiguin AE (2004) Real-time evolution using the density matrix renormalization group. Phys Rev Lett 93:076401
57. Schmitteckert P (2004) Nonequilibrium electron transport using the density matrix renormalization group method. Phys Rev B 70:121302
58. Al-Hassanieh KA, Feiguin AE, Riera JA, Büsser CA, Dagotto E (2006) Adaptive time-dependent density-matrix renormalization-group technique for calculating the conductance of strongly correlated nanostructures. Phys Rev B 73:195304
59. Dias da Silva LGGV, Heidrich-Meisner F, Feiguin AE, Büsser CA, Martins GB, Anda EV, Dagotto E (2008) Transport properties and Kondo correlations in nanostructures: time-dependent dmrg method applied to quantum dots coupled to Wilson chains. Phys Rev B 78:195317
60. Weichselbaum A, Verstraete F, Schollwöck U, Cirac JI, von Delft J (2009) Variational matrix-product-state approach to quantum impurity models. Phys Rev B 80:165117
61. Heidrich-Meisner F, Feiguin AE, Dagotto E (2009) Real-time simulations of nonequilibrium transport in the single-impurity Anderson model. Phys Rev B 79:235336
62. Heidrich-Meisner F, González I, Al-Hassanieh KA, Feiguin AE, Rozenberg MJ, Dagotto E (2010) Nonequilibrium electronic transport in a one-dimensional Mott insulator. Phys Rev B 82:205110
63. Nghiem HTM, Costi TA (2017) Time evolution of the Kondo resonance in response to a quench. Phys Rev Lett 119:156601
64. Anders FB, Schiller A (2005) Real-time dynamics in quantum-impurity systems: a time-dependent numerical renormalization-group approach. Phys Rev Lett 95:196801
65. Anders FB, Schiller A (2006) Spin precession and real-time dynamics in the Kondo model: time-dependent numerical renormalization-group study. Phys Rev B 74:245113
66. Anders FB, Bulla R, Vojta M (2007) Equilibrium and nonequilibrium dynamics of the sub-ohmic spin-boson model. Phys Rev Lett 98:210402
67. Anders FB (2008) Steady-state currents through nanodevices: a scattering-states numerical renormalization-group approach to open quantum systems. Phys Rev Lett 101:066804
68. Roosen D, Wegewijs MR, Hofstetter W (2008) Nonequilibrium dynamics of anisotropic large spins in the Kondo regime: time-dependent numerical renormalization group analysis. Phys Rev Lett 100:087201
69. Eckel J, Heidrich-Meisner F, Jakobs SG, Thorwart M, Pletyukhov M, Egger R (2010) Comparative study of theoretical methods for non-equilibrium quantum transport. New J Phys 12:043042

70. Lechtenberg B, Anders FB (2014) Spatial and temporal propagation of Kondo correlations. Phys Rev B 90:045117
71. Nuss M, Ganahl M, Arrigoni E, von der Linden W, Evertz HG (2015) Nonequilibrium spatiotemporal formation of the Kondo screening cloud on a lattice. Phys Rev B 91:085127
72. Dóra B, Werner MA, Moca CP (2017) Information scrambling at an impurity quantum critical point. Phys Rev B 96:155116
73. Lesage F, Saleur H, Skorik S (1996) Time correlations in 1D quantum impurity problems. Phys Rev Lett 76:3388–3391
74. Lesage F, Saleur H (1998) Boundary interaction changing operators and dynamical correlations in quantum impurity problems. Phys Rev Lett 80:4370–4373
75. Schiller A, Hershfield S (1998) Toulouse limit for the nonequilibrium Kondo impurity: currents, noise spectra, and magnetic properties. Phys Rev B 58:14978–15010
76. Lobaskin D, Kehrein S (2005) Crossover from nonequilibrium to equilibrium behavior in the time-dependent Kondo model. Phys Rev B 71:193303
77. Vasseur R, Trinh K, Haas S, Saleur H (2013) Crossover physics in the nonequilibrium dynamics of quenched quantum impurity systems. Phys Rev Lett 110:240601
78. Ghosh S, Ribeiro P, Haque M (2014) Real-space structure of the impurity screening cloud in the resonant level model. J Stat Mech Theor Exp 2014:P04011
79. Medvedyeva M, Hoffmann A, Kehrein S (2013) Spatiotemporal buildup of the Kondo screening cloud. Phys Rev B 88:094306
80. Bolech CJ, Shah N (2016) Consistent bosonization-debosonization. ii. the two-lead Kondo problem and the fate of its nonequilibrium toulouse point. Phys Rev B 93:085441
81. Schollwöck U (2011) The density-matrix renormalization group in the age of matrix product states. Ann Phys 326:96–192
82. Rosch A (2012) Wilson chains are not thermal reservoirs. Eur Phys J B 85:6
83. Leggett AJ (2001) Bose-Einstein condensation in the alkali gases: some fundamental concepts. Rev Mod Phys 73:307–356
84. Verstraete F, Murg V, Cirac JI (2008) Matrix product states, projected entangled pair states, and variational renormalization group methods for quantum spin systems. Adv Phys 57:143–224
85. Ashida Y, Shi T, Bañuls MC, Cirac JI, Demler E (2018) Variational principle for quantum impurity systems in and out of equilibrium: application to Kondo problems. Phys Rev B 98:024103
86. Tomonaga S (1947) On the effect of the field reactions on the interaction of mesotrons and nuclear particles. III. Prog Theor Phys 2:6–24
87. Lee TD, Low FE, Pines D (1953) The motion of slow electrons in a polar crystal. Phys Rev 90:297–302
88. Devreese JT, Alexandrov AS (2009) Fröhlich polaron and bipolaron: recent developments. Rep Prog Phys 72:066501
89. Devreese JT (2010) Fröhlich polarons. lecture course including detailed theoretical derivations. arXiv:1012.4576
90. Alexandrov AS (2007) Polarons in advanced materials. Canopus Pub
91. Feynman RP (1955) Slow electrons in a polar crystal. Phys Rev 97:660–665
92. Bardeen J, Baym G, Pines D (1967) Effective interaction of He 3 atoms in dilute solutions of He 3 in He 4 at low temperatures. Phys Rev 156:207
93. Nagy P (1990) The polaron and squeezed states. J Phys Cond Matt 2:10573
94. Zhang W-M, Feng DH, Gilmore R (1990) Coherent states: theory and some applications. Rev Mod Phys 62:867–927
95. Altanhan T, Kandemir BS (1993) A squeezed state approach for the large polarons. J Phys Cond Matt 5:6729
96. Tempere J, Casteels W, Oberthaler MK, Knoop S, Timmermans E, Devreese JT (2009) Feynman path-integral treatment of the BEC-impurity polaron. Phys Rev B 80:184504
97. Novikov A, Ovchinnikov M (2010) Variational approach to the ground state of an impurity in a Bose-Einstein condensate. J Phys B 43:105301

98. Casteels W, Van Cauteren T, Tempere J, Devreese JT (2011) Strong coupling treatment of the polaronic system consisting of an impurity in a condensate. Laser Phys 21:1480–1485
99. Casteels W, Tempere J, Devreese JT (2011) Many-polaron description of impurities in a Bose-Einstein condensate in the weak-coupling regime. Phys Rev A 84:063612
100. Rath SP, Schmidt R (2013) Field-theoretical study of the Bose polaron. Phys Rev A 88:053632
101. Vlietinck J, Ryckebusch J, Van Houcke K (2013) Quasiparticle properties of an impurity in a Fermi gas. Phys Rev B 87:1–11
102. Vlietinck J, Casteels W, Van Houcke K, Tempere J, Ryckebusch J, Devreese JT (2014) Diagrammatic Monte Carlo study of the acoustic and the BEC polaron. New J Phys:9
103. Li W, Das Sarma S (2014) Variational study of polarons in Bose-Einstein condensates. Phys Rev A 90:013618
104. Grusdt F, Shchadilova YE, Rubtsov AN, Demler E (2015) Renormalization group approach to the Fröhlich polaron model: application to impurity-BEC problem. Sci Rep 5:12124
105. Shchadilova YE, Grusdt F, Rubtsov AN, Demler E (2016) Polaronic mass renormalization of impurities in Bose-Einstein condensates: correlated Gaussian-wave-function approach. Phys Rev A 93:043606
106. Shchadilova YE, Schmidt R, Grusdt F, Demler E (2016) Quantum dynamics of ultracold Bose polarons. Phys Rev Lett 117:113002
107. Ashida Y, Schmidt R, Tarruell L, Demler E (2018) Many-body interferometry of magnetic polaron dynamics. Phys Rev B 97:060302
108. Konik RM, Saleur H, Ludwig A (2002) Transport in quantum dots from the integrability of the Anderson model. Phys Rev B 66:125304
109. Bendkowsky V, Butscher B, Nipper J, Shaffer JP, Löw R, Pfau T (2009) Observation of ultralong-range Rydberg molecules. Nature 458:1005–1008
110. Gaj A, Krupp AT, Balewski JB, Löw R, Hofferberth S, Pfau T (2014) From molecular spectra to a density shift in dense Rydberg gases. Nat Commun 5:4546
111. Böttcher F, Gaj A, Westphal KM, Schlagmüller M, Kleinbach KS, Löw R, Liebisch TC, Pfau T, Hofferberth S (2016) Observation of mixed singlet-triplet Rb_2 Rydberg molecules. Phys Rev A 93:032512
112. Schliemann J, Khaetskii A, Loss D (2003) Electron spin dynamics in quantum dots and related nanostructures due to hyperfine interaction with nuclei. J Phys Condens Matter 15:R1809
113. Zaránd G, von Delft J (2000) Analytical calculation of the finite-size crossover spectrum of the anisotropic two-channel Kondo model. Phys Rev B 61:6918–6933
114. Sala P, Shi T, Kühn S, Bañuls MC, Demler E, Cirac JI (2018) Variational study of U(1) and SU(2) lattice gauge theories with Gaussian states in 1+1 dimensions. arXiv:1805.05190
115. Weedbrook C, Pirandola S, García-Patrón R, Cerf NJ, Ralph TC, Shapiro JH, Lloyd S (2012) Gaussian quantum information. Rev Mod Phys 84:621–669
116. Mitroy J, Bubin S, Horiuchi W, Suzuki Y, Adamowicz L, Cencek W, Szalewicz K, Komasa J, Blume D, Varga K (2013) Theory and application of explicitly correlated Gaussians. Rev Mod Phys 85:693–749
117. Kraus CV, Cirac JI (2010) Generalized Hartree-Fock theory for interacting fermions in lattices: numerical methods. New J Phys 12:113004
118. Jackiw R, Kerman A (1979) Time-dependent variational principle and the effective action. Phys Lett A 71:1–5
119. Kramer P (2008) A review of the time-dependent variational principle. J Phys Conf Ser 99:012009
120. Shi T, Demler E, Cirac JI (2018) Variational study of fermionic and bosonic systems with non-Gaussian states: theory and applications. Ann Phys 390:245–302
121. Petersen KB, Petersen MS (2006) The matrix cookbook. Version 20051003
122. Weigmann PB (1980) JETP Lett 31:364
123. Andrei N (1980) Diagonalization of the Kondo hamiltonian. Phys Rev Lett 45:379–382
124. Andrei N, Lowenstein JH (1981) Scales and scaling in the Kondo model. Phys Rev Lett 46:356–360
125. Yosida K (1966) Bound state due to the $s - d$ exchange interaction. Phys Rev 147:223–227

References

126. Varma CM, Yafet Y (1976) Magnetic susceptibility of mixed-valence rare-earth compounds. Phys Rev B 13:2950–2954
127. Bergmann G, Zhang L (2007) Compact approximate solution to the Kondo problem. Phys Rev B 76:064401
128. Yang C, Feiguin AE (2017) Unveiling the internal entanglement structure of the Kondo singlet. Phys Rev B 95:115106
129. White SR (1992) Density matrix formulation for quantum renormalization groups. Phys Rev Lett 69:2863–2866
130. Verstraete F, Porras D, Cirac JI (2004) Density matrix renormalization group and periodic boundary conditions: a quantum information perspective. Phys Rev Lett 93:227205
131. Vidal G (2003) Efficient classical simulation of slightly entangled quantum computations. Phys Rev Lett 91:147902
132. Vidal G (2004) Efficient simulation of one-dimensional quantum many-body systems. Phys Rev Lett 93:040502
133. Daley AJ, Kollath C, Schollwöck U, Vidal G (2004) Time-dependent density-matrix renormalization-group using adaptive effective Hilbert spaces. J Stat Mech Theor Exp 2004:P04005
134. Pirvu B, Murg V, Cirac JI, Verstraete F (2010) Matrix product operator representations. New J Phys 12:025012
135. Trotter HF (1959) On the product of semi-groups of operators. Proc Amer Math Soc 10:545–551
136. Suzuki M (1985) Decomposition formulas of exponential operators and lie exponentials with some applications to quantum mechanics and statistical physics. J Math Phys 26:601–612
137. Verstraete F, García-Ripoll JJ, Cirac JI (2004) Matrix product density operators: simulation of finite-temperature and dissipative systems. Phys Rev Lett 93:207204
138. Ashida Y, Shi T, Bañuls MC, Cirac JI, Demler E (2018) Solving quantum impurity problems in and out of equilibrium with the variational approach. Phys Rev Lett 121:026805
139. Holzner A, McCulloch IP, Schollwöck U, von Delft J, Heidrich-Meisner F (2009) Kondo screening cloud in the single-impurity Anderson model: a density matrix renormalization group study. Phys Rev B 80:205114
140. Ishii H (1978) Spin correlation in dilute magnetic alloys. J Low Temp Phys 32:457–467
141. Barzykin V, Affleck I (1998) Screening cloud in the k-channel Kondo model: perturbative and large-k results. Phys Rev B 57:432–448
142. Hand T, Kroha J, Monien H (2006) Spin correlations and finite-size effects in the one-dimensional Kondo box. Phys Rev Lett 97:136604
143. Endres M, Cheneau M, Fukuhara T, Weitenberg C, Schauß P, Gross C, Mazza L, Bañuls MC, Pollet L, Bloch I, Kuhr S (2011) Observation of correlated particle-hole pairs and string order in low-dimensional Mott insulators. Science 334:200–203
144. Cheneau M, Barmettler P, Poletti D, Endres M, Schauss P, Fukuhara T, Gross C, Bloch I, Kollath C, Kuhr S (2012) Light-cone-like spreading of correlations in a quantum many-body system. Nature 481:484–487
145. Fukuhara T, Hild S, Zeiher J, Schauß P, Bloch I, Endres M, Gross C (2015) Spatially resolved detection of a spin-entanglement wave in a Bose-Hubbard chain. Phys Rev Lett 115:035302
146. Kaufman AM, Tai ME, Lukin A, Rispoli M, Schittko R, Preiss PM, Greiner M (2016) Quantum thermalization through entanglement in an isolated many-body system. Science 353:794–800
147. Kanász-Nagy M, Ashida Y, Shi T, Moca CP, Ikeda TN, Fölling S, Cirac JI, Zaránd G, Demler EA (2018) Exploring the anisotropic Kondo model in and out of equilibrium with alkaline-earth atoms. Phys Rev B 97:155156
148. Basov DN, Averitt RD, van der Marel D, Dressel M, Haule K (2011) Electrodynamics of correlated electron materials. Rev Mod Phys 83:471–541
149. Kormos M, Collura M, Takács G, Calabrese P (2017) Real-time confinement following a quantum quench to a non-integrable model. Nat Phys 13:246–249
150. Rosch A, Kroha J, Wölfle P (2001) Kondo effect in quantum dots at high voltage: universality and scaling. Phys Rev Lett 87:156802

151. Hewson AC, Bauer J, Oguri A (2005) Non-equilibrium differential conductance through a quantum dot in a magnetic field. J Phys Cond Matt 17:5413
152. Sela E, Malecki J (2009) Nonequilibrium conductance of asymmetric nanodevices in the Kondo regime. Phys Rev B 80:233103
153. Mora C, Vitushinsky P, Leyronas X, Clerk AA, Le Hur K (2009) Theory of nonequilibrium transport in the SU(n) Kondo regime. Phys Rev B 80:155322
154. Mora C, Moca CP, von Delft J, Zaránd G (2015) Fermi-liquid theory for the single-impurity Anderson model. Phys Rev B 92:075120
155. Filippone M, Moca CP, von Delft J, Mora C (2017) At which magnetic field, exactly, does the Kondo resonance begin to split? A fermi liquid description of the low-energy properties of the Anderson model. Phys Rev B 95:165404
156. Oguri A, Hewson AC (2018) Higher-order fermi-liquid corrections for an Anderson impurity away from half filling. Phys Rev Lett 120:126802
157. Oguri A, Hewson AC (2018) Higher-order fermi-liquid corrections for an Anderson impurity away from half filling: equilibrium properties. Phys Rev B 97:045406
158. Shah N, Bolech CJ (2016) Consistent bosonization-debosonization. I resolution of the nonequilibrium transport puzzle. Phys Rev B 93:085440
159. Schmidt R, Sadeghpour HR, Demler E (2016) Mesoscopic Rydberg impurity in an atomic quantum gas. Phys Rev Lett 116:105302
160. Fermi E (1934) Sopra lo spostamento per pressione delle righe elevate delle serie spettrali. II Nuovo Cimento 11:157
161. Ashida Y, Shi T, Schmidt R, Sadeghpour HR, Cirac JI, Demler E (2019) Quantum rydberg central spin model. Phys Rev Lett 123:183001
162. Khaetskii A, Loss D, Glazman L (2003) Electron spin evolution induced by interaction with nuclei in a quantum dot. Phys Rev B 67:195329
163. Bortz M, Stolze J (2007) Exact dynamics in the inhomogeneous central-spin model. Phys Rev B 76:014304
164. Coish WA, Loss D (2004) Hyperfine interaction in a quantum dot: non-markovian electron spin dynamics. Phys Rev B 70:195340
165. Hanson R, Dobrovitski VV, Feiguin AE, Gywat O, Awschalom DD (2008) Coherent dynamics of a single spin interacting with an adjustable spin bath. Science 320:352–355
166. Witzel WM, Carroll MS, Morello A, Cywiński L, Das Sarma S (2010) Electron spin decoherence in isotope-enriched silicon. Phys Rev Lett 105:187602
167. Tyryshkin AM, Tojo S, Morton JJL, Riemann H, Abrosimov NV, Becker P, Pohl H-J, Schenkel T, Thewalt MLW, Itoh KM et al (2012) Electron spin coherence exceeding seconds in high-purity silicon. Nat Mater 11:143
168. Ashida Y, Shi T, Schmidt R, Sadeghpour HR, Cirac JI, Demler E (2019) Efficient variational approach to dynamics of a spatially extended bosonic kondo model. Phys Rev A 100:043618
169. Miranda M, Inoue R, Okuyama Y, Nakamoto A, Kozuma M (2015) Site-resolved imaging of ytterbium atoms in a two-dimensional optical lattice. Phys Rev A 91:063414
170. Ashida Y, Ueda M (2015) Diffraction-unlimited position measurement of ultracold atoms in an optical lattice. Phys Rev Lett 115:095301
171. Yamamoto R, Kobayashi J, Kuno T, Kato K, Takahashi Y (2016) An ytterbium quantum gas microscope with narrow-line laser cooling. New J Phys 18:023016
172. Recati A, Fedichev PO, Zwerger W, von Delft J, Zoller P (2005) Atomic quantum dots coupled to a reservoir of a superfluid Bose-Einstein condensate. Phys Rev Lett 94:040404
173. Pekker D, Babadi M, Sensarma R, Zinner N, Pollet L, Zwierlein MW, Demler E (2011) Competition between pairing and ferromagnetic instabilities in ultracold fermi gases near feshbach resonances. Phys Rev Lett 106:050402
174. Bauer J, Salomon C, Demler E (2013) Realizing a Kondo-correlated state with ultracold atoms. Phys Rev Lett 111:215304
175. Nishida Y (2013) SU(3) orbital Kondo effect with ultracold atoms. Phys Rev Lett 111:135301
176. Nishida Y (2016) Transport measurement of the orbital Kondo effect with ultracold atoms. Phys Rev A 93:011606

177. Nakagawa M, Kawakami N (2015) Laser-induced Kondo effect in ultracold alkaline-earth fermions. Phys Rev Lett 115:165303
178. Zhang R, Cheng Y, Zhai H, Zhang P (2015) Orbital feshbach resonance in alkali-earth atoms. Phys Rev Lett 115:135301
179. Zhang R, Zhang D, Cheng Y, Chen W, Zhang P, Zhai H (2016) Kondo effect in alkaline-earth-metal atomic gases with confinement-induced resonances. Phys Rev A 93:043601
180. Miranda E, Dobrosavljevic V, Kotliar G (1996) Kondo disorder: a possible route towards non-fermi-liquid behaviour. J Phys Cond Matt 8:9871
181. Tsunetsugu H, Sigrist M, Ueda K (1997) The ground-state phase diagram of the one-dimensional Kondo lattice model. Rev Mod Phys 69:809–864
182. Bao Z-Q, Zhang F (2017) Topological majorana two-channel Kondo effect. Phys Rev Lett 119:187701
183. Kleinbach KS, Meinert F, Engel F, Kwon WJ, Löw R, Pfau T, Raithel G (2017) Photoassociation of trilobite Rydberg molecules via resonant spin-orbit coupling. Phys Rev Lett 118:223001
184. Camargo F, Schmidt R, Whalen JD, Ding R, Woehl G Jr, Yoshida S, Burgdörfer J, Dunning FB, Sadeghpour HR, Demler E, Killian TC (2017) Creation of Rydberg Polarons in a Bose Gas. arXiv:1706.03717
185. Romeo F, Citro R (2009) Adiabatic pumping in a double quantum dot structure with strong spin-orbit interaction. Phys Rev B 80:165311
186. Peng Y, Vinkler-Aviv Y, Brouwer PW, Glazman LI, von Oppen F (2016) Parity anomaly and spin transmutation in quantum spin Hall Josephson junctions. Phys Rev Lett 117:267001
187. Gutman DB, Gefen Y, Mirlin AD (2010) Bosonization of one-dimensional fermions out of equilibrium. Phys Rev B 81:085436
188. Levitov LS, Lee H, Lesovik GB (1996) Electron counting statistics and coherent states of electric current. J Math Phys 37:4845–4866
189. Maldacena J, Shenker SH, Stanford D (2016) A bound on chaos. J High Energy Phys 2016:106
190. Bentsen G, Potirniche I-D, Bulchandani VB, Scaffidi T, Cao X, Qi X-L, Schleier-Smith M, Altman E (2019) Integrable and chaotic dynamics of spins coupled to an optical cavity. Phys Rev X 9:041011
191. Cui J, Cirac JI, Bañuls MC (2015) Variational matrix product operators for the steady state of dissipative quantum systems. Phys Rev Lett 114:220601

Chapter 6
Quantum Particle in a Magnetic Environment

Abstract In this Chapter, we study yet another fundamental paradigm of a quantum impurity, a mobile spinless particle interacting with a many-particle environment. The ultimate building block of such a system is the formation of a polarization cloud of collective excitations around an impurity, which is known as a polaron cloud. Yet, ever since the original paper by Landau and Pekar [1], it has long remained a fundamental challenge to measure it in an unambiguous manner due to the elusive nature of the polaron cloud. We here present a novel platform using ultracold atoms to overcome the obstacle and allow one to directly probe the polaron-cloud formation in real time. We reveal the emergence of rich nonequilibrium dynamics of the polaron cloud in the strong-coupling regime that is not readily attainable in solid-state materials. To our knowledge, our work suggests the first concrete possibility for a direct real-time measurement of the polaron-cloud formation.

Keywords Polaron · Quantum impurity · Magnetic polaron · Ultracold atoms

6.1 Introduction: New Frontiers in Polaron Physics

In the previous Chapter, we have studied the strongly correlated physics in open quantum systems by focusing on a localized spin-impurity system. Here, we study yet another fundamental paradigm of quantum impurity problems, a mobile spinless quantum particle interacting with a many-particle environment. There, a "dress" of collective excitations around the impurity, which is known as a polaron cloud [2], will be formed. A concept of a polaron was originally suggested by Landau and Pekar [1] and has played a central role in determining thermodynamic properties of various condensed matter systems [3–7]. Recent experiments in ultracold atoms have allowed one to study polaron physics in strongly interacting regimes that are not attainable in solid-state materials. These systems consist of imbalanced mixtures of atomic gases; the density of one type of atoms is much lower than that of the other. Thus, the minority atoms act as impurities while the majority atoms form a many-particle environment interacting with the impurities. The impurities are dressed by the collective excitations in the environment and constitute Fermi or Bose polarons depending on species of the majority atoms. The major advantages in ultracold atomic physics include a high controllability of physical parameters and the ability to precisely measure dynamical properties, which are made possible by rich experimental tool box in atomic, molecular and optical (AMO) physics such as radio-frequency absorption measurements [8–15] and interferometric measurements [16–19]. Until now, most studies of polaron physics in ultracold atoms have focused on impurities immersed in a single-component Bose-Einstein condensate (BEC) [13–15, 20–43] or Fermi gas of atoms [8–12, 44–52]. These studies have revealed new aspects of polaron physics beyond the Anderson orthogonality catastrophe and the conventional Fröhlich paradigm [16, 17, 19, 53, 54]. Nevertheless, revealing real-time dynamics of the polaron cloud has still remained a great challenge in both ultracold atomic gases and solid-state materials. Here, a subtle density change in the environment associated with the formation of the polaron cloud poses a severe challenge for its direct observation.

The aim of this Chapter is to show that the use of the Ramsey interferometry performed on the environment can overcome this challenge and allow one to directly measure the polaron-cloud formation in real time, which would provide an unambiguous signature of the system-environment entanglement. In previous setups [13, 20–31, 33–41], only phonon excitations are coupled to the impurity; this fact has made a direct observation of the polaron cloud extremely difficult due to the minuscule nature of density modifications. We here consider a magnetic polaron [55] that is dressed by spin-wave excitations in addition to phonon excitations. The Ramsey interferometry can then be used to directly measure the polaron cloud via this magnetic dressing.

In Sect. 6.2, applying this idea to impurity atoms immersed in a two-species BEC (see Fig. 6.1a), we analyze impurities interacting with a synthetic magnetic environment and derive an effective Hamiltonian addressed in this Chapter. In Sect. 6.3, we show how one can use the Ramsey interferometry to directly measure the polaron-cloud dynamics in real time to reveal its rich out-of-equilibrium physics. In Sect. 6.4, we employ the time-dependent variational principle and study nonequilibrium dynamics of the polaron cloud. We show that, in the strong-coupling regime, the polaron cloud forms many-body bound states, which is a nontrivial feature that is not readily attainable in solid-state systems. The formations of many-body bound states lead to the dynamical 'phase diagram' of the polaron cloud (see Fig. 6.1b), which corresponds to its distinct oscillatory dynamics (see Fig. 6.1c). One of the main advantages in our scheme is that it can effectively enhance observable signals from the impurity. The reason is that, for each impurity, a multiple number of excitations in the bath are generated and they are subsequently detected via the interferometric measurement acting on the environment. We mention that this protocol can be applied to other experimental systems [57–63], where interferometric techniques are already available. In Sect. 6.5, we discuss a concrete experimental setup to realize our proposal. Finally, we conclude this Chapter with an outlook in Sect. 6.6.

6.2 System: A Mobile Particle in a Synthetic Magnetic Environment

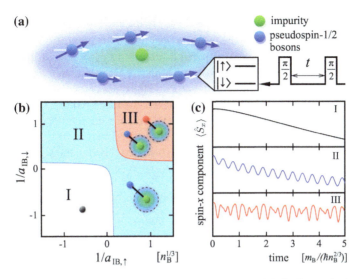

Fig. 6.1 a A schematic figure of a mobile spinless quantum particle in the synthetic magnetic environment. **b** The dynamical phase diagram. The system has zero, one, and two many-body bound states depending on the impurity-boson scattering lengths $a_{\text{IB},\uparrow\downarrow}$. We use $a_{\text{BB}} n_{\text{B}}^{1/3} = 0.05$ and $m_{\text{I}}/m_{\text{B}} = 0.95$ as appropriate for the ^{41}K-^{39}K mixture. **c** The dynamics of environment spins in each regime of the dynamical phase diagram. Reproduced from Fig. 1 of Ref. [56]. Copyright © 2018 by the American Physical Society

6.2 System: A Mobile Particle in a Synthetic Magnetic Environment

To be concrete, we consider an impurity atom immersed in a weakly interacting two-component BEC (see Fig. 6.1a). The impurity is dressed by spin-wave excitations of a synthetic ferromagnetic environment and forms a magnetic polaron. The total Hamiltonian is

$$\hat{H} = \hat{H}_{\text{B}} + \hat{V}_{\text{IB}} + \hat{H}_{\text{I}}, \quad (6.1)$$

where

$$\hat{H}_{\text{B}} = \sum_{\mathbf{k}\sigma} \epsilon_{\mathbf{k}} \hat{a}^\dagger_{\mathbf{k}\sigma} \hat{a}_{\mathbf{k}\sigma} + \frac{g_{\text{BB}}}{2V} \sum_{\mathbf{k}\mathbf{k}'\mathbf{q}\sigma\sigma'} \hat{a}^\dagger_{\mathbf{k}+\mathbf{q}\sigma} \hat{a}^\dagger_{\mathbf{k}'-\mathbf{q}\sigma'} \hat{a}_{\mathbf{k}'\sigma'} \hat{a}_{\mathbf{k}\sigma} \quad (6.2)$$

is the bath Hamiltonian of the background BEC of density n_{B}. The interaction between the impurity and the environment is given by

$$\hat{V}_{\text{IB}} = \frac{1}{V} \sum_{\mathbf{k}\mathbf{q}\sigma} g_{\text{IB},\sigma} \hat{a}^\dagger_{\mathbf{k}+\mathbf{q}\sigma} \hat{a}_{\mathbf{k}\sigma} e^{i\mathbf{q}\hat{\mathbf{R}}}. \quad (6.3)$$

An operator $\hat{a}_{\mathbf{k}\sigma}$ ($\hat{a}^\dagger_{\mathbf{k}\sigma}$) annihilates (creates) an environmental boson with a wavenumber $\mathbf{k}$ and a spin $\sigma = \uparrow, \downarrow$, $\hat{\mathbf{R}}$ is the position operator of the impurity, V is the volume of the system, and we denote a dispersion relation as $\epsilon_\mathbf{k} = \hbar^2 \mathbf{k}^2/(2m_\mathrm{B})$ with m_B being the mass of surrounding bosons. The impurity Hamiltonian is $\hat{H}_\mathrm{I} = \hat{\mathbf{P}}^2/(2m_\mathrm{I})$ with $\hat{\mathbf{P}}$ and m_I being the momentum operator and the mass of the impurity. We assume that the interaction term of the environment bosons (the second term in Eq. (6.2)) is spin-independent and thus characterized by a single parameter g_BB as realized for many atoms [64, 65].[1] In contrast, the impurity-bath interaction is spin-dependent and characterized by two parameters $g_{\mathrm{IB},\sigma}$. They are related to the scattering lengths $a_{\mathrm{IB},\sigma}$ via the Lippmann-Schwinger equation:

$$\frac{1}{g_{\mathrm{IB},\sigma}} = \frac{m_\mathrm{red}}{2\pi a_{\mathrm{IB},\sigma}} - \frac{1}{V}\sum_\mathbf{k}^\Lambda \frac{2m_\mathrm{red}}{\mathbf{k}^2}, \quad (6.4)$$

where $m_\mathrm{red} = m_\mathrm{I} m_\mathrm{B}/(m_\mathrm{I} + m_\mathrm{B})$ is the reduced mass. We note that these expressions are fully regularized and the momentum cutoff Λ introduced in Eq. (6.4) can be taken to infinity.

To realize a synthetic ferromagnetic medium, we start from a superposition state of the pseudospin-1/2 BEC: $|\Psi_\mathrm{BEC}\rangle \propto (\hat{a}^\dagger_{\mathbf{0}\uparrow} + \hat{a}^\dagger_{\mathbf{0}\downarrow})^{N_\mathrm{B}}|0\rangle$, where N_B is the total number of environment particles. Since the bath Hamiltonian $\hat{H}_\mathrm{B}$ possesses the SU(2) symmetry, it does not cause decoherence of the spin dynamics. In contrast, the impurity-bath interaction breaks this symmetry and leads to the dephasing of environment spins.

We can simplify the environment Hamiltonian following the standard procedure. To take into account an initial macroscopic population of the environment bosons in the $\mathbf{k} = 0$ mode, we expand $\hat{a}_{\mathbf{0}\sigma}$ around $\langle \hat{a}_{\mathbf{0}\sigma}\rangle = \sqrt{N_\mathrm{B}/2}$. Here the factor of $1/2$ accounts for the fact that the bosons are prepared in a superposition of $\uparrow$ and $\downarrow$ states. We then diagonalize the bath Hamiltonian (6.2) using the Bogoliubov transformation:

$$\hat{a}_{\mathbf{k},\uparrow} = \frac{1}{\sqrt{2}}\left(\hat{\gamma}^s_\mathbf{k} + u_\mathbf{k}\hat{\gamma}^c_\mathbf{k} - v_\mathbf{k}\hat{\gamma}^{\dagger c}_{-\mathbf{k}}\right), \quad \hat{a}_{\mathbf{k},\downarrow} = \frac{1}{\sqrt{2}}\left(-\hat{\gamma}^s_\mathbf{k} + u_\mathbf{k}\hat{\gamma}^c_\mathbf{k} - v_\mathbf{k}\hat{\gamma}^{\dagger c}_{-\mathbf{k}}\right), (6.5)$$

where the coefficients are given by

$$u_\mathbf{k} = \sqrt{\frac{\epsilon_\mathbf{k} + g_\mathrm{BB} n_\mathrm{B}}{2\epsilon^c_\mathbf{k}} + \frac{1}{2}}, \quad v_\mathbf{k} = \sqrt{\frac{\epsilon_\mathbf{k} + g_\mathrm{BB} n_\mathrm{B}}{2\epsilon^c_\mathbf{k}} - \frac{1}{2}} \quad (6.6)$$

and the dispersion relation of the charge (i.e., phonon) excitations in the environment is

$$\epsilon^c_\mathbf{k} = \sqrt{\epsilon_\mathbf{k}(\epsilon_\mathbf{k} + 2g_\mathrm{BB} n_\mathrm{B})}. \quad (6.7)$$

---

[1]This condition can be relaxed. See Sect. 6.5 for discussions on a possible breaking of the SU(2) symmetry in the environment interactions.

6.2 System: A Mobile Particle in a Synthetic Magnetic Environment

The operators $\hat{\gamma}_{\mathbf{k}}^{c,s}$ and $\hat{\gamma}_{\mathbf{k}}^{\dagger c,s}$ represent the creation and annihilation operators of the charge and spin excitations and satisfy the commutation relations

$$[\hat{\gamma}_{\mathbf{k}}^{\xi}, \hat{\gamma}_{\mathbf{k}'}^{\eta}] = [\hat{\gamma}_{\mathbf{k}}^{\dagger\xi}, \hat{\gamma}_{\mathbf{k}'}^{\dagger\eta}] = 0, \quad [\hat{\gamma}_{\mathbf{k}}^{\xi}, \hat{\gamma}_{\mathbf{k}'}^{\dagger\eta}] = \delta_{\xi,\eta}\delta_{\mathbf{k},\mathbf{k}'} \quad (\xi, \eta = s, c). \tag{6.8}$$

The resulting expression for the total Hamiltonian $\hat{H}$ is

$$\hat{H} = g_{\mathrm{IB}}^+ n_{\mathrm{B}} + \frac{\hat{\mathbf{P}}^2}{2m_{\mathrm{I}}} + \sum_{\mathbf{k}}\left(\epsilon_{\mathbf{k}}^c\hat{\gamma}_{\mathbf{k}}^{\dagger c}\hat{\gamma}_{\mathbf{k}}^c + \epsilon_{\mathbf{k}}^s\hat{\gamma}_{\mathbf{k}}^{\dagger s}\hat{\gamma}_{\mathbf{k}}^s\right) + \sqrt{\frac{n_{\mathrm{B}}}{V}}\sum_{\mathbf{k}}\left[g_{\mathrm{IB}}^+ W_{\mathbf{k}}\left(\hat{\gamma}_{\mathbf{k}}^c + \hat{\gamma}_{-\mathbf{k}}^{\dagger c}\right) + g_{\mathrm{IB}}^-\left(\hat{\gamma}_{\mathbf{k}}^s + \hat{\gamma}_{-\mathbf{k}}^{\dagger s}\right)\right]e^{-i\mathbf{k}\hat{\mathbf{R}}}$$

$$+ \frac{g_{\mathrm{IB}}^+}{2V}\sum_{\mathbf{k},\mathbf{k}'}\left(V_{\mathbf{k}\mathbf{k}'}^{(1)}\hat{\gamma}_{\mathbf{k}'}^{\dagger c}\hat{\gamma}_{\mathbf{k}}^c e^{i(\mathbf{k}-\mathbf{k}')\hat{\mathbf{R}}} + \hat{\gamma}_{\mathbf{k}'}^{\dagger s}\hat{\gamma}_{\mathbf{k}}^s e^{i(\mathbf{k}-\mathbf{k}')\hat{\mathbf{R}}} + V_{\mathbf{k}\mathbf{k}'}^{(2)}\hat{\gamma}_{\mathbf{k}'}^{\dagger c}\hat{\gamma}_{\mathbf{k}}^{\dagger c} e^{i(\mathbf{k}+\mathbf{k}')\hat{\mathbf{R}}} + \mathrm{H.c.}\right)$$

$$+ \frac{g_{\mathrm{IB}}^-}{V}\sum_{\mathbf{k},\mathbf{k}'}\left(u_{\mathbf{k}}\hat{\gamma}_{\mathbf{k}'}^{\dagger s}\hat{\gamma}_{\mathbf{k}}^c e^{i(\mathbf{k}'-\mathbf{k})\hat{\mathbf{R}}} - v_{\mathbf{k}}\hat{\gamma}_{\mathbf{k}'}^{\dagger s}\hat{\gamma}_{\mathbf{k}}^{\dagger c} e^{i(\mathbf{k}'+\mathbf{k})\hat{\mathbf{R}}} + \mathrm{H.c.}\right), \tag{6.9}$$

where $\epsilon_{\mathbf{k}}^s = \hbar^2 \mathbf{k}^2/(2m_{\mathrm{B}})$ is the dispersion relation of the spin-wave excitations in the environment, $g_{\mathrm{IB}}^{\pm} = (g_{\mathrm{IB},\uparrow} \pm g_{\mathrm{IB},\downarrow})/2$ are the average and difference of the impurity-boson interaction parameters, $W_{\mathbf{k}} = \sqrt{\epsilon_{\mathbf{k}}/\epsilon_{\mathbf{k}}^c}$ and $V_{\mathbf{k}\mathbf{k}'}^{(1)} \pm V_{\mathbf{k}\mathbf{k}'}^{(2)} = (W_{\mathbf{k}} W_{\mathbf{k}'})^{\pm 1}$ are the vertices.[2] When $g_{\mathrm{IB}}^- \neq 0$, spin waves are generated due to the imbalance of the impurity-boson couplings.

To simplify the problem, we go onto the comoving frame of the impurity via

$$\hat{U}_{\mathrm{LLP}} = e^{i\hat{\mathbf{R}}\hat{\mathbf{P}}_{\mathrm{B}}}, \tag{6.10}$$

which is known as the Lee-Low-Pines (LLP) transformation [66]. Here, we introduce the total momentum of the environment by $\hat{\mathbf{P}}_{\mathrm{B}} = \sum_{\mathbf{k}}\mathbf{k}(\hat{\gamma}_{\mathbf{k}}^{\dagger c}\hat{\gamma}_{\mathbf{k}}^c + \hat{\gamma}_{\mathbf{k}}^{\dagger s}\hat{\gamma}_{\mathbf{k}}^s)$ (we set $\hbar = 1$). This leads to the transformed Hamiltonian $\hat{\mathcal{H}} = \hat{U}_{\mathrm{LLP}}^{\dagger}\hat{H}\hat{U}_{\mathrm{LLP}}$, which commutes with the impurity momentum $\hat{\mathbf{P}}$. This can be inferred from the fact that the total momentum $\hat{\mathbf{P}} + \hat{\mathbf{P}}_{\mathrm{B}}$ of the impurity and the environment is conserved in the original frame while it is mapped to $\hat{\mathbf{P}}$ under the transformation $\hat{U}_{\mathrm{LLP}}$:

$$\hat{U}_{\mathrm{LLP}}^{\dagger}(\hat{\mathbf{P}} + \hat{\mathbf{P}}_{\mathrm{B}})\hat{U}_{\mathrm{LLP}} = \hat{\mathbf{P}}. \tag{6.11}$$

Thus, $\hat{\mathbf{P}}$ can be taken as a classical number $\mathbf{P}$ in the transformed frame and the impurity is completely decoupled from the environmental degrees of freedom. The explicit form of the transformed Hamiltonian $\hat{\mathcal{H}}$ can be obtained as

---

[2] We note that the vertex functions that couple to the spin sector acquire an additional momentum dependence when the SU(2) symmetry of the bath is broken.

$$\hat{\mathcal{H}} = g_{\mathrm{IB}}^+ n_{\mathrm{B}} + \frac{(\mathbf{P} - \hat{\mathbf{P}}_{\mathrm{B}})^2}{2m_{\mathrm{I}}} + \sum_{\mathbf{k}} \left(\epsilon_{\mathbf{k}}^c \hat{\gamma}_{\mathbf{k}}^{\dagger c} \hat{\gamma}_{\mathbf{k}}^c + \epsilon_{\mathbf{k}}^s \hat{\gamma}_{\mathbf{k}}^{\dagger s} \hat{\gamma}_{\mathbf{k}}^s \right) + \sqrt{\frac{n_{\mathrm{B}}}{V}} \sum_{\mathbf{k}} \left[g_{\mathrm{IB}}^+ W_{\mathbf{k}} \left(\hat{\gamma}_{\mathbf{k}}^c + \hat{\gamma}_{-\mathbf{k}}^{\dagger c} \right) + g_{\mathrm{IB}}^- \left(\hat{\gamma}_{\mathbf{k}}^s + \hat{\gamma}_{-\mathbf{k}}^{\dagger s} \right) \right]$$

$$+ \frac{g_{\mathrm{IB}}^+}{2V} \sum_{\mathbf{k},\mathbf{k}'} \left(V_{\mathbf{k}\mathbf{k}'}^{(1)} \hat{\gamma}_{\mathbf{k}}^{\dagger c} \hat{\gamma}_{\mathbf{k}'}^c + \hat{\gamma}_{\mathbf{k}}^{\dagger s} \hat{\gamma}_{\mathbf{k}'}^s + V_{\mathbf{k}\mathbf{k}'}^{(2)} \hat{\gamma}_{\mathbf{k}}^{\dagger c} \hat{\gamma}_{\mathbf{k}'}^{\dagger c} + \mathrm{H.c.} \right)$$

$$+ \frac{g_{\mathrm{IB}}^-}{V} \sum_{\mathbf{k},\mathbf{k}'} \left(u_{\mathbf{k}} \hat{\gamma}_{\mathbf{k}'}^{\dagger s} \hat{\gamma}_{\mathbf{k}}^c - v_{\mathbf{k}} \hat{\gamma}_{\mathbf{k}'}^{\dagger s} \hat{\gamma}_{\mathbf{k}}^{\dagger c} + \mathrm{H.c.} \right). \tag{6.12}$$

Hereafter, we consider the sector $\mathbf{P} = 0$ in the transformed frame, i.e., we assume that the total momentum of the impurity plus the environment in the original frame is zero. We also assume that the initial state in the original frame is an eigenstate of the momentum operator of the environment with the zero eigenvalue $\hat{\mathbf{P}}_{\mathrm{B}} = 0$. Since $\hat{\mathbf{P}}_{\mathrm{B}}$ is invariant under $\hat{U}_{\mathrm{LLP}}$, we can take $\hat{\mathbf{P}}_{\mathrm{B}} = 0$ at the initial time also in the transformed frame. Then, in the sector $\mathbf{P} = 0$, one can show that $\hat{\mathbf{P}}_{\mathrm{B}}$ in the transformed frame remains to be zero in the course of the time evolution owing to the symmetry of the Hamiltonian $\hat{\mathcal{H}}$ under $\hat{\mathbf{P}}_{\mathrm{B}} \to -\hat{\mathbf{P}}_{\mathrm{B}}$. Thus, the second term in the first line on the right-hand side in Eq. (6.12) plays no role in our consideration. The second line in Eq. (6.12) describes the contributions beyond the Fröhlich paradigm [67], leading to new types of nonequilibrium polaron physics as detailed below.

6.3 Many-Body Interferometry Acting on the Environment

To directly probe the dynamics of the polaron cloud in real time, we propose employing the Ramsey interference acting on the environment atoms. We here outline its specific protocol.

One starts from the environment atoms prepared in the ↑ state and then performs a $\pi/2$ pulse, resulting in a superposition of ↑ and ↓ states as described in the previous section. After letting the system evolve in time t, one again applies a $\pi/2$ pulse and measures the number $N_\uparrow$ of environment atoms in the ↑ state, which directly provides the number $N^s(t) = \sum_{\mathbf{k}} \langle \hat{\gamma}_{\mathbf{k}}^{\dagger s} \hat{\gamma}_{\mathbf{k}}^s \rangle$ of spin-wave excitations. Although we treat the system as a single-impurity problem, our results are still valid for a system with a finite density of impurities if, as realized in typical experimental setups, the impurity density is so low that one can neglect impurity-impurity interactions. In such a case, the interferometric signal $N^s = N_\uparrow$ linearly scales with the number of impurities and can be enhanced by increasing the density of impurities. Moreover, the signal N^s can easily exceed the number of impurities since a single impurity can generate a multiple number of spin-wave excitations as shown below. The Ramsey interferometry then allows one to make a precise measurement of the enhanced signals N^s, thus enabling a direct observation of the polaron cloud. This has been challenging to achieve in conventional setups [13, 20–31, 33–41] due to the difficulty of detecting minuscule modifications of densities induced by phonon excitations.

6.4 Out-of-Equilibrium Dynamics of the Magnetic Polaron

6.4.1 Time-Dependent Variational Principle

The polaron cloud exhibits distinct dynamics depending on the impurity-bath couplings. To analyze out-of-equilibrium dynamics of the magnetic polaron, we use the time-dependent variational approach [68]. We consider the following family of variational states

$$|\Psi(t)\rangle = e^{\sum_{\mathbf{k}}(\alpha_{\mathbf{k}}^c(t)\hat{\gamma}_{\mathbf{k}}^c + \alpha_{\mathbf{k}}^s(t)\hat{\gamma}_{\mathbf{k}}^s - \text{h.c.})}|0\rangle, \quad (6.13)$$

where $|0\rangle$ is the vacuum of the charge and spin-wave excitations and $\alpha_{\mathbf{k}}^{c,s}(t)$ represent their amplitudes. We remark that this variational state becomes exact in the limit of $m_{\mathrm{I}} \to \infty$.

The variational principle $\delta[\langle\Psi|i\hbar\partial_t - \hat{\mathcal{H}}|\Psi\rangle] = 0$ provides the equations of motion for $\alpha_{\mathbf{k}}^{c,s}$, which are the following coupled integral equations:

$$i\dot{\alpha}_{\mathbf{k}}^c = \Omega_{\mathbf{k}}^c \alpha_{\mathbf{k}}^c - \frac{\mathbf{k}\cdot(\mathbf{P} - \mathbf{P}_{\mathrm{B}}[\alpha^{c,s}])}{m_{\mathrm{I}}}\alpha_{\mathbf{k}}^c + g_{\mathrm{IB}}^+\left(\sum_{\mathbf{k}'}V_{\mathbf{k}\mathbf{k}'}^{(1)}\alpha_{\mathbf{k}'}^c + \sum_{\mathbf{k}'}V_{\mathbf{k}\mathbf{k}'}^{(2)}\alpha_{\mathbf{k}'}^{*c}\right)$$

$$+ g_{\mathrm{IB}}^-\left(u_{\mathbf{k}}\sum_{\mathbf{k}'}\alpha_{\mathbf{k}'}^s - v_{\mathbf{k}}\sum_{\mathbf{k}'}\alpha_{\mathbf{k}'}^{*s}\right) + g_{\mathrm{IB}}^+\sqrt{n_{\mathrm{B}}}W_{\mathbf{k}}, \quad (6.14)$$

$$i\dot{\alpha}_{\mathbf{k}}^s = \Omega_{\mathbf{k}}^s \alpha_{\mathbf{k}}^s - \frac{\mathbf{k}\cdot(\mathbf{P} - \mathbf{P}_{\mathrm{B}}[\alpha^{c,s}])}{m_{\mathrm{I}}}\alpha_{\mathbf{k}}^s + g_{\mathrm{IB}}^+\sum_{\mathbf{k}'}\alpha_{\mathbf{k}'}^s + g_{\mathrm{IB}}^-\left[\sum_{\mathbf{k}'}(u_{\mathbf{k}'}\alpha_{\mathbf{k}'}^c - v_{\mathbf{k}'}\alpha_{\mathbf{k}'}^{*c})\right] + g_{\mathrm{IB}}^-\sqrt{n_{\mathrm{B}}}, \quad (6.15)$$

where $\Omega_{\mathbf{k}}^{s,c} = \hbar^2\mathbf{k}^2/(2m_{\mathrm{I}}) + \epsilon_{\mathbf{k}}^{s,c}$ and $\mathbf{P}_{\mathrm{B}} = \sum_{\mathbf{k}}\mathbf{k}(|\alpha_{\mathbf{k}}^c|^2 + |\alpha_{\mathbf{k}}^s|^2)$. As we mentioned before, for our choice of the initial condition, we can set $\mathbf{P} = \mathbf{P}_{\mathrm{B}} = 0$ in the course of time evolution. Thus, the above equations reduce to linear inhomogeneous equations for $\alpha_{\mathbf{k}}^{c,s}$, where the last terms on the right-hand sides of Eqs. (6.14) and (6.15) describe driving forces.

To begin with, we analyze the equilibrium solution $\bar{\alpha}_{\mathbf{k}}^{c,s}$ which can be derived by setting the left-hand sides of Eqs. (6.14) and (6.15) to be zero. From the conditions that the real and imaginary parts of the right-hand sides of Eqs. (6.14) and (6.15) must vanish independently, it follows that the imaginary parts of $\bar{\alpha}_{\mathbf{k}}^{c,s}$ are zero. Solving the coupled integral equations for the remaining real parts of $\bar{\alpha}_{\mathbf{k}}^{c,s}$, we obtain the stationary solution:

$$\bar{\alpha}_{\mathbf{k}}^c = -\frac{\sqrt{n_{\mathrm{B}}}W_{\mathbf{k}}}{\Omega_{\mathbf{k}}^c}\frac{g_{\mathrm{IB}}^+(1 + g_{\mathrm{IB}}^+ B) - (g_{\mathrm{IB}}^-)^2 B}{D}, \quad \bar{\alpha}_{\mathbf{k}}^s = -\frac{\sqrt{n_{\mathrm{B}}}}{\Omega_{\mathbf{k}}^s}\frac{g_{\mathrm{IB}}^-}{D}. \quad (6.16)$$

Here we introduce the coefficients as

$$D = (1 + g_{\mathrm{IB}}^+ A)(1 + g_{\mathrm{IB}}^+ B) - (g_{\mathrm{IB}}^-)^2 AB, \quad A = \sum_{\mathbf{k}}\frac{W_{\mathbf{k}}^2}{\Omega_{\mathbf{k}}^c}, \quad B = \sum_{\mathbf{k}}\frac{1}{\Omega_{\mathbf{k}}^s}. \quad (6.17)$$

When there exists an imbalance in the impurity-bath couplings, i.e., $g_{IB}^- \neq 0$, the stationary solution for the spin part $\bar{\alpha}_\mathbf{k}^s$ is nonvanishing and the magnetic polaron is formed.

The energy of the equilibrium magnetic polaron is given by the expectation value $E_{mpol} = \langle \Psi_{mpol} | \hat{\mathcal{H}} | \Psi_{mpol} \rangle$ with respect to the stationary state $|\Psi_{mpol}\rangle = \exp\left[\sum_\mathbf{k} \left(\bar{\alpha}_\mathbf{k}^c \hat{\gamma}_\mathbf{k}^c + \bar{\alpha}_\mathbf{k}^s \hat{\gamma}_\mathbf{k}^s - \text{h.c.}\right)\right]|0\rangle$. Using the solution (6.16) and expressing the interaction strengths $g_{IB,\sigma}$ in terms of the scattering lengths $a_{IB,\sigma}$ by the Lippmann-Schwinger equation (6.4), we obtain

$$E_{mpol} = \frac{2\pi n_B}{m_{red}\left(1/a_{IB}^+ - 1/l_0\right)}, \tag{6.18}$$

where we define

$$a_{IB}^+ = \frac{a_{IB,\uparrow} + a_{IB,\downarrow}}{2}, \quad l_0 = \left(4\pi \sum_\mathbf{k} \frac{1}{\mathbf{k}^2} - \frac{2\pi}{m_{red}} \sum_\mathbf{k} \frac{W_\mathbf{k}^2}{\Omega_\mathbf{k}^c}\right)^{-1}. \tag{6.19}$$

6.4.2 Quantum Dynamics of the Environment

We now analyze out-of-equilibrium dynamics of the polaron cloud by integrating the time-evolution equations (6.14) and (6.15). Figure 6.2 plots the time evolutions of $N^{s,c}(t)$ at different impurity-bath couplings $a_{IB,\sigma}$, which correspond to the regions I, II, and III in which zero, one, and two bound states are present, respectively (see Fig. 6.1b). As shown in panel I in Fig. 6.2, $N^c(t)$ saturates while $N^s(t)$ grows as $\propto \sqrt{t}$ in the absence of bound states. Thus, the latter can easily surpass one, indicating that the observable signal $N^s = N_\uparrow$ at a finite impurity density can exceed the number of impurities. In contrast, the number of impurities strictly sets an upper bound on that of detectable signals in the conventional approaches [8–18] that utilize the impurity also as a probe. Our proposed approach using environment atoms as interferometric probes thus allow one to enhance experimental signals of the impurities.

The $\sqrt{t}$ growth of $N^s(t)$ can be understood from the quadratic dispersion of the magnon excitation as follows. For the sake of simplicity, let us neglect interactions between different momentum modes and consider the simplified Hamiltonian where different sectors of momentum $\mathbf{k}$ are decoupled: $\hat{H}_\mathbf{k} = \omega_\mathbf{k} \hat{\gamma}_\mathbf{k}^\dagger \hat{\gamma}_\mathbf{k} + V_\mathbf{k}(\hat{\gamma}_\mathbf{k} + \hat{\gamma}_\mathbf{k}^\dagger)$. Here $\hat{\gamma}_\mathbf{k}$ denotes a spin or charge annihilation operator at momentum $\mathbf{k}$ and we assume $V_{-\mathbf{k}} = V_\mathbf{k}$. In the nonequilibrium interacting problem such as the one studied here, the number of excitations $n_\mathbf{k}$ in the mode $\mathbf{k}$ will in general oscillate. However, for times longer than $t > 1/\omega_\mathbf{k}$, one can give a simple scaling argument for the behavior of $n_\mathbf{k}$. At such times, the occupation of excitations have a scaling $\langle \hat{\gamma}_\mathbf{k} \rangle \propto V_\mathbf{k}/\omega_\mathbf{k}$ and thus $n_\mathbf{k} \propto (V_\mathbf{k}/\omega_\mathbf{k})^2$ as can be seen from an inspection of the equation of motion. Let us then consider the total number of excitations at time t. We identify all modes $\mathbf{k}$ satisfying $\omega_\mathbf{k} t > 1$, i.e., the modes $|\mathbf{k}| > k^*$ where k^* is determined by $\omega_{k^*} t = 1$, as

6.4 Out-of-Equilibrium Dynamics of the Magnetic Polaron

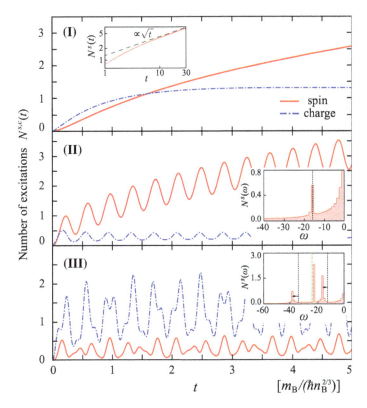

Fig. 6.2 Time evolutions of the number of charge (blue dashed curves) and spin-wave (red solid curves) excitations. We use the parameters $(n_B^{1/3} a_{IB,\uparrow}, n_B^{1/3} a_{IB,\downarrow}) = (-1, -10)$ in I, $(3.5, -5)$ in II, and $(3.5, 5)$ in III. The Fourier spectra of $N^s(t)$ are plotted in insets of panels II and III, where the black dotted lines correspond to the bound-state energies obtained from Eq. (6.26) while their difference is shown as a green dashed line. Reproduced from Fig. 2 of Ref. [56]. Copyright © 2018 by the American Physical Society

contributions to the excitations. Using the estimate $n_\mathbf{k} \propto (V_\mathbf{k}/\omega_\mathbf{k})^2$, we then integrate over these modes to find the scaling of the spin-wave excitations

$$N^s(t) = \int d^3\mathbf{k}\, n_\mathbf{k}^s \propto \int_{k^*}^\infty k^2 dk\, \frac{1}{k^4} \propto \sqrt{t}. \quad (6.20)$$

Here we use the fact that for the spin sector $V_\mathbf{k} = $ const., and the magnon dispersion relation scales as $\omega_\mathbf{k} \propto k^2$, leading to $k^* \propto 1/\sqrt{t}$.

We note that the $\sqrt{t}$ behavior remains observable also in the presence of a small imbalance in the boson-boson scattering length (i.e., a broken SU(2) symmetry). In this case, the magnon dispersion relation has a linear low-energy contribution and the number of spin excitations will ultimately saturate. However, since the imbalance in the scattering lengths is typically very small, we expect a large time window for which the $\sqrt{t}$ behavior remains valid before saturation. Specifically, let $t = t^*$ be the time scale at which k^* (determined by $\omega_{k^*}t = 1$) reaches a small value such that the magnon dispersion becomes linear for $k < k^*$. Under this condition, the scaling argument given in Eq. (6.20) remains valid as long as $t < t^*$. Importantly, because a quadratic dispersion is relevant to collective excitations in various setups such as dipolar or Rydberg gases [61–63], multi-component BEC [59, 60], and fermionic gases [57, 58], our protocol to enhance the impurity signal via the interferometric tools can also be realized in these systems.

6.4.3 Many-Body Bound States

As shown in panels II and III in Fig. 6.2, the presence of bound states induce the oscillatory dynamics in the bath excitations. We can understand the single- and multi-frequency oscillations as the formation and coupling of many-body bound states, respectively. To give physical insights, let us analyze the bound states based on the following variational state:

$$|\Psi_b(t)\rangle = \sum_{\mathbf{k}} \left(\psi^c_{\mathbf{k}}(t) \hat{\gamma}^{\dagger c}_{\mathbf{k}} + \psi^s_{\mathbf{k}}(t) \hat{\gamma}^{\dagger s}_{\mathbf{k}} \right) |\Psi_{\mathrm{mpol}}\rangle, \qquad (6.21)$$

which describes the magnetic polaron bound to a single spin and charge excitations. The equations of motion for $\psi^{c,s}_{\mathbf{k}}$ are derived from the time-dependent variational principle $\delta[\langle\Psi_b|i\partial_t - \hat{\mathcal{H}}|\Psi_b\rangle] = 0$. They are given by

$$i\dot{\psi}^c_{\mathbf{k}} = (E_{\mathrm{mpol}} + \Omega^c_{\mathbf{k}})\psi^c_{\mathbf{k}} + g^+_{\mathrm{IB}} \sum_{\mathbf{k}'} V^{(1)}_{\mathbf{k}\mathbf{k}'} \psi^c_{\mathbf{k}'} + g^-_{\mathrm{IB}} u_{\mathbf{k}} \sum_{\mathbf{k}'} \psi^s_{\mathbf{k}'}, \qquad (6.22)$$

$$i\dot{\psi}^s_{\mathbf{k}} = (E_{\mathrm{mpol}} + \Omega^s_{\mathbf{k}})\psi^s_{\mathbf{k}} + g^+_{\mathrm{IB}} \sum_{\mathbf{k}'} \psi^s_{\mathbf{k}'} + g^-_{\mathrm{IB}} \sum_{\mathbf{k}'} u_{\mathbf{k}'} \psi^c_{\mathbf{k}'}. \qquad (6.23)$$

In order to find the eigenmodes of these equations, we assume the solutions of the form $\psi^{c,s}_{\mathbf{k}} \propto e^{-i\omega t}$ which oscillate in time with frequency ω. Substituting this ansatz into Eqs. (6.22) and (6.23), we obtain the equation

$$\begin{pmatrix} 1 - \frac{g^+_{\mathrm{IB}}}{2}\Pi_{w^2} & -\frac{g^+_{\mathrm{IB}}}{2}\Pi & 0 & -g^-_{\mathrm{IB}}\Pi_{uw} \\ -\frac{g^+_{\mathrm{IB}}}{2}\Pi & 1 - \frac{g^+_{\mathrm{IB}}}{2}\Pi_{w^{-2}} & 0 & -g^-_{\mathrm{IB}}\Pi_{uw^{-1}} \\ -\frac{g^+_{\mathrm{IB}}}{2}\Pi_{uw} & -\frac{g^+_{\mathrm{IB}}}{2}\Pi_{uw^{-1}} & 1 & -g^-_{\mathrm{IB}}\Pi_{u^2} \\ 0 & 0 & -g^-_{\mathrm{IB}}\Pi_s & 1 - g^+_{\mathrm{IB}}\Pi_s \end{pmatrix} \begin{pmatrix} \sum_{\mathbf{k}} W_{\mathbf{k}} \psi^c_{\mathbf{k}} \\ \sum_{\mathbf{k}} W^{-1}_{\mathbf{k}} \psi^c_{\mathbf{k}} \\ \sum_{\mathbf{k}} u_{\mathbf{k}} \psi^c_{\mathbf{k}} \\ \sum_{\mathbf{k}} \psi^s_{\mathbf{k}} \end{pmatrix} = \begin{pmatrix} 0 \\ 0 \\ 0 \\ 0 \end{pmatrix}, \qquad (6.24)$$

where we define

$$\Pi_{w^{\pm 2}} = \sum_{\mathbf{k}} \frac{W^{\pm 2}_{\mathbf{k}}}{\omega - E_{\mathrm{mpol}} - \Omega^c_{\mathbf{k}}}, \quad \Pi = \sum_{\mathbf{k}} \frac{1}{\omega - E_{\mathrm{mpol}} - \Omega^c_{\mathbf{k}}}, \quad \Pi_{uw^{\pm 1}} = \sum_{\mathbf{k}} \frac{u_{\mathbf{k}} W^{\pm 1}_{\mathbf{k}}}{\omega - E_{\mathrm{mpol}} - \Omega^c_{\mathbf{k}}},$$

6.4 Out-of-Equilibrium Dynamics of the Magnetic Polaron

$$\Pi_{u^2} = \sum_{\mathbf{k}} \frac{u_{\mathbf{k}}^2}{\omega - E_{\text{mpol}} - \Omega_{\mathbf{k}}^c}, \quad \Pi_s = \sum_{\mathbf{k}} \frac{1}{\omega - E_{\text{mpol}} - \Omega_{\mathbf{k}}^s}. \tag{6.25}$$

Equation (6.24) has nontrivial solutions only if the determinant of the matrix on the left-hand side vanishes. Expressing the interaction strengths $g_{\text{IB},\sigma}$ in terms of the scattering lengths $a_{\text{IB},\sigma}$ via Eq. (6.4) and collecting the leading terms in the limit of $\Lambda \to \infty$, we obtain the equation

$$\left[a_{\text{IB}}^+ - \left(4\pi \sum_{\mathbf{k}}^{\Lambda} \frac{1}{\mathbf{k}^2} + \frac{2\pi}{m_{\text{red}}} \Pi_s \right)^{-1} \right] \left[a_{\text{IB}}^+ - \left(4\pi \sum_{\mathbf{k}}^{\Lambda} \frac{1}{\mathbf{k}^2} + \frac{2\pi}{m_{\text{red}}} \frac{\Pi_{w^2} + \Pi_{w^{-2}}}{2} \right)^{-1} \right] = (a_{\text{IB}}^-)^2. \tag{6.26}$$

Solving this equation for ω gives the bound-state energy ω_{bound}. The number of solutions determine the dynamical phase diagram in Fig. 6.1b.

In the two-particle problem with the impurity and a single bath atom, for each positive scattering length $a_{\text{IB},\sigma}$, there exists a bound state whose energy is $\epsilon_{\text{dim}} = \hbar^2/(2m_{\text{red}}a_{\text{IB},\sigma}^2)$. Thus, the corresponding diagram (cf. Fig. 6.1b) has four distinct regions depending on signs of $a_{\text{IB},\sigma}$. In contrast, the many-body phase diagram in Fig. 6.1b shows a unified region II as a consequence of the hybridization of the bound states due to the exchange of spin-wave excitations.

The presence of a single bound state in this region II leads to the oscillatory spin dynamics whose frequency agrees with the bound-state energy calculated from Eq. (6.26) (see inset of panel II in Fig. 6.2). In region III, there are two bound states that lead to the multi-frequency oscillations in the spin dynamics. As shown in inset of panel III in Fig. 6.2, the coupling between the bound states induces shifts in the oscillation frequencies from the bare bound-state energies obtained from Eq. (6.26) (black dotted lines). The coupling also leads to a large peak lying at the difference between two energies (green dashed line). Physically, we can understand these shifts as a consequence of the magnon-mediated polaronic nonlinearity [69].

The coupling of the two bound states in region III weakens as we depart from the strongly interacting regime. The oscillation frequencies in the spin dynamics then eventually converge to the bare bound-state energies obtained from Eq. (6.26). To clarify this point further, we show the spin dynamics for varying scattering lengths in Fig. 6.3. When both states are weakly bound and close in energy (i.e., both scattering lengths are large and take on similar values), the coupling between the bound states can be strong enough to induce shifts in the oscillation frequencies as shown in Fig. 6.3a. In this regime, a large peak placed at the difference of the two oscillatory modes indicates a strong coupling of the two bound states. As the strength of one of the scattering lengths is decreased, while keeping the other unchanged, the coupling of the bound states becomes weak and the oscillation frequencies eventually converge to the bare bound-state energies indicated by the dashed black lines (see Fig. 6.3b–d). As the energy gap between both states increases, the gradual decoupling of the two bound states can also be seen as a decrease in the peak height at the difference between the two oscillatory mode energies.

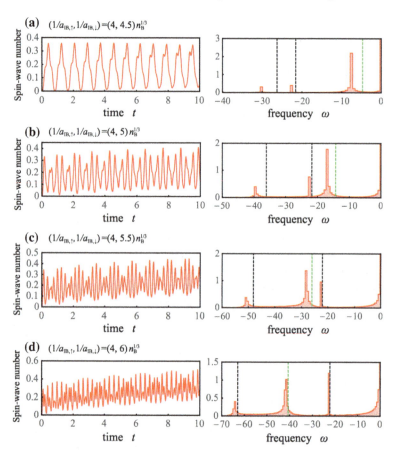

Fig. 6.3 Dynamics of environment spins and their Fourier spectra in the presence of two bound states. One of the scattering lengths is set to be $1/(a_{\mathrm{IB},\uparrow} n_B^{1/3}) = 4$, while the other $1/(a_{\mathrm{IB},\downarrow} n_B^{1/3})$ is varied over the range from 4.5 to 6 in (**a**)–(**d**). The black dashed lines in the Fourier spectra indicate the eigenenergies ω_{bound} of the magnetic bound states calculated from Eq. (6.26), while the green dashed line shows the difference between these energies. We use the parameters $a_{\mathrm{BB}} n_B^{1/3} = 0.05$ and $m_\mathrm{I}/m_\mathrm{B} = 0.95$. Time and frequency are shown in units of $m_\mathrm{B}/(\hbar n_B^{2/3})$. Reproduced from Fig. S1 of Ref. [56]. Copyright © 2018 by the American Physical Society

6.5 Experimental Implementation in Ultracold Atomic Gases

A variety of imbalanced mixtures of atomic gases allow one to experimentally implement our proposal. To be concrete, we here consider the ^{41}K-^{39}K mixture. Two miscible states $|\uparrow\rangle = |F = 1, m_F = 1\rangle$ and $|\downarrow\rangle = |F = 1, m_F = 0\rangle$ of ^{41}K can be identified as the environment bosons and $|i\rangle = |F = 1, m_F = 1\rangle$ of ^{39}K plays a role

6.5 Experimental Implementation in Ultracold Atomic Gases

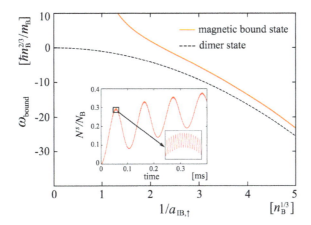

Fig. 6.4 The bound-state energy ω_{bound} (solid curve) obtained from Eq. (6.26) and the dimer energy $\epsilon_{\text{dim}} = \hbar^2/(2m_{\text{red}}a_{\text{IB},\uparrow}^2)$ (dashed curve). The inset shows the spin dynamics at $1/(a_{\text{IB},\uparrow}n_{\text{B}}^{1/3}) = 2$. We choose the densities as $n_{\text{B}} = 10^{14}\,\text{cm}^{-3}$ and $n_{\text{I}}/n_{\text{B}} = 0.1$. We use the parameters $a_{\text{BB}}n_{\text{B}}^{1/3} = 0.05$, $1/(a_{\text{IB},\downarrow}n_{\text{B}}^{1/3}) = 15$ and $m_{\text{I}}/m_{\text{B}} = 0.95$ as realized in a ^{41}K-^{39}K mixture. Reproduced from Fig. 3 of Ref. [56]. Copyright © 2018 by the American Physical Society

as the impurity.[3] The impurity-bath interaction can be controlled via a Feshbach resonance at 500 G. The bound-state energy ω_{bound} obtained from Eq. (6.26) is plotted in Fig. 6.4 in the vicinity of a Feshbach resonance. In this regime, $a_{\text{IB},\downarrow}$ takes a small, positive value while $a_{\text{IB},\uparrow}$ takes a large positive one. The imbalanced scattering lengths lead to a generation of large spin-wave excitations $N^s(t)$ (see the inset of Fig. 6.4). The shallow bound state induces a relatively slow oscillation with ∼10 kHz, which can be measured by current techniques [17]. While the oscillation frequency corresponds to a temperature scale $T/k_B \simeq 500$ nK that has already been achieved in several experiments [17, 71, 72], the predicted results should be accessible in higher temperatures by performing local measurements [54] or by localizing the impurities around the center of the system [17].

We have so far assumed that the scattering length a_{BB} between environment bosons is independent of the spin components $\sigma = \uparrow, \downarrow$. However, in practice, there may exist a small imbalance in the boson-boson interactions. For example, identifying the hyperfine states $|F = 1, m_F = 1\rangle$ and $|F = 1, m_F = 0\rangle$ as $\uparrow$-and $\downarrow$-state, respectively, the imbalance in scattering lengths is ∼0.4% for ^{41}K [73] and ∼0.5% for ^{87}Rb atoms [64]. In general, this weak symmetry breaking causes spin decoherence that is additional to the one induced by the impurity. To estimate the size of such a contribution, we consider the Hamiltonian of a two-component gas of host bosons in the absence of the impurity

---

[3] As the background scattering length for $|i\rangle = |F = 1, m_F = 1\rangle$ of ^{39}K is negative [70], this component should be used as the minority atoms (i.e., impurities) in the mixture.

$$\hat{H}_B = \sum_{k\sigma} \epsilon_k \hat{a}^\dagger_{k\sigma} \hat{a}_{k\sigma} + \frac{1}{2V} \sum_{k,k',q} \bigg(g_{BB\uparrow\uparrow} \hat{a}^\dagger_{k+q,\uparrow} \hat{a}^\dagger_{k'-q,\uparrow} \hat{a}_{k,\uparrow} \hat{a}_{k',\uparrow}$$
$$+ 2 g_{BB\uparrow\downarrow} \hat{a}^\dagger_{k+q,\uparrow} \hat{a}^\dagger_{k'-q,\downarrow} \hat{a}_{k,\uparrow} \hat{a}_{k',\downarrow} + g_{BB\downarrow\downarrow} \hat{a}^\dagger_{k+q,\downarrow} \hat{a}^\dagger_{k'-q,\downarrow} \hat{a}_{k,\downarrow} \hat{a}_{k',\downarrow} \bigg). \quad (6.27)$$

Here $g_{BB\sigma\sigma'}$ denotes the interaction strength between host bosons of spin component σ and σ'. Let us introduce imbalance parameters $r_{1,2}$ by

$$g_{BB\uparrow\uparrow} = g_{BB\uparrow\downarrow}(1+r_1), \quad g_{BB\downarrow\downarrow} = g_{BB\uparrow\downarrow}(1+r_2), \quad g_{BB\uparrow\downarrow} \equiv g_{BB}. \quad (6.28)$$

Then, by following a similar procedure as outlined in Sect. 6.2, we can diagonalize the Hamiltonian to obtain

$$\hat{H}_B = \sum_k \left(\tilde{\epsilon}^s_k \hat{\tilde{\gamma}}^{\dagger s}_k \hat{\tilde{\gamma}}^s_k + \tilde{\epsilon}^c_k \hat{\tilde{\gamma}}^{\dagger c}_k \hat{\tilde{\gamma}}^c_k \right), \quad (6.29)$$

where the dispersion relations are given by

$$\tilde{\epsilon}^{s,c}_k = \sqrt{\left(\frac{k^2}{2m_B}\right)^2 + \frac{g n_B k^2}{2m_B}\left[\left(1 + \frac{r_1+r_2}{2}\right) \mp \sqrt{1+\left(\frac{r_1-r_2}{2}\right)^2}\right]}. \quad (6.30)$$

To ensure that the energies are real, we require that the parameters satisfy the miscible condition:

$$(1+r_1)(1+r_2) > 1 \iff g_{BB\uparrow\uparrow} g_{BB\downarrow\downarrow} > g^2_{BB\uparrow\downarrow}. \quad (6.31)$$

The operators $\hat{\tilde{\gamma}}^{s,c}_k$ are related to $\hat{a}_{k,\sigma}$ by the Bogoliubov transformation. As an example, we show the expressions in the case of $r_1 = r_2 = r$:

$$\begin{cases} \hat{a}_{k\uparrow} = \frac{1}{\sqrt{2}} \left(\tilde{u}^s_k \hat{\tilde{\gamma}}^s_k - \tilde{v}^s_{-k} \hat{\tilde{\gamma}}^{\dagger s}_{-k} + \tilde{u}^c_k \hat{\tilde{\gamma}}^c_k - \tilde{v}^c_{-k} \hat{\tilde{\gamma}}^{\dagger c}_{-k} \right); \\ \hat{a}_{k\downarrow} = \frac{1}{\sqrt{2}} \left(-\tilde{u}^s_k \hat{\tilde{\gamma}}^s_k + \tilde{v}^s_{-k} \hat{\tilde{\gamma}}^{\dagger s}_{-k} + \tilde{u}^c_k \hat{\tilde{\gamma}}^c_k - \tilde{v}^c_{-k} \hat{\tilde{\gamma}}^{\dagger c}_{-k} \right), \end{cases} \quad (6.32)$$

where

$$\tilde{u}^s_k = \sqrt{\frac{\frac{k^2}{2m_B} + \frac{g n_B}{2} r}{2\sqrt{\left(\frac{k^2}{2m_B}\right)^2 + \frac{g n_B k^2}{2m_B} r}} + \frac{1}{2}},$$

$$\tilde{v}^s_k = \frac{g n_B r}{2\sqrt{2}} \left[\left(\frac{k^2}{2m_B}\right)^2 + \frac{g n_B k^2}{2m_B} r + \left(\frac{k^2}{2m_B} + \frac{g n_B}{2} r\right) \sqrt{\left(\frac{k^2}{2m_B}\right)^2 + \frac{g n_B k^2}{2m_B} r} \right]^{-1/2}. \quad (6.33)$$

For the sake of simplicity, let us focus on this particular case.[4] In order to study how the initially prepared superposition state $|\Psi_{\text{BEC}}\rangle \propto (\hat{a}_{0\uparrow}^\dagger + \hat{a}_{0\downarrow}^\dagger)^{N_B}|0\rangle$ dephases due to the imbalance in the spin-dependent boson-boson scattering lengths, we consider the initial state in terms of the operators $\hat{\tilde{\gamma}}_{\mathbf{k}}^{s,c}$:

$$|\Psi_{\text{BEC}}(0)\rangle \propto \exp\left[\frac{1}{2}\sum_{\mathbf{k}\neq 0}\left(\frac{\tilde{v}_{-\mathbf{k}}^s}{\tilde{u}_{\mathbf{k}}^s}\hat{\tilde{\gamma}}_{\mathbf{k}}^{\dagger s}\hat{\tilde{\gamma}}_{-\mathbf{k}}^{\dagger s} + \frac{\tilde{v}_{-\mathbf{k}}^c}{\tilde{u}_{\mathbf{k}}^c}\hat{\tilde{\gamma}}_{\mathbf{k}}^{\dagger c}\hat{\tilde{\gamma}}_{-\mathbf{k}}^{\dagger c}\right)\right]|0\rangle_\gamma, \quad (6.34)$$

which satisfies $\hat{a}_{\mathbf{k}\sigma}|\Psi_{\text{BEC}}(0)\rangle = 0$ for $\mathbf{k} \neq 0$, where $|0\rangle_\gamma$ denotes the vacuum of the $\hat{\tilde{\gamma}}_{\mathbf{k}}^{s,c}$ operators. From the Hamiltonian (6.29), the time evolution of the quantum state follows

$$|\Psi_{\text{BEC}}(t)\rangle \propto \exp\left[\frac{1}{2}\sum_{\mathbf{k}\neq 0}\left(\frac{\tilde{v}_{-\mathbf{k}}^s e^{-2i\tilde{\epsilon}_{\mathbf{k}}^s t}}{\tilde{u}_{\mathbf{k}}^s}\hat{\tilde{\gamma}}_{\mathbf{k}}^{\dagger s}\hat{\tilde{\gamma}}_{-\mathbf{k}}^{\dagger s} + \frac{\tilde{v}_{-\mathbf{k}}^c e^{-2i\tilde{\epsilon}_{\mathbf{k}}^c t}}{\tilde{u}_{\mathbf{k}}^c}\hat{\tilde{\gamma}}_{\mathbf{k}}^{\dagger c}\hat{\tilde{\gamma}}_{-\mathbf{k}}^{\dagger c}\right)\right]|0\rangle_\gamma. \quad (6.35)$$

Then, by denoting $\langle \cdots \rangle$ as the expectation value with respect to $|\Psi_{\text{BEC}}(t)\rangle$, the time evolution of the spin operator becomes

$$\frac{\langle \hat{S}_x \rangle}{N_B} = \left\langle \frac{1}{2N_B}\sum_{\mathbf{k}}\left(\hat{a}_{\mathbf{k},\uparrow}^\dagger \hat{a}_{\mathbf{k},\downarrow} + \hat{a}_{\mathbf{k},\downarrow}^\dagger \hat{a}_{\mathbf{k},\uparrow}\right)\right\rangle \quad (6.36)$$

$$= \frac{1}{2} - \frac{1}{n_B}\int \frac{d^3\mathbf{k}}{(2\pi)^3} 2(\tilde{u}_{\mathbf{k}}^s)^2(\tilde{v}_{\mathbf{k}}^s)^2(1 - \cos(2\tilde{\epsilon}_{\mathbf{k}}^s t)), \quad (6.37)$$

where we used the expressions of the Bogoliubov transformations (6.32). The second term in Eq. (6.37) represents the decoherence factor induced by the spin-dependent internal interactions between the host bosons. The integral over $(\tilde{u}_{\mathbf{k}}^s)^2$ roughly equals the number of excited particles, which is typically less than 1%, and $\tilde{v}_{\mathbf{k}}^s$ is on the order of r. Thus, the decoherence factor can be estimated by the multiplication of these two factors.

As an example, we assume an imbalance $r = 0.01$. Then the total decoherence factor induced by the internal dynamics (6.37) is about $<10^{-6}$ which is negligible compared to the dephasing induced by the impurity. Figure 6.5 shows the time evolution of the decoherence given by Eq. (6.37) for the imbalance parameters $r = 0.01$, and 0.1. Our numerical findings support the above estimation of the decoherence factor. In particular, the decoherence is still greatly suppressed even for an imbalance in the boson-boson scattering lengths of about 10%. Thus, our predictions on the magnetic polaron dynamics should be detectable also using a miscible pair of hyperfine states of ^{23}Na by identifying, for instance, $|F=1, m_F=1\rangle$ as the $\uparrow$-state and $|F=1, m_F=0\rangle$ as the $\downarrow$-state, leading to an imbalance $\sim 8\%$ [65]. For this choice, ^{40}K [13] will be the most promising candidate for the impurity atoms.

---

[4] A generalization to the case of $r_1 \neq r_2$ is straightforward and leads to a similar conclusion.

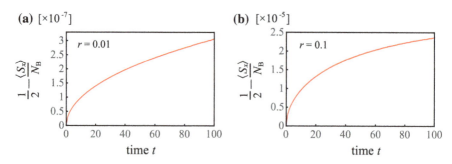

Fig. 6.5 Decoherence induced by an imbalance in the interactions between the different components of the environment bosons. The time evolutions are plotted for **a** $r = 0.01$ and **b** $r = 0.1$. The boson-boson interaction is set to $a_{BB} n_B^{1/3} = 0.1$ and we plot time in units of $m_B/(\hbar n_B^{2/3})$. Reproduced from Fig. S2 of Ref. [56]. Copyright © 2018 by the American Physical Society

6.6 Conclusions and Outlook

We have studied out-of-equilibrium dynamics of yet another paradigmatic open-quantum system, a mobile spinless particle interacting with a many-particle environment. In particular, we have shown that the formation of a polaron cloud can be directly measured in real time by using the Ramsey interferometry acting on the synthetic magnetic environment. We have demonstrated that the key signature of the magnetic polaron is the generation of spin excitations in the environment atoms. In the strong-coupling regime that is attainable in ultracold atomic systems (but not in solid-state materials), we found that the many-body bound states can induce oscillatory spin dynamics whose frequencies are characterized by the bound-state energies. Finally, we discussed a concrete experimental implementation by the state-of-the-art techniques in ultracold atoms.

Our work suggests several new directions. Firstly, combining our interferometric approach with in-situ imaging techniques such as quantum gas microscopy [74–78], one can directly observe the real-space structure of the polaron cloud. In particular, it will be intriguing to reveal how the entanglement generated by the impurity-environment coupling propagates in real space and time. Secondly, a generalization of our approach to large-spin spinor BECs [79] presents an intriguing direction where the formation of unconventional magnetic polarons is expected. Thirdly, while we considered the physics close to a broad Feshbach resonance where range corrections are negligible, it will be interesting to see how they affect the spin and charge dynamics. Finally, it remains an important open question how the nature of magnetic polarons changes as the density of impurities is increased to a degree at which a magnon-mediated interaction between polarons [80] becomes important, potentially leading to a superfluid instability of a fermionic polaron gas.

References

1. Landau L, Pekar S (1948) Effective mass of the polaron. J Exp Theor Phys 423:71–74
2. Mahan G (2000) Many-particle physics. Kluwer Academic/Plenum Publishers, New York
3. Jungwirth T, Sinova J, Mašek J, Kučera J, MacDonald AH (2006) Theory of ferromagnetic (III, Mn)V semiconductors. Rev Mod Phys 78:809–864
4. Majumdar P, Littlewood PB (1998) Dependence of magnetoresistivity on charge-carrier density in metallic ferromagnets and doped magnetic semiconductors. Nature 395:479–481
5. Salje EKH, Alexandrov AS, Liang WY (eds) (2005) Polarons and bipolarons in high temperature superconductors and related materials. Cambridge University Press, Cambridge
6. Bardeen J, Baym G, Pines D (1967) Effective interaction of He 3 atoms in dilute solutions of He 3 in He 4 at low temperatures. Phys Rev 156:207
7. De Teresa JM, Ibarra MR, Algarabel PA, Ritter C, Marquina C, Blasco J, Garcia J, del Moral A, Arnold Z (1997) Evidence for magnetic polarons in the magnetoresistive perovskites. Nature 386:256–259
8. Schirotzek A, Wu C-H, Sommer A, Zwierlein MW (2009) Observation of fermi polarons in a tunable fermi liquid of ultracold atoms. Phys Rev Lett 102:230402
9. Nascimbène S, Navon N, Jiang KJ, Tarruell L, Teichmann M, McKeever J, Chevy F, Salomon C (2009) Collective oscillations of an imbalanced fermi gas: axial compression modes and polaron effective mass. Phys Rev Lett 103:170402
10. Koschorreck M, Pertot D, Vogt E, Frohlich B, Feld M, Köhl M (2012) Attractive and repulsive Fermi polarons in two dimensions. Nature 485:619–622
11. Kohstall C, Zaccanti M, Jag M, Trenkwalder A, Massignan P, Bruun GM, Schreck F, Grimm R (2012) Metastability and coherence of repulsive polarons in a strongly interacting Fermi mixture. Nature 485:615–618
12. Zhang Y, Ong W, Arakelyan I, Thomas JE (2012) Polaron-to-polaron transitions in the radio-frequency spectrum of a quasi-two-dimensional fermi gas. Phys Rev Lett 108:235302
13. Wu C-H, Park JW, Ahmadi P, Will S, Zwierlein MW (2012) Ultracold fermionic feshbach molecules of ^{23}Na^{40}K. Phys Rev Lett 109:085301
14. Hu M-G, Van de Graaff MJ, Kedar D, Corson JP, Cornell EA, Jin DS (2016) Bose polarons in the strongly interacting regime. Phys Rev Lett 117:055301
15. Jørgensen NB, Wacker L, Skalmstang KT, Parish MM, Levinsen J, Christensen RS, Bruun GM, Arlt JJ (2016) Observation of attractive and repulsive polarons in a Bose-Einstein condensate. Phys Rev Lett 117:055302
16. Cetina M, Jag M, Lous RS, Walraven JTM, Grimm R, Christensen RS, Bruun GM (2015) Decoherence of impurities in a fermi sea of ultracold atoms. Phys Rev Lett 115:135302
17. Cetina M, Jag M, Lous RS, Fritsche I, Walraven JTM, Grimm R, Levinsen J, Parish MM, Schmidt R, Knap M, Demler E (2016) Ultrafast many-body interferometry of impurities coupled to a fermi sea. Science 354:96–99
18. Parish MM, Levinsen J (2016) Quantum dynamics of impurities coupled to a fermi sea. Phys Rev B 94:184303
19. Schmidt R, Knap M, Ivanov DA, You J-S, Cetina M, Demler E (2018) Universal many-body response of heavy impurities coupled to a fermi sea: a review of recent progress. Rep Prog Phys 81:024401
20. Mathey L, Wang DW, Hofstetter W, Lukin MD, Demler E (2004) Luttinger liquid of polarons in one-dimensional boson-fermion mixtures. Phys Rev Lett 93:120404
21. Palzer S, Zipkes C, Sias C, Köhl M (2009) Quantum transport through a Tonks-Girardeau gas. Phys Rev Lett 103:150601
22. Cucchietti FM, Timmermans E (2006) Strong-coupling polarons in dilute gas Bose-Einstein condensates. Phys Rev Lett 96:210401
23. Klein A, Bruderer M, Clark SR, Jaksch D (2007) Dynamics, dephasing and clustering of impurity atoms in Bose-Einstein condensates. New J Phys 9:411
24. Tempere J, Casteels W, Oberthaler MK, Knoop S, Timmermans E, Devreese JT (2009) Feynman path-integral treatment of the BEC-impurity polaron. Phys Rev B 80:184504

25. Casteels W, Tempere J, Devreese JT (2011) Many-polaron description of impurities in a Bose-Einstein condensate in the weak-coupling regime. Phys Rev A 84:063612
26. Spethmann N, Kindermann F, John S, Weber C, Meschede D, Widera A (2012) Dynamics of single neutral impurity atoms immersed in an ultracold gas. Phys Rev Lett 109:235301
27. Casteels W, Tempere J, Devreese JT (2013) Bipolarons and multipolarons consisting of impurity atoms in a Bose-Einstein condensate. Phys Rev A 88:013613
28. Rath SP, Schmidt R (2013) Field-theoretical study of the Bose polaron. Phys Rev A 88:053632
29. Li W, Das Sarma S (2014) Variational study of polarons in Bose-Einstein condensates. Phys Rev A 90:013618
30. Christensen RS, Levinsen J, Bruun GM (2015) Quasiparticle properties of a mobile impurity in a Bose-Einstein condensate. Phys Rev Lett 115:160401
31. Ardila LAP, Giorgini S (2015) Impurity in a Bose-Einstein condensate: study of the attractive and repulsive branch using quantum Monte Carlo methods. Phys Rev A 92:033612
32. Ardila LAP, Giorgini S (2016) Bose polaron problem: effect of mass imbalance on binding energy. Phys Rev A 94:063640
33. Levinsen J, Parish MM, Bruun GM (2015) Impurity in a Bose-Einstein condensate and the Efimov effect. Phys Rev Lett 115:125302
34. Shchadilova YE, Schmidt R, Grusdt F, Demler E (2016) Quantum dynamics of ultracold Bose polarons. Phys Rev Lett 117:113002
35. Grusdt F, Shchadilova YE, Rubtsov AN, Demler E (2015) Renormalization group approach to the Fröhlich polaron model: application to impurity-BEC problem. Sci Rep 5:12124
36. Vlietinck J, Casteels W, Van Houcke K, Tempere J, Ryckebusch J, Devreese JT (2014) Diagrammatic Monte Carlo study of the acoustic and the BEC polaron. New J Phys, 9
37. Schmidt R, Lemeshko M (2015) Rotation of quantum impurities in the presence of a many-body environment. Phys Rev Lett 114:203001
38. Volosniev AG, Hammer H-W, Zinner NT (2015) Real-time dynamics of an impurity in an ideal Bose gas in a trap. Phys Rev A 92:023623
39. Shchadilova YE, Grusdt F, Rubtsov AN, Demler E (2016) Polaronic mass renormalization of impurities in Bose-Einstein condensates: correlated Gaussian-wave-function approach. Phys Rev A 93:043606
40. Schmidt R, Lemeshko M (2016) Deformation of a quantum many-particle system by a rotating impurity. Phys Rev X 6:011012
41. Schmidt R, Sadeghpour HR, Demler E (2016) Mesoscopic Rydberg impurity in an atomic quantum gas. Phys Rev Lett 116:105302
42. Bellotti FF, Frederico T, Yamashita MT, Fedorov DV, Jensen AS, Zinner NT (2016) Three-body bound states of two bosonic impurities immersed in a fermi sea in 2d. New J Phys 18:043023
43. Midya B, Tomza M, Schmidt R, Lemeshko M (2016) Rotation of cold molecular ions inside a Bose-Einstein condensate. Phys Rev A 94:041601
44. Cui X, Zhai H (2010) Stability of a fully magnetized ferromagnetic state in repulsively interacting ultracold fermi gases. Phys Rev A 81:041602
45. Schmidt R, Enss T (2011) Excitation spectra and rf response near the polaron-to-molecule transition from the functional renormalization group 83:063620
46. Massignan P, Bruun GM (2011) Repulsive polarons and itinerant ferromagnetism in strongly polarized fermi gases. Eur Phys J D 65:83–89
47. Schmidt R, Enss T, Pietilä V, Demler E (2012) Fermi polarons in two dimensions. Phys Rev A 85:021602
48. Mathy CJM, Zvonarev MB, Demler E (2012) Quantum flutter of supersonic particles in one-dimensional quantum liquids. Nat Phys 8:881–886
49. Massignan P, Zaccanti M, Bruun GM (2014) Polarons, dressed molecules and itinerant ferromagnetism in ultracold fermi gases. Rep Prog Phys 77:034401
50. Yi W, Cui X (2015) Polarons in ultracold fermi superfluids. Phys Rev A 92:013620
51. Ong W, Cheng C, Arakelyan I, Thomas JE (2015) Spin-imbalanced quasi-two-dimensional fermi gases. Phys Rev Lett 114:110403

References

52. Meinert F, Knap M, Kirilov E, Jag-Lauber K, Zvonarev MB, Demler E, Nägerl H-C (2017) Bloch oscillations in the absence of a lattice. Science 356:945–948
53. Knap M, Kantian A, Giamarchi T, Bloch I, Lukin MD, Demler E (2013) Probing real-space and time-resolved correlation functions with many-body ramsey interferometry. Phys Rev Lett 111:147205
54. Scazza F, Valtolina G, Massignan P, Recati A, Amico A, Burchianti A, Fort C, Inguscio M, Zaccanti M, Roati G (2017) Repulsive fermi polarons in a resonant mixture of ultracold ^{6}Li atoms. Phys Rev Lett 118:083602
55. de Gennes PG (1960) Effects of double exchange in magnetic crystals. Phys Rev 118:141–154
56. Ashida Y, Schmidt R, Tarruell L, Demler E (2018) Many-body interferometry of magnetic polaron dynamics. Phys Rev B 97:060302
57. Mazurenko A, Chiu CS, Ji G, Parsons MF, Kanász-Nagy M, Schmidt R, Grusdt F, Demler E, Greif D, Greiner M (2017) A cold-atom Fermi-Hubbard antiferromagnet. Nature 545:462–466
58. Valtolina G, Scazza F, Amico A, Burchianti A, Recati A, Enss T, Inguscio M, Zaccanti M, Roati G (2017) Exploring the ferromagnetic behaviour of a repulsive fermi gas through spin dynamics. Nat Phys 13:704–709
59. Gorshkov AV, Hermele M, Gurarie V, Xu C, Julienne PS, Ye J, Zoller P, Demler E, Lukin MD, Rey AM (2010) Two-orbital SU(N) magnetism with ultracold alkaline-earth atoms. Nat Phys 6:289–295
60. Marti GE, MacRae A, Olf R, Lourette S, Fang F, Stamper-Kurn DM (2014) Coherent magnon optics in a ferromagnetic spinor Bose-Einstein condensate. Phys Rev Lett 113:155302
61. Zeiher J, Van Bijnen R, Schauß P, Hild S, Choi J-Y, Pohl T, Bloch I, Gross C (2016) Many-body interferometry of a Rydberg-dressed spin lattice. Nat Phys 12:1095–1099
62. Böttcher F, Gaj A, Westphal KM, Schlagmüller M, Kleinbach KS, Löw R, Liebisch TC, Pfau T, Hofferberth S (2016) Observation of mixed singlet-triplet Rb$_2$ Rydberg molecules. Phys Rev A 93:032512
63. Petrosyan D (2017) Dipolar exchange induced transparency with Rydberg atoms. New J Phys 19:033001
64. van Kempen EGM, Kokkelmans SJJMF, Heinzen DJ, Verhaar BJ (2002) Interisotope determination of ultracold rubidium interactions from three high-precision experiments. Phys Rev Lett 88:093201
65. Samuelis C, Tiesinga E, Laue T, Elbs M, Knöckel H, Tiemann E (2000) Cold atomic collisions studied by molecular spectroscopy. Phys Rev A 63:012710
66. Lee TD, Low FE, Pines D (1953) The motion of slow electrons in a polar crystal. Phys Rev 90:297–302
67. Fröhlich H (1954) Electrons in lattice fields. Adv Phys 3:325–361
68. Jackiw R, Kerman A (1979) Time-dependent variational principle and the effective action. Phys Lett A 71:1–5
69. Aubry S (1997) Breathers in nonlinear lattices: existence, linear stability and quantization. Phys D 103:201–250
70. D'Errico C, Zaccanti M, Fattori M, Roati G, Inguscio M, Modugno G, Simoni A (2007) Feshbach resonances in ultracold 39 k. New J Phys 9:223
71. Fang F, Olf R, Wu S, Kadau H, Stamper-Kurn DM (2016) Condensing magnons in a degenerate ferromagnetic spinor Bose gas. Phys Rev Lett 116:095301
72. Fletcher RJ, Lopes R, Man J, Navon N, Smith RP, Zwierlein MW, Hadzibabic Z (2017) Two- and three-body contacts in the unitary Bose gas. Science 355:377–380
73. Tanzi L, Cabrera CR, Sanz J, Cheiney P, Tomza M, Tarruell L (2018) Feshbach resonances in potassium Bose-Bose mixtures. arXiv:1810.12453
74. Bakr WS, Gillen JI, Peng A, Fölling S, Greiner M (2009) A quantum gas microscope for detecting single atoms in a Hubbard-regime optical lattice. Nature 462:74–77
75. Sherson JF, Weitenberg C, Endres M, Cheneau M, Bloch I, Kuhr S (2010) Single-atom-resolved fluorescence imaging of an atomic Mott insulator. Nature 467:68–72
76. Ashida Y, Ueda M (2015) Diffraction-unlimited position measurement of ultracold atoms in an optical lattice. Phys Rev Lett 115:095301

77. Ashida Y, Ueda M (2016) Precise multi-emitter localization method for fast super-resolution imaging. Opt Lett 41:72–75
78. Alberti A, Robens C, Alt W, Brakhane S, Karski M, Reimann R, Widera A, Meschede D (2016) Super-resolution microscopy of single atoms in optical lattices. New J Phys 18:053010
79. Stamper-Kurn DM, Ueda M (2013) Spinor Bose gases: symmetries, magnetism, and quantum dynamics. Rev Mod Phys 85:1191–1244
80. Naidon P (2018) Two impurities in a Bose-Einstein condensate: from Yukawa to Efimov attracted polarons. J Phys Soc Jpn 87:043002

Chapter 7
Conclusions and Outlook

Abstract We provide summary and outlooks of the studies presented in this thesis. The first part of this thesis deals with the influence from an external observer on quantum many-body phenomena. We discuss several intriguing open questions such as possible thermalization of integrable or many-body localized systems under measurement or dissipation, and effects of quantum jumps on many-body dynamics in long-time regimes. The second part is concerned with open quantum systems correlated with an external many-body environment. We discuss how the versatile variational approaches developed in this thesis can be applied to address important open questions in various fields such as condensed matter physics and AMO physics.

Keywords Open quantum systems · Quantum many-body systems · Quantum criticality · Nonequilibrium dynamics · Quantum impurity

In this Thesis, we have studied the fundamental aspects of many-body physics in quantum systems open to an external world, with particular focus on their quantum criticality, out-of-equilibrium dynamics, and entanglement structures. The interactions with an external observer or environment fundamentally alter the underlying physics and can give rise to new types of quantum many-body phenomena beyond the conventional paradigm.

The first part of this Thesis is devoted to elucidating the influence of measurement backaction on quantum many-body phenomena. First, by analyzing a universal low-energy behavior of a one-dimensional many-body system, we have identified that continuous observation leads to two possible types of relevant perturbations to an effective Hamiltonian. Extending the Tomonaga-Luttinger liquid theory to non-Hermitian cases, we have revealed new types of critical phenomena beyond the conventional paradigm. Analyzing the Bose-Hubbard model under continuous observation, we found that the measurement backaction can shift the quantum critical point. Our formalism and analyses can readily be extended to other types of models with different symmetries or dimensions. It will be intriguing to further search previously unexplored phenomena that have no analogues in Hermitian many-body systems. Second, we have studied how the measurement backaction qualitatively alters out-of-equilibrium dynamics of many-particle systems. We have developed the notion of the full-counting dynamics, in which we found peculiar dynamics fundamentally different from the unconditional or unitary dynamics. Most strikingly, we have revealed the propagation of correlations beyond the conventional maximal speed limit known as the Lieb-Robinson bound at the expense of probabilistic nature of quantum measurement. While we have focused on the solvable noninteracting model, it is intriguing to explore similar unconventional phenomena also in interacting many-body systems. Employing the idea of the eigenstate-thermalization hypothesis, we have also addressed the thermalization and heating in generic (nonintegrable) many-body systems under continuous observation. We find that the nonunitary nature of quantum measurement leads to several unique features in thermalization mechanism that are unseen in closed systems. We leave it as an interesting open question to elucidate thermalization of an integrable many-body system under measurement. Taking the diffusive limit of continuous measurement, we have obtained the time-evolution equation for indistinguishable particles under a weak spatial observation. We expect that the derived equation should be useful to analyze the backaction from a variety of minimally destructive in-situ measurements of quantum systems since specific spatial profiles of measurement processes will disappear after taking the diffusive limit of measurement.

The second part of this Thesis is devoted to studies of strongly correlated open quantum systems, where the system-environment entanglement plays an essential role. We have focused on a quantum impurity as their most fundamental paradigm, and developed a versatile and efficient theoretical approach to studying its in- and out-of-equilibrium physics. This approach has almost achieved the state-of-the-art accuracy realized by the matrix-product-state ansatz with several orders of magnitude fewer variational parameters. Moreover, it has allowed one to explore new types of nonequilibrium phenomena that have been difficult to study through previous approaches. Our theoretical approach and, in particular, the newly constructed canonical transformations can readily be generalized to other types of many-body problems. We expect that they should play pivotal roles in the future studies of solving challenging problems in strongly correlated systems. We have also analyzed a mobile spinless particle strongly coupled to the magnetic environment and found new types of out-of-equilibrium dynamics in the regime that is not attainable in solid-state materials. Our idea of employing the Ramsey interferometry for the environment to reveal the impurity dynamics should have a broad range of applicability other than the model we discuss. Together with quantum gas microscopy, this possibility will open a way to reveal how the system-environment entanglement develops in real time and space.

In short, this Thesis has been devoted to advancing our understanding of open and out-of-equilibrium physics in quantum many-body systems. I hope that the research presented in this Thesis will stimulate further studies in this new frontier.

Curriculum Vitae

Yuto Ashida

Address (Current affiliation):
Department of Applied Physics, The University of Tokyo
7-3-1 Hongo, Bunkyo-ku, Tokyo, Japan
Postal code: 113-8656
Tel.: +81-3-5841-6814
e-mail: ashida@ap.t.u-tokyo.ac.jp
Web: https://sites.google.com/site/yutoashida/home

Education

Department of Physics, The University of Tokyo
Degree: M.Sc. Mar. 2016, Ph.D. Mar. 2019
Thesis advisor: Masahito Ueda
Thesis topic: Quantum many-body physics in open systems

Department of Physics, The University of Tokyo
Degree: Mar. 2014, B.Sc.

Major Honors and Awards

- President's award from the University of Tokyo, 2019
- 9-th JSPS Ikushi Prize, 2019
- Research Fellow of Japan Society for the Promotion of Science (DC1), 2016–2019
- Harvard ITAMP postdoctoral fellowship, 2019 (offered)

- Kyoto University Hakubi research fellow, 2019 (offered)
- The Research Award for Graduate students of Faculty of Science, 2016 and 2019
- 40-th International Physics Olympiad Gold Medal, 2009

Research Interests

I am interested in an interdisciplinary field of condensed matter physics, AMO physics and statistical physics. Recently, my studies have particularly focused on open and out-of-equilibrium physics of quantum many-body systems, where the interplay between many-body interactions and influences of external environments, meausrements, and manipulations exhibits rich and novel phenomena beyond the conventional paradigms of closed, equilibrium systems.

Research Experience:

Apr. 2019–present. Department of Applied Physics, The University of Tokyo
Assistant Professor (or Research associate without students)

Aug. 2016–present. Department of Physics, Harvard University
Visiting Researcher

Apr. 2016–Mar. 2019. Department of Physics, The University of Tokyo
Research fellow of JSPS (DC1)

Apr. 2014–Mar. 2019. Department of Physics, The University of Tokyo
Research assistant of Advanced Leading Graduate Course for Photon Science (ALPS).

CPSIA information can be obtained
at www.ICGtesting.com
Printed in the USA
LVHW010843120121
676227LV00006B/13